GROUP INVERSES OF M-MATRICES AND THEIR APPLICATIONS

CHAPMAN & HALL/CRC APPLIED MATHEMATICS AND NONLINEAR SCIENCE SERIES

Series Editor *H. T. Banks*

Published Titles

Advanced Differential Quadrature Methods, Zhi Zong and Yingyan Zhang
Computing with hp-ADAPTIVE FINITE ELEMENTS, Volume 1, One and Two Dimensional Elliptic and Maxwell Problems, Leszek Demkowicz
Computing with hp-ADAPTIVE FINITE ELEMENTS, Volume 2, Frontiers: Three Dimensional Elliptic and Maxwell Problems with Applications, Leszek Demkowicz, Jason Kurtz, David Pardo, Maciej Paszyński, Waldemar Rachowicz, and Adam Zdunek
CRC Standard Curves and Surfaces with Mathematica®*: Second Edition,* David H. von Seggern
Discovering Evolution Equations with Applications: Volume 1-Deterministic Equations, Mark A. McKibben
Discovering Evolution Equations with Applications: Volume 2-Stochastic Equations, Mark A. McKibben
Exact Solutions and Invariant Subspaces of Nonlinear Partial Differential Equations in Mechanics and Physics, Victor A. Galaktionov and Sergey R. Svirshchevskii
Fourier Series in Several Variables with Applications to Partial Differential Equations, Victor L. Shapiro
Geometric Sturmian Theory of Nonlinear Parabolic Equations and Applications, Victor A. Galaktionov
Green's Functions and Linear Differential Equations: Theory, Applications, and Computation, Prem K. Kythe
Group Inverses of M-Matrices and Their Applications, Stephen J. Kirkland and Michael Neumann
Introduction to Fuzzy Systems, Guanrong Chen and Trung Tat Pham
Introduction to non-Kerr Law Optical Solitons, Anjan Biswas and Swapan Konar
Introduction to Partial Differential Equations with MATLAB®, Matthew P. Coleman
Introduction to Quantum Control and Dynamics, Domenico D'Alessandro
Mathematical Methods in Physics and Engineering with Mathematica, Ferdinand F. Cap
Mathematical Theory of Quantum Computation, Goong Chen and Zijian Diao
Mathematics of Quantum Computation and Quantum Technology, Goong Chen, Louis Kauffman, and Samuel J. Lomonaco
Mixed Boundary Value Problems, Dean G. Duffy
Modeling and Control in Vibrational and Structural Dynamics, Peng-Fei Yao
Multi-Resolution Methods for Modeling and Control of Dynamical Systems, Puneet Singla and John L. Junkins
Nonlinear Optimal Control Theory, Leonard D. Berkovitz and Negash G. Medhin
Optimal Estimation of Dynamic Systems, Second Edition, John L. Crassidis and John L. Junkins
Quantum Computing Devices: Principles, Designs, and Analysis, Goong Chen, David A. Church, Berthold-Georg Englert, Carsten Henkel, Bernd Rohwedder, Marlan O. Scully, and M. Suhail Zubairy
A Shock-Fitting Primer, Manuel D. Salas
Stochastic Partial Differential Equations, Pao-Liu Chow

CHAPMAN & HALL/CRC APPLIED MATHEMATICS
AND NONLINEAR SCIENCE SERIES

GROUP INVERSES OF M-MATRICES AND THEIR APPLICATIONS

Stephen J. Kirkland
Michael Neumann

CRC Press is an imprint of the
Taylor & Francis Group, an **informa** business
A CHAPMAN & HALL BOOK

CRC Press
Taylor & Francis Group
6000 Broken Sound Parkway NW, Suite 300
Boca Raton, FL 33487-2742

First issued in paperback 2019

ISBN-13: 978-1-4398-8858-2 (hbk)
ISBN-13: 978-0-367-38048-9 (pbk)

Visit the Taylor & Francis Web site at
http://www.taylorandfrancis.com

and the CRC Press Web site at
http://www.crcpress.com

To Seema

–S.J.K.

To Helen

–M.N.

Contents

List of Figures

Preface

Nonnegative matrices and M-matrices have become a staple in contemporary linear algebra, and they arise frequently in its applications. Such matrices are encountered not only in matrix analysis, but also in stochastic processes, graph theory, electrical networks, and demographic models. We, the authors have worked for many years investigating these topics, and have found group inverses for singular M-matrices to be a recurring and useful tool.

The aim of this monograph is to bring together a diverse collection of results on the group inverses of M-matrices, highlighting their importance and utility in this body of research. By presenting these results in a single volume, we hope to emphasise the connections between problems arising in Markov chains, Perron eigenvalue analysis, spectral graph theory, and the like, and to show how insight into each of these areas can be gained through the use of group inverses.

While we have tried to make this book as self-contained as possible, we assume that the reader is familiar with the basics on the theory of nonnegative matrices, directed and undirected graphs, and Markov chains. There are many excellent texts covering these subjects, and we will refer the reader to these as needed.

The structure of the book is as follows. As a way of motivating the questions considered in the remainder of the book, in Chapter 1 we pose some sample problems associated with Leslie matrices (which arise in a demographic model) and stochastic matrices (which are central in the theory of Markov chains). In Chapter 2, we develop the basic algebraic and spectral properties of the group inverse of a general matrix, with a specific focus on the case that the matrix in question is a singular and irreducible M-matrix. In Chapter 3, we consider the Perron value and vector of a nonnegative matrix as a function of the entries in that matrix, and derive formulas for their derivatives in terms of the group inverse of an associated M-matrix. These formulas are then applied in Chapter 4 to several classes of matrices, including the Leslie matrices mentioned above. The elasticity of the Perron value, which measures the contribution of an individual entry to the Perron value, is also considered, and the group inverse approach is used to analyse the derivatives of the elasticity with respect to the entries in the matrix.

In Chapters 5 and 6, we turn our attention to Markov chains. In Chapter 5, the group inverse of an appropriate M-matrix is used not only in the perturbation analysis of the stationary distribution vector, but also in deriving bounds on the asymptotic rate of convergence of the underlying Markov chain. In Chapter 6, we will show how the group inverse can be used to compute and analyse the mean first passage matrix for a Markov chain. An analogous approach is also used to discuss the Kemeny constant for a Markov chain.

Chapter 7 has a combinatorial flavour, and is devoted to the Laplacian matrix for an undirected graph. Here we use the group inverse of the Laplacian matrix in order to discuss the Weiner and Kirchhoff indices, as well as the algebraic connectivity of the graph. Applications to electrical networks are also considered.

Chapter 8 concludes the book with a discussion on computing the group inverse, and on the stability of doing so. Several approaches are considered and compared.

In writing this book, we were informed and inspired by the work of many. In particular, we benefited from the books of Adi Ben–Israel and Thomas Greville, Abraham Berman and Robert Plemmons, Stephen Campbell and Carl Meyer, and Hal Caswell. We should also single out the 1975 paper of Meyer, which serves as the starting point for many of the developments reported in this book. Over the years, we have had the pleasure of working with many talented co-authors, and a number of collaborative results appear in this book. As a show of our gratitude, let us list those collaborators here: Mahmud Akelbek, Minerva Catral, Yonghong Chen, Emeric Deutsch, Ludwig Elsner, Ilse Ipsen, Charles Johnson, Srinivasa Mohan, Jason Molitierno, Nic Ormes, K.G. Ramamurthy, Bryan Shader, Nun–Sing Sze, and Jianhong Xu.

S.J.K. and M.N., March, 2011.

The text above was written while Michael Neumann (Miki, to his friends) was visiting me in March of 2011. He and I had been working on the monograph long-distance for about seven months, and during his visit with me, we solidified the contents of the book, developed preliminary drafts of the first few chapters, and mapped out a plan for writing the remaining chapters. In April of 2011, Miki died, of natural causes, but very suddenly. The news of his passing came as a shock to me. We had been collaborators for some twenty years—more than that, we were good friends.

I resolved to complete the monograph on my own, attempting to stay true to both the content and the spirit of the project that Miki and I had envisioned together. What follows is the result, though it must be said that that the book would have been stronger had Miki been able to continue his work on it. The experience of completing this monograph on my own has been bittersweet for me: it has made me acutely aware of Miki's absence, but has also served as an extended farewell to my close friend.

While I was writing this book, my research was supported in part by the Science Foundation Ireland, under Grant No. SFI/07/SK/I1216b. Thanks are also due to Iarnród Éireann, as much of the book was written while I was availing myself of its service.

S.J.K., May, 2012.

Author Bios

Stephen Kirkland received a Ph.D. in Mathematics from the University of Toronto in 1989, having previously taken an M.Sc. in Mathematics from the University of Toronto (1985), and a B.Sc. (Honours) in Mathematics from the University of British Columbia (1984). He held postdoctoral positions at Queen's University at Kingston (1989–1991), and the University of Minnesota (1991–1992), and was a faculty member at the University of Regina from 1992 until 2009. He then moved to Ireland, and is currently a Stokes Professor at the National University of Ireland Maynooth. He is an editor-in-chief of the journal *Linear and Multilinear Algebra*, and serves on the editorial boards of several other journals. Kirkland's research interests are primarily in matrix theory and graph theory, with an emphasis on the interconnections between those two areas.

Michael Neumann received a B.Sc. in Mathematics and Statistics from Tel Aviv University in 1970, and a Ph.D. in Computer Science from London University in 1972. He subsequently held positions at the University of Reading (1972–1973), the Technion (1973–1975), the University of Nottingham (1975–1980) and the University of South Carolina (1980–1985). In 1985 he moved to the University of Connecticut, where he remained as a faculty member until his death in 2011. Over the course of his career, Neumann published more than 160 mathematical papers, primarily in matrix theory, numerical linear algebra and numerical analysis. He was elected as a member of the Connecticut Academy of Arts and Sciences in 2007, was named a Board of Trustees Distinguished Professor in 2007, and was appointed as the Stuart and Joan Sidney Professor of Mathematics in 2010.

1

Motivation and Examples

Nonnegative matrices arise naturally in a number of models of physical and biological phenomena. The reason for this is transparent: many of the quantities that come up in those settings are necessarily nonnegative. In this chapter, we present two examples of nonnegative matrices; one arises in mathematical demography, the other in stochastic processes. In both instances, we indicate how the group inverse of an appropriate M-matrix can be used to analyse the model. Our aim is to help motivate the kinds of questions raised in later chapters, and to give the reader a sense of the techniques upon which we will focus.

1.1 An example from population modelling

An important context in which nonnegative matrices appear is in the modelling of populations. The book of Caswell [17] provides a wealth of examples of models of that type. One of the simplest such models is the so-called Leslie model for age-classified populations.

In the Leslie model, we consider a population closed to migration that has been classified into n age groups, each of equal length. The population is modelled in discrete time, where the unit of time under consideration coincides with the length of a single age class. Let $u_i(k)$ denote the number of individuals in the i–th age group at time k. Let f_i, $i = 1, \ldots, n$, denote the fecundity (i.e., birth rate) of each individual in the i-th age group and let p_i, $i = 1, \ldots, n-1$, denote the proportion of individuals at age i who survive to age $i+1$. Assume that both the fecundity and survival proportions are independent of the time k. Then, as can be readily ascertained, the population vector at time k can be described by the matrix–vector relation:

$$u(k) = Au(k-1),$$

where $u(k) = \begin{bmatrix} u_1(k) & \dots & u_n(k) \end{bmatrix}^t$, for all $k \geq 0$, and where

$$A = \begin{bmatrix} f_1 & f_2 & & \dots & f_{n-1} & f_n \\ p_1 & 0 & & \dots & 0 & 0 \\ 0 & p_2 & 0 & \dots & 0 & 0 \\ \vdots & & \ddots & & & \vdots \\ 0 & \dots & & 0 & p_{n-1} & 0 \end{bmatrix}$$

is the *Leslie matrix* corresponding to the population under consideration. Evidently we have $u(k) = A^k u(0)$, so that the sequence of population vectors is a realisation of the power method (see, for example, Stewart's book [109, section 7.2]) applied to $u(0)$.

Denote the all ones vector in $\mathbb{R}^n$ by $\mathbf{1}$. The *age distribution vector at time* k is given by $u(k)/\mathbf{1}^t u(k)$, while the *total size of the population* at time k is $\mathbf{1}^t u(k)$. Suppose that our Leslie matrix A is primitive, i.e., some power of A has all positive entries; this is often the case in applications. Then as $k \to \infty$, we find that the population size is asymptotically growing like $r(A)^k$, where $r(A)$ is the dominant eigenvalue (i.e., the Perron value) of A, while the age distribution vector converges to an appropriately scaled eigenvector of A corresponding to $r(A)$. In particular, the eigenvalue $r(A)$ has demographic significance, as it is interpreted as the asymptotic growth rate for the population under consideration.

How does this asymptotic growth rate $r(A)$ behave when considered as a function of the fecundity rates $f_1, \dots, f_n$? It is intuitively obvious that $r(A)$ is an increasing function of each f_i, and this intuition is indeed confirmed by fundamental results in Perron–Frobenius theory. However, as the following (somewhat artificial) example illustrates, the behaviour of $r(A)$ as a function of f_i varies with different choices of the index i.

EXAMPLE 1.1.1 Here we consider a 7×7 Leslie matrix of the following form:

$$A = \begin{bmatrix} 0 & 0 & 0 & 0 & 0 & .2 & .8 \\ 1 & 0 & 0 & 0 & 0 & 0 & 0 \\ 0 & 1 & 0 & 0 & 0 & 0 & 0 \\ 0 & 0 & 1 & 0 & 0 & 0 & 0 \\ 0 & 0 & 0 & 1 & 0 & 0 & 0 \\ 0 & 0 & 0 & 0 & 1 & 0 & 0 \\ 0 & 0 & 0 & 0 & 0 & 1 & 0 \end{bmatrix}.$$

Next, we consider the effect on the Perron value (i.e., asymptotic

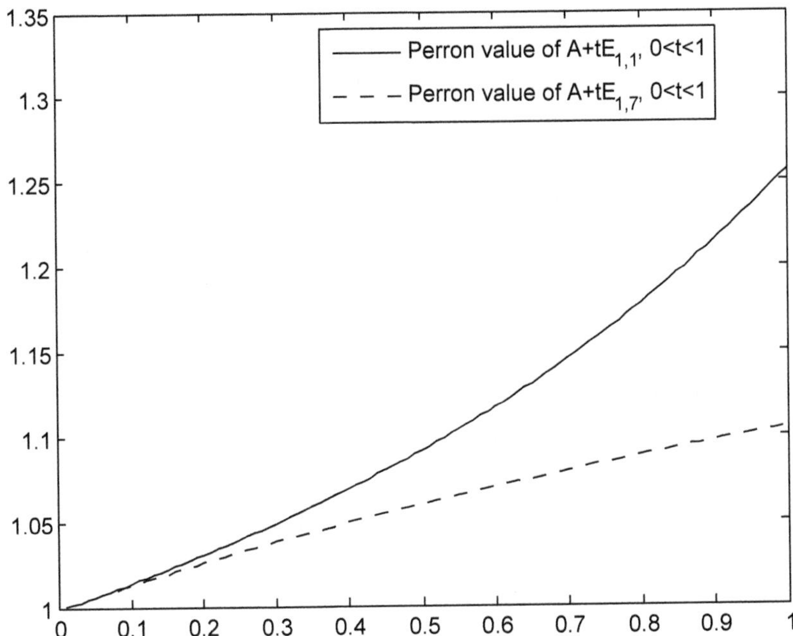

FIGURE 1.1
Perron value as a function of $a_{1,1}$ and $a_{1,7}$

growth rate) of changing the fecundity rates corresponding to the first and last age groups in the population. We increase the fecundity rate for the first age group from 0 to 1, and, in a separate computation, we increase the fecundity rate for the last age group from .8 to 1.8. This is equivalent to computing $r(A + tE_{1,1})$ and $r(A + tE_{1,7})$ for $t \in [0, 1]$, where $E_{i,j}$ denotes the $(0, 1)$ matrix with a one in the (i, j) position and zeros elsewhere. Figure 1.1 plots the resulting curves.

As expected, each of the curves in Figure 1.1 is increasing, but note that the curve corresponding to changes for age group one is concave up, while the curve corresponding to changes for age group seven is concave down. Consequently, the asymptotic growth rate is rapidly increasing as a function of the fecundity rate for age one, but more slowly increasing as a function of the fecundity rate for age seven.

In order to understand this phenomenon mathematically, we must develop a tool that will help to analyse the second derivatives of the dominant eigenvalue with respect the entries in the top row of A. In Chapter 3 (sections 3.2 and 3.3), we provide formulas and analysis for the second derivatives of the Perron value for a general nonnegative matrix, while in Chapter 4, we focus on the special case of Leslie matrices. In particular, we show that for a Leslie matrix, the dominant eigenvalue is a concave up function of f_i for lower age groups, a concave down function of f_i for higher age groups, with the switch from concave up to concave down taking place no later than age group $\frac{n}{2}$. As we will see, the convexity or concavity of the Perron eigenvalue as a function of a particular entry in A can be read off from the signs of the entries of a certain group generalised inverse associated with A.

1.2 An example from Markov chains

Recall that a square matrix T is *stochastic* if it is nonnegative with all row sums equal to 1. Assuming that T is irreducible, we find from Perron–Frobenius theory that the dominant eigenvalue of T is 1. The *stationary distribution vector (or stationary vector)* for T is the left eigenvector w of T corresponding to the eigenvalue 1, normalised so that the entries of w sum to 1. Stochastic matrices are central to the theory of discrete time, finite state space, time homogeneous Markov chains, and thorough overviews of that subject can be found in the books of Kemeny and Snell [66] and Seneta [107]. A recurring theme in the analysis of discrete time, homogeneous ergodic Markov chains, is that of determining how the stationary vector changes as the entries in the corresponding transition matrix are perturbed to yield another stochastic matrix. The following example illustrates how small perturbations can affect the entries in the stationary distribution vector.

EXAMPLE 1.2.1 Consider the following 20×20 transition matrix:

$$T = \begin{bmatrix} 0 & 1 & 0 & 0 & \dots & 0 & 0 & 0 \\ \frac{1}{2} & 0 & \frac{1}{2} & 0 & 0 & \dots & 0 & 0 \\ 0 & \frac{1}{2} & 0 & \frac{1}{2} & 0 & 0 & \dots & 0 \\ & & & & & & & \\ \vdots & & \ddots & \ddots & \ddots & & & \vdots \\ & & & & & & & \\ 0 & 0 & \dots & 0 & 0 & \frac{1}{2} & 0 & \frac{1}{2} \\ 0 & 0 & 0 & \dots & 0 & 0 & 1 & 0 \end{bmatrix}.$$

It is straightforward to determine that the stationary vector w for T

satisfies

$$w^t = \frac{1}{38}\left[\begin{array}{ccccccc} 1 & 2 & 2 & \dots & 2 & 2 & 1 \end{array}\right].$$

Next, we consider the 20×20 matrix B given by

$$B = \begin{bmatrix} 0 & -1 & 0 & 0 & \dots & 0 & 1 & 0 \\ 0 & 0 & -1 & 0 & \dots & 0 & 1 & 0 \\ 0 & 0 & 0 & -1 & 0 & \dots & 1 & 0 \\ & & & & & & & \\ \vdots & & \ddots & \ddots & \ddots & & & \vdots \\ & & & & & & & \\ 0 & 0 & 0 & \dots & 0 & -1 & 1 & 0 \\ 0 & 0 & 0 & \dots & 0 & 0 & 0 & 0 \\ 0 & 0 & 0 & \dots & 0 & 0 & 1 & -1 \\ 0 & 0 & 0 & \dots & 0 & 0 & 0 & 0 \end{bmatrix}.$$

Observe that for each $s \in [0, \frac{1}{2})$, the matrix $T + sB$ is irreducible and stochastic, with stationary vector say $w(s)$. Certainly we expect that $w(s) \to w$ as $s \to 0^+$, but the plots below shows that different entries in $w(s)$ can behave in very different ways for values of s near 0. The plots in Figure 1.2 exhibit the graphs for

$$\frac{w_{19}(s) - w_{19}}{s} \text{ and } \frac{w_{12}(s) - w_{12}}{s}$$

for $s \in [0, 0.1]$. From the plots, we see that for s near 0, $w_{19}(s)$ is significantly more sensitive to changes in s than $w_{12}(s)$.

How can we predict when one or more entries in the stationary vector will be sensitive to changes in the transition matrix? How can we quantify that sensitivity? Again, a group generalised inverse provides us with the tools needed to answer these questions. As we will see in Chapter 3, this group inverse is used to find the derivative of a Perron vector with respect to the entries in a nonnegative matrix, while in Chapter 5, we provide an extensive discussion of how the group inverse serves to measure error and conditioning properties of the stationary distribution vector under perturbation of the corresponding transition matrix.

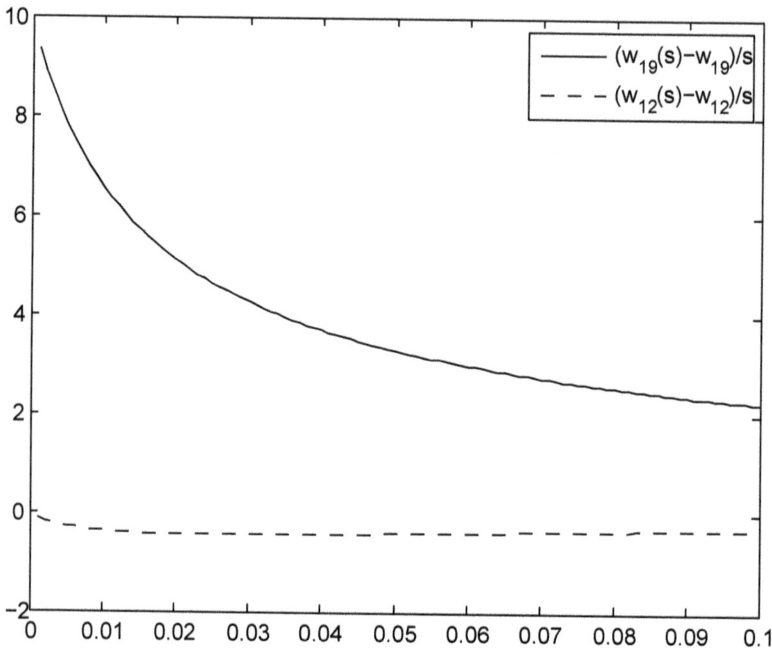

FIGURE 1.2
$\frac{w_{19}(s)-w_{19}}{s}$ and $\frac{w_{12}(s)-w_{12}}{s}$, for $s \in [0, 0.1]$

2

The Group Inverse

It is suggested in Chapter 1 that a certain generalised inverse, called the *group inverse*, plays a key role in the development of results on nonnegative matrices. In this chapter we develop the background material and basic properties of the group inverse. We will pay particular attention to the properties of group generalised inverses when the matrices under discussion are singular and irreducible M-matrices.

We remark that much of the material present in this chapter can be found in well–known books on generalised inverses, such as those by Ben–Israel and Greville [10] and by Campbell and Meyer [16].

2.1 Definition and general properties of the group inverse

Suppose that we have an $n \times n$ complex singular matrix B and that its eigenvalue 0 of B is *semisimple*; that is, its algebraic and geometric multiplicities coincide. When this is the case, the *group inverse* of B, denoted $B^{\#}$, is the unique $X \in \mathbb{C}^{n,n}$ satisfying the following three matrix equations:

$$\text{(i) } BXB = B; \quad \text{(ii) } XBX = X; \quad \text{(iii) } BX = XB. \tag{2.1}$$

We remark that below we will see why the semisimplicity of 0 as an eigenvalue of B implies the existence and uniqueness of such a matrix X. We further mention that the name "group inverse" derives from the fact that there is a multiplicative group G of matrices in $\mathbb{C}^{n,n}$ such that $B \in G$, and $B^{\#}$ serves as the multiplicative inverse of B in G.

We now proceed to construct the group inverse. Indeed, suppose that 0 is a semisimple eigenvalue of B and that B has rank r. Then by the Jordan canonical form we know that, via similarity, B admits the representation

$$B = P \left[\begin{array}{c|c} C & 0 \\ \hline 0 & 0 \end{array} \right] P^{-1}, \tag{2.2}$$

where P is an $n \times n$ invertible matrix and where $C \in \mathbb{C}^{r,r}$ is invertible. Evidently there are many choices for P and C; for instance, one might choose C to be a direct sum of the Jordan blocks of B corresponding to the nonzero eigenvalues of B. It is easy to see that the matrix

$$X_0 = P \left[\begin{array}{c|c} C^{-1} & 0 \\ \hline 0 & 0 \end{array} \right] P^{-1} \tag{2.3}$$

then satisfies (i), (ii), and (iii) of (2.1). Thus, the existence of at least one solution to (2.1) is established.

Next, we turn our attention to the uniqueness of the solution to (2.1). Suppose that X satisfies (i), (ii) and (iii) of (2.1). Then, in particular,

$$(BX)(BX) = B(XBX) = BX \tag{2.4}$$

so that BX is a projection matrix. Denoting the range space and null space of a matrix M by $\mathcal{R}(M)$ and $\mathcal{N}(M)$, respectively, we find that since $\mathcal{R}(BX) \subseteq \mathcal{R}(B)$ and

$$\text{rank}(B) = \text{rank}(BXB) \leq \text{rank}(BX),$$

it must be the case that $\mathcal{R}(B) = \mathcal{R}(BX)$. Using similar arguments we can show that

$$\mathcal{R}(X) = \mathcal{R}(XB) = \mathcal{R}(BX) = \mathcal{R}(B) \tag{2.5}$$

and that

$$\mathcal{N}(X) = \mathcal{N}(XB) = \mathcal{N}(BX) = \mathcal{N}(B). \tag{2.6}$$

From the above relations we can readily prove the uniqueness of the solution to equations (i), (ii), and (iii) of (2.1), as follows. Letting $P_{\mathcal{R}(B),\mathcal{N}(B)}$ denote the oblique projector with range $\mathcal{R}(B)$ and null space $\mathcal{N}(B)$, it follows that if X_1 and X_2 both satisfy (2.1), then we have:

$$\begin{aligned} X_1 = X_1(BX_1) = X_1 P_{\mathcal{R}(B),\mathcal{N}(B)} = X_1(BX_2) \\ = (X_1B)X_2 = P_{\mathcal{R}(B),\mathcal{N}(B)} X_2 = (X_2B)X_2 = X_2, \end{aligned}$$

as desired. As noted above, there is a multiplicative group of matrices in which B and $B^{\#}$ are multiplicative inverses. Indeed there are many such groups; the group $G = \{B, B^{\#}, BB^{\#}\}$ where the unit element is

$BB^{\#}$, is one such example.

The notion of the group inverse extends to a more general setting in which the eigenvalue 0 is not constrained to be semisimple. Given any square singular matrix B, the *index of* B is the smallest $k \in \mathbb{N}$ such that $\text{rank}(B^k) = \text{rank}(B^{k+1})$. For B of index k, the *Drazin inverse* of B, denoted B^D, is the unique matrix $X \in \mathbb{R}^{n,n}$ such that

$$\text{(i}'\text{)}\ XBX = X, \quad \text{(ii}'\text{)}\ B^{k+1}X = B^k, \quad \text{(iii}'\text{)}\ BX = XB.$$

The interested reader is referred to [10] and [16] for further details on the Drazin inverse. Evidently in the case that the eigenvalue 0 of B is semisimple, the index of B is 1, and $B^{\#}$ coincides with B^D.

2.2 Spectral properties of the group inverse

Let us continue now with the assumption that 0 is a semisimple eigenvalue of B and list several spectral properties of $B^{\#}$ of which we will make repeated use in this book.

As mentioned in the previous section, one of the forms to which B may be transformed by similarity is (2.2), where $C = \tilde{J}$, viz.,

$$B = P \left[\begin{array}{c|c} \tilde{J} & 0 \\ \hline 0 & 0 \end{array} \right] P^{-1}, \tag{2.7}$$

where $\tilde{J}$ is a direct sum of all Jordan blocks of B corresponding to the nonzero eigenvalues of B. Each such block is thus of the form:

$$\begin{bmatrix} \lambda & 1 & 0 & \dots & & 0 \\ 0 & \lambda & 1 & 0 & \dots & 0 \\ \vdots & & \ddots & \ddots & & \vdots \\ & & & & & \\ 0 & & \dots & & 0 & \lambda \end{bmatrix} \in \mathbb{C}^{k,k}, \tag{2.8}$$

for some $1 \le k \le n-1$, where $\lambda \neq 0$. Then from (2.2)–(2.6), and by identifying $B^{\#}$ with the matrix X in (2.3), we find the following:

i) $B^{\#}$ has 0 as a semisimple eigenvalue and its multiplicity coincides with the multiplicity of 0 as an eigenvalue of B.
ii) A vector x is a right null vector for B if and only if it is a right null

vector for $B^\#$. An analogous statement holds for left null vectors.
iii) The number $\lambda \neq 0$ is an eigenvalue of B of algebraic multiplicity m if and only if $1/\lambda$ is an eigenvalue of $B^\#$ of algebraic multiplicity m.
iv) A vector x is a right eigenvector for B corresponding to the eigenvalue $\lambda \neq 0$ if and only if it is a right eigenvector for $B^\#$ corresponding to the eigenvalue $1/\lambda$. An analogous statement holds for left eigenvectors.
v) Suppose that $\lambda \neq 0$ is an eigenvalue of B and that $v_1, \ldots, v_k$ is a Jordan chain of generalised λ–eigenvectors for B, that is,

$$Bv_1 = \lambda v_1, \text{ and } Bv_j = \lambda v_j + v_{j-1}, \quad j = 2, \ldots, k.$$

Set

$$u_1 \; = \; v_1, \text{ and } u_p \; = \; (-1)^{p-1}\lambda^p \sum_{j=0}^{p-2} \binom{p-2}{j} \lambda^j v_{j+2}, \quad p = 2, \ldots, k.$$

Then

$$B^\# u_1 = \frac{1}{\lambda} u_1, \text{ and } B^\# u_p = \frac{1}{\lambda} u_p + u_{p-1}, \quad p = 2, \ldots, k,$$

so that $u_1, \ldots, u_k$ is a Jordan chain of generalised $1/\lambda$–eigenvectors for $B^\#$.
vi) We know that the subspaces $\mathcal{N}(B)$ and $\mathcal{R}(B)$ are complementary subspaces and that, in fact, $BB^\# = P_{\mathcal{R}(B),\mathcal{N}(B)}$. From (2.7) and (2.8), we find that $\mathcal{R}(B)$ is just the direct sum of all (generalised) eigenspaces of B corresponding to the eigenvalues of B other than 0. Thus $BB^\#$ is the eigenprojection of B onto the direct sum of all its (generalised) eigenspaces other than those corresponding to the eigenvalue 0. Then denoting the identity matrix by I, the matrix $I - BB^\#$ is the eigenprojection of B onto the eigenspace of B corresponding to 0. In particular, if 0 is a simple eigenvalue of B, with right and left null vectors x and y^t, respectively, normalised so that $y^t x = 1$, then $BB^\# = B^\# B = I - xy^t$.

2.3 Expressions for the group inverse

From the developments in the previous section we see that if we know a Jordan basis for B, then we can compute $B^\#$. However, we may not have a Jordan basis for B at our disposal, which compels us to find other ways to compute $B^\#$. It turns out that there are several methods.

One method is via a *full rank factorisation.*

PROPOSITION 2.3.1 *Let $B \in \mathbb{C}^{n,n}$ be a nonzero singular matrix with rank r and suppose that 0 is a semisimple eigenvalue of B. Let EF be a full rank factorisation of B; that is, $E \in \mathbb{C}^{n,r}, F \in \mathbb{C}^{r,n}$, and $B = EF$. Then FE is invertible and $B^{\#} = E(FE)^{-2}F$.*

<u>**Proof**</u>: Since 0 is a semisimple eigenvalue of B, we know that B has r nonzero eigenvalues (including multiplicities). Since the matrices FE and $EF(= B)$ have the same nonzero eigenvalues (see the book of Horn and Johnson [57, Theorem 1.3.20]), we see that FE is invertible. Let $X = E(FE)^{-2}F$, and note that $BX = XB$, $BXB = B$, and $XBX = X$. We conclude from (2.1) that $B^{\#} = X$. □

Next, we apply the technique of Rose in [106] for computing the Drazin inverse in order to express $B^{\#}$ as a polynomial in B when the characteristic polynomial of the latter matrix is available.

THEOREM 2.3.2 *Let $B \in \mathbb{C}^{n,n}$ be a nonzero singular matrix with nullity r and suppose that 0 is a semisimple eigenvalue of B. Suppose that the characteristic polynomial for B is given by*

$$\lambda^r \left(\lambda^{n-r} + \sum_{j=0}^{n-r-1} b_j \lambda^j \right).$$

Then $B^{\#}$ admits the following representation as a polynomial in B:

$$B^{\#} = \frac{b_1}{b_0^2} B^{n-r} + \left(\frac{b_1 b_{n-r-1}}{b_0^2} - \frac{1}{b_0} \right) B^{n-r-1} + \sum_{j=1}^{n-r-2} \left(\frac{b_1 b_j}{b_0^2} - \frac{b_{j+1}}{b_0} \right) B^j. \tag{2.9}$$

<u>**Proof**</u>: Since B has the semisimple eigenvalue 0 with multiplicity r, there is an invertible matrix P of order n and an invertible matrix C of order $n-r$ such that B admits the representation in (2.2), in which case $B^{\#}$ is given by the right hand side of (2.3). Note further that as the multiplicity of 0 as an eigenvalue of B is r, the characteristic polynomial of C is given by:

$$\lambda^{n-r} + \sum_{j=0}^{n-r-1} b_j \lambda^j.$$

By the Cayley–Hamilton theorem (see [57, Theorem 2.4.2]), we have $C^{n-r} + \sum_{j=0}^{n-r-1} b_j C^j = 0$ and, since C is invertible, we may rearrange this last equation to yield

$$C^{-1} = -\frac{1}{b_0}\left(C^{n-r-1} + \sum_{j=0}^{n-r-2} b_{j+1}C^j\right). \tag{2.10}$$

Multiplying (2.10) by C^{-1} now gives

$$C^{-2} = -\frac{1}{b_0}\left(C^{n-r-2} + \sum_{j=1}^{n-r-2} b_{j+1}C^{j-1} + b_1C^{-1}\right). \tag{2.11}$$

Finally, substituting (2.10) into (2.11) and simplifying yields

$$\begin{aligned} C^{-2} = {} & \frac{b_1}{b_0^2}C^{n-r-1} + \left(\frac{b_1b_{n-r-1}}{b_0^2} - \frac{1}{b_0}\right)C^{n-r-2} \\ & + \sum_{j=0}^{n-r-3}\left(\frac{b_1b_{j+1}}{b_0^2} - \frac{b_{j+2}}{b_0}\right)C^j. \end{aligned} \tag{2.12}$$

Next, consider the matrix E given by the expression

$$E = \frac{b_1}{b_0^2}C^{n-r} + \left(\frac{b_1b_{n-r-1}}{b_0^2} - \frac{1}{b_0}\right)C^{n-r-1} + \sum_{j=1}^{n-r-2}\left(\frac{b_1b_j}{b_0^2} - \frac{b_{j+1}}{b_0}\right)C^j.$$

Recall that

$$B = P\left[\begin{array}{c|c} C & 0 \\ \hline 0 & 0 \end{array}\right]P^{-1}.$$

It follows by checking that

$$\frac{b_1}{b_0^2}B^{n-r} + \left(\frac{b_1b_{n-r-1}}{b_0^2} - \frac{1}{b_0}\right)B^{n-r-1} + \sum_{j=1}^{n-r-2}\left(\frac{b_1b_j}{b_0^2} - \frac{b_{j+1}}{b_0}\right)B^j$$

$$= P\left[\begin{array}{c|c} E & 0 \\ \hline 0 & 0 \end{array}\right]P^{-1} = P\left[\begin{array}{c|c} CC^{-2} & 0 \\ \hline 0 & 0 \end{array}\right]P^{-1} = P\left[\begin{array}{c|c} C^{-1} & 0 \\ \hline 0 & 0 \end{array}\right]P^{-1}.$$

The conclusion follows. □

Our next example considers a certain circulant matrix, and illustrates Theorem 2.3.2 . We refer the reader to [57, section 0.9.6] for basics on circulant matrices.

EXAMPLE 2.3.3 Let $G \in \mathbb{R}^{n,n}$ be the circulant matrix whose elements are specified by

$$g_{i,j} = \begin{cases} 1, & j = i+1,\ i = 1, \ldots, n-1 \\ 1, & i = n,\ j = 1, \\ 0, & \text{otherwise.} \end{cases}$$

Next, we consider the matrix $B = I - G$. It is readily seen that the characteristic polynomial for B is given by

$$(\lambda - 1)^n + (-1)^{n-1} = \lambda \left[\lambda^{n-1} + \sum_{j=0}^{n-2} (-1)^{n-j-1} \binom{n}{j+1} \lambda^j \right].$$

Thus, in the notation of Theorem 2.3.2, we have that $r = 1$ and that

$$b_j = (-1)^{n-j-1} \binom{n}{j+1}, \quad j = 0, \ldots, n-2.$$

On applying Theorem 2.3.2, we find that

$$\begin{aligned} B^{\#} &= \frac{(-1)^{n-2}\binom{n}{2}}{n^2} B^{n-1} + \left[\frac{(-1)^{n-2}\binom{n}{2}(-n)}{n^2} - \frac{1}{(-1)^{n-1} n} \right] B^{n-2} \\ &+ \sum_{j=1}^{n-3} \left[\frac{(-1)^{n-2}\binom{n}{2}(-1)^{n-j-1}\binom{n}{j+1}}{n^2} - \frac{(-1)^{n-j-2}\binom{n}{j+2}}{(-1)^{n-1} n} \right] B^j. \end{aligned}$$

After a few steps of simplification, we find from (2.13) that

$$B^{\#} = \sum_{j=1}^{n-1} (-1)^{j+1} \frac{j}{2n} \binom{n+1}{j+2} B^j.$$

In section 5.2, using another method, we will find an explicit expression for $B^{\#}$ in terms of the matrix G.

In section 2.5, we derive a formula for the group inverse when B is a singular and irreducible M-matrix. It is actually a special case of the following observation, which essentially appears in a paper of Meyer [94], in which we will use the fact that right and left eigenvectors corresponding to a simple eigenvalue of a matrix cannot be orthogonal to each other.

OBSERVATION 2.3.4 *Suppose that 0 is an algebraically simple eigenvalue of $B \in \mathbb{C}^{n,n}$. Let x and y denote right and left null vectors for B, respectively. Assume that neither x_n nor y_n is zero. Denote the leading principal submatrix of order $n-1$ of B by $B_{1,1}$ and let $\bar{x}$ and $\bar{y}$ be the vectors constructed from x and y by deleting their last entries, respectively. Then:*

$$B^{\#} = \frac{\bar{y}^t B_{1,1}^{-1} \bar{x}}{y^t x} x y^t + \left[\begin{array}{c|c} B_{1,1}^{-1} - \frac{1}{y^t x} B_{1,1}^{-1} \bar{x} \bar{y}^t - \frac{1}{y^t x} \bar{x} \bar{y}^t B_{1,1}^{-1} & -\frac{y_n}{y^t x} B_{1,1}^{-1} \bar{x} \\ \hline -\frac{x_n}{y^t x} \bar{y}^t B_{1,1}^{-1} & 0 \end{array} \right]. \tag{2.13}$$

<u>**Proof**</u>: Partition B as

$$\left[\begin{array}{c|c} B_{1,1} & B_{1,2} \\ \hline B_{2,1} & B_{2,2} \end{array} \right].$$

From this partitioning and the facts that $Bx = 0$ and $y^t B = 0^t$, we find that

$$B_{1,2} = -\frac{1}{x_n} B_{1,1} \bar{x} \text{ and } B_{2,1} = -\frac{1}{y_n} \bar{y}^t B_{1,1}.$$

On setting

$$E = \left[\begin{array}{c} B_{1,1} \\ \hline -\frac{1}{y_n} \bar{y}^t B_{1,1} \end{array} \right] \text{ and } F = \left[\begin{array}{c|c} I & -\frac{1}{x_n} \bar{x} \end{array} \right],$$

we observe that EF is a full rank factorisation of B. The expression (2.13) now follows by using the formula for $B^{\#}$ given in Proposition 2.3.1. □

We note that the hypothesis that neither x_n nor y_n is 0 in Observation 2.3.4 is not much of a restriction. Since $y^t x$ cannot be zero, there is at least one index j such that $x_j y_j \neq 0$, so we may always perform a permutation similarity on our matrix B so that it satisfies the hypotheses of the observation.

2.4 Group inverse versus Moore–Penrose inverse

Recall that the *Moore–Penrose generalised inverse* of a matrix $B \in \mathbb{C}^{m,n}$, denoted by $B^{\dagger}$, is the unique matrix $X \in \mathbb{C}^{n,m}$ which satisfies the four matrix equations:

$$(\tilde{\text{i}})\ BXB=B;\ (\widetilde{\text{ii}})\ XBX=X;\ (\widetilde{\text{iii}})\ BX=(BX)^*;\ (\widetilde{\text{iv}})\ XB=(XB)^*. \tag{2.14}$$

In view of the resemblance of the first two conditions of (2.14) to the first two conditions of (2.1), it is natural to consider the relationship between the group inverse and the Moore–Penrose inverse for a square matrix B. The fact that the Moore–Penrose inverse always exists while the group inverse may not, suggests that even when $m = n$, the two generalised inverses do not coincide in general.

It is known that given the linear system $Bx = c$, the vector $y = B^{\dagger}c$ is the unique least-squares solution of minimal 2-norm to the system. Further, if one happens to know a singular value decomposition of B, then $B^{\dagger}$ is readily computed from that information. On the other hand, as we have seen in the previous section, the properties of the group inverse are closely associated with spectral properties of a matrix, which typically are distinct from either least-squares or singular value properties. So this is another aspect which sets the two inverses apart. Still, the question arises: under what circumstances do the two generalised inverses coincide?

To answer the above question we assume (as we must) that $m = n$ and that B has 0 as a semisimple eigenvalue. Just as we have done with the group inverse, we can use properties $(\tilde{\text{i}})$–$(\widetilde{\text{iv}})$ to show that for the Moore–Penrose inverse, $BB^{\dagger}$ and $B^{\dagger}B$ are (in fact orthogonal) projection matrices and that the range and null space of $B^{\dagger}$ also satisfy

$$\mathcal{R}(B^{\dagger}) = \mathcal{R}(B^{\dagger}B) = \mathcal{R}(B^*) \tag{2.15}$$

and

$$\mathcal{N}(B^{\dagger}) = \mathcal{N}(BB^{\dagger}) = \mathcal{N}(B^*). \tag{2.16}$$

On comparing (2.5) and (2.6) with (2.15) and (2.16), respectively, we see that if $B^{\#} = B^{\dagger}$, then necessarily it must be the case that $\mathcal{N}(B) = \mathcal{N}(B^*)$ (and $\mathcal{R}(B) = \mathcal{R}(B^*)$).

The sufficiency of the condition $\mathcal{N}(B) = \mathcal{N}(B^*)$ for $B^{\#}$ to equal $B^{\dagger}$ can be seen as follows. Suppose that B has rank r with 0 as a semisimple eigenvalue, and that $\mathcal{N}(B) = \mathcal{N}(B^*)$. It follows that we can find an orthonormal basis of eigenvectors of $\mathcal{N}(B)$, say $u_1, \ldots, u_{n-r}$, and further that $B^{\#}B = BB^{\#} = I - \sum_{j=1}^{n-r} u_j u_j^*$. In this case, and as can be checked, we find that $B^{\#}$ satisfies the four equations of (2.14), so that necessarily $B^{\#} = B^{\dagger}$.

Thus, in particular, we see that if B is Hermitian, then $B^{\#} = B^{\dagger}$. The following simple example further illustrates the distinction between $B^{\#}$ and $B^{\dagger}$.

EXAMPLE 2.4.1 Suppose that we have vectors $u, v \in \mathbb{R}^n$ such that $u^t v \neq 0$, and consider the rank 1 matrix $B = uv^t$. Then 0 is a semisimple eigenvalue of B of multiplicity $n-1$. Appealing to Proposition 2.3.1, we find that

$$B^{\#} = \frac{1}{(u^t v)^2} uv^t.$$

On the other hand, it is straightforward to verify that the Moore–Penrose inverse for B is given by

$$B^{\dagger} = \frac{1}{(u^t u)(v^t v)} vu^t.$$

In particular, we see that $B^{\#} = B^{\dagger}$ if and only if u is a scalar multiple of v.

2.5 The group inverse associated with an M-matrix

In this section we outline some basic properties of the group inverse of a singular and irreducible M-matrix. We begin by giving a formula for the group inverse of such a matrix in partitioned form which is a special case of Observation 2.3.4. Recall that a square matrix T is *stochastic* provided that T is entrywise nonnegative and $T\mathbf{1} = \mathbf{1}$, where $\mathbf{1}$ denotes an all ones vector of the appropriate order.

PROPOSITION 2.5.1 ([94]) *Let $T \in \mathbb{R}^{n,n}$ be an irreducible stochastic matrix and partition T as*

$$T = \left[\begin{array}{c|c} T_{1,1} & T_{1,2} \\ \hline T_{2,1} & T_{2,2} \end{array}\right],$$

where $T_{1,1} \in \mathbb{R}^{n-1,n-1}$. Then $(I-T)^{\#}$ can be written in partitioned form as

$$\frac{T_{2,1}(I-T_{1,1})^{-2}\mathbf{1}}{(1+T_{2,1}(I-T_{1,1})^{-1}\mathbf{1})^2}\mathbf{1}\left[\begin{array}{c|c} T_{2,1}(I-T_{1,1})^{-1} & 1 \end{array}\right] + \left[\begin{array}{c|c} U_{1,1} & U_{1,2} \\ \hline U_{2,1} & 0 \end{array}\right], \tag{2.17}$$

where

$$\begin{cases} U_{1,1} = (I - T_{1,1})^{-1} - \frac{1}{1 + T_{2,1}(I - T_{1,1})^{-1}\mathbf{1}}\mathbf{1}T_{2,1}(I - T_{1,1})^{-2}, \\ \quad -\frac{1}{1 + T_{2,1}(I - T_{1,1})^{-1}\mathbf{1}}(I - T_{1,1})^{-1}\mathbf{1}T_{2,1}(I - T_{1,1})^{-1} \\ U_{1,2} = \frac{-1}{1 + T_{2,1}(I - T_{1,1})^{-1}\mathbf{1}}(I - T_{1,1})^{-1}\mathbf{1}, \\ U_{2,1} = \frac{-1}{1 + T_{2,1}(I - T_{1,1})^{-1}\mathbf{1}}T_{2,1}(I - T_{1,1})^{-2}. \end{cases}$$

Proof: First we note that the right null vector for $I-T$ can be taken to be $x = \mathbf{1}$. But then, on letting y be the left null vector for $I - T$ normalised so that $y^t x = 1$, it follows that y^t can be written in partitioned form as

$$y^t = \frac{1}{1 + T_{2,1}(I - T_{1,1})^{-1}\mathbf{1}} \left[\; T_{2,1}(I - T_{1,1})^{-1} \mid 1 \; \right].$$

The conclusion now follows by applying Observation 2.3.4. □

REMARK 2.5.2 It follows from Proposition 2.5.1 that if T is an $n \times n$ irreducible and stochastic matrix, then for each $j = 1, \ldots, n$, $(I-T)^{\#}_{j,j} > (I - T)^{\#}_{i,j}$ for each index $i = 1, \ldots, n$ with $i \neq j$. Thus, in each column of $(I - T)^{\#}$, the maximum entry is found on the diagonal. However, it is possible that the entry of maximum absolute value in a particular column of $(I - T)^{\#}$ is not found on the diagonal. For example, if we let T be given by

$$T = \begin{bmatrix} 0 & 1 & 0 & 0 \\ \frac{1}{2} & 0 & \frac{1}{2} & 0 \\ 0 & \frac{1}{2} & 0 & \frac{1}{2} \\ 0 & 0 & 1 & 0 \end{bmatrix},$$

then

$$(I - T)^{\#} = \frac{1}{36}\begin{bmatrix} 35 & 10 & -26 & -19 \\ 5 & 22 & -14 & -13 \\ -13 & -14 & 22 & 5 \\ -19 & -26 & 10 & 35 \end{bmatrix}.$$

Observe then that $|(I-T)^{\#}_{4,2}| > (I-T)^{\#}_{2,2}$ and $|(I-T)^{\#}_{1,3}| > (I-T)^{\#}_{3,3}$.

In the course of this book, there are a number of instances in which it will be convenient to frame our arguments for a general singular and irreducible M-matrix in terms of a related M–matrix $I - T$, where T is irreducible and stochastic. The following remark establishes the correspondence between these two types of matrices.

REMARK 2.5.3 Starting with a singular and irreducible M–matrix Q, write $Q = rI - A$, where A is an irreducible nonnegative matrix with Perron value r. Let x denote a right Perron vector for A. For any vector v we define diag(v) to be the diagonal matrix whose diagonal entries are given by the corresponding entries of v. Setting $X = \text{diag}(x)$, we find that the matrix $T = \frac{1}{r}X^{-1}AX$ is irreducible and stochastic. Consider the matrix $\tilde{Q} = I - T$; since $Q = rX\tilde{Q}X^{-1}$, it follows readily that

$$Q^{\#} = \frac{1}{r}X\tilde{Q}^{\#}X^{-1}.$$

Let $A \in \mathbb{R}^{n,n}$ be a nonnegative and irreducible matrix, and denote its Perron value by $r(A)$. Evidently for any $s \geq r(A)$,

$$Q = sI - A$$

is an M–matrix. When $s > r(A)$, Q is a nonsingular M-matrix and as explored in Chapter 6 of the book of Berman and Plemmons [12], Q is endowed with many different types of properties. In particular nonsingular M-matrices, and hence their inverses as well, have the property that all of their principal minors (i.e., the determinants of the principal submatrices) are positive. This fact seems to have been first proved by Fiedler and Pták in [41]. The class of all real matrices $n \times n$ which possess the property that all their principal minors are positive is denoted by $\mathcal{P}^n$.

Because the principal submatrices of a singular and irreducible M-matrix of all orders less than n are invertible, it follows, by continuity arguments, that all the proper principal minors of a singular and irreducible M-matrix are positive. Let $\mathcal{P}_0^n$ be the class of all $n \times n$ real singular matrices whose proper principal minors are positive.

Our next goal is to prove that if Q is a singular and irreducible M-matrix, then $Q^{\#} \in \mathcal{P}_0^n$. The following result of Fiedler and Ptak in [42] will be useful in that endeavour.

THEOREM 2.5.4 *Suppose that $M \in \mathbb{R}^{n,n}$ and that $M + M^t$ is a positive semidefinite matrix. Then M is in the class $\mathcal{P}_0^n$.*

Recall that a matrix $M \in \mathbb{R}^{n,n}$ is *row diagonally dominant* if $|m_{i,i}| \geq \sum_{j \neq i} |m_{i,j}|, i = 1, \ldots, n$; an analogous collection of inequalities defines the term *column diagonally dominant.* These notions will be useful in showing next that $Q^{\#} \in \mathcal{P}_0^n$. We employ a technique of Deutsch and Neumann used in [30].

THEOREM 2.5.5 *Let Q be a singular irreducible M-matrix of order n. Then $Q^\#$ is in the class $\mathcal{P}_0^n$.*

Proof: Suppose that x and y are positive right and left null vectors for Q, respectively, normalised so that $y^t x = 1$. Let $X = \text{diag}(x)$, $Y = \text{diag}(y)$, and consider the following two matrices: $G = YQ^\#X$ and $B = YQX$. Observe that since G is formed from $Q^\#$ by multiplication by positive diagonal matrices, it suffices to show that G is in the class $\mathcal{P}_0^n$. Note also that $B\mathbf{1} = 0$ and $\mathbf{1}^t B = 0^t$, from which we find that B is both row and column diagonally dominant. Consequently, applying Gershgorin's theorem ([57, section 6.1]), we find that $B + B^t$ is a positive semidefinite matrix; in particular we find that for any vector $v \in \mathbb{R}^n$, $v^t Bv \geq 0$.

We claim that for any vector $u \in \mathbb{R}^n$, there is a vector v such that $u^t G^t u = v^t Bv$. To see the claim, suppose that u has been given and let $v = X^{-1}Q^\# Xu$. We then have

$$\begin{aligned} v^t Bv &= (u^t X(Q^\#)^t X^{-1})YQX(X^{-1}Q^\# Xu) \\ &= u^t X(Q^\#)^t X^{-1} Y(I - xy^t)Xu \\ &= u^t X(Q^\#)^t YX^{-1}(I - xy^t)Xu \\ &= u^t G^t u - u^t X(Q^\#)^t yy^t Xu = u^t G^t u, \end{aligned}$$

the third equality holding since diagonal matrices commute. Hence $u^t G^t u = v^t Bv$, as claimed.

It now follows that for any $u \in \mathbb{R}^n$, $u^t(G + G^t)u \geq 0$, so that $G + G^t$ is a positive semidefinite matrix. Applying Theorem 2.5.4, we find that G is in the class $\mathcal{P}_0^n$, and hence so is $Q^\#$. $\square$

One of the characterising properties of matrix in $B \in \mathcal{P}^n$ is that B does not reverse the sign of any vector in $\mathbb{R}^n$. That is, if $v = \begin{bmatrix} v_1 & \dots & v_n \end{bmatrix}^t$ is a nonzero vector in $\mathbb{R}^n$ and $w = \begin{bmatrix} w_1 & \dots & w_n \end{bmatrix}^t = Av$, then there exists at least one index $1 \leq i \leq n$ such that $w_i v_i > 0$. Let us take an example. Consider the matrix:

$$C = \begin{bmatrix} 1 & 1 \\ 1 & 2 \end{bmatrix} = \begin{bmatrix} 2 & -1 \\ -1 & 1 \end{bmatrix}^{-1}.$$

If we let $v = \begin{bmatrix} -1 & 2 \end{bmatrix}^t$, then $w = \begin{bmatrix} 1 & 3 \end{bmatrix}^t$. Thus, $w_2 v_2 > 0$, so that for the index $i = 2$, the sign of the corresponding entry in v is not reversed in w. Note however that the only negative entry of v is mapped to a positive entry in w. So, as we pass from v to w, while one entry has its sign preserved, for this example it is not the case that both a positive entry and a negative entry have their signs preserved. As we will see below, multiplying a vector by the group inverse of a singular and irreducible M-matrix results in some subtler behaviour.

The following result is taken from Mohan, Neumann, and Ramamurthy's paper [98].

LEMMA 2.5.6 *Let $Q \in \mathbb{R}^{n,n}$ be a singular and irreducible M-matrix and suppose that $v = \begin{bmatrix} v_1 & \dots & v_n \end{bmatrix}^t \notin \mathcal{N}(Q)$. Let $v_{\min}$ and $v_{\max}$ denote the minimal and maximal entries of v, respectively. Then:*

(i) *If $v_{\min} \le 0$ and $v_{\max} \ge 0$, then there exist indices $1 \le r, s \le n$ such that*

$$v_r \le 0 \text{ and } (Qv)_r < 0 \tag{2.18}$$

and

$$v_s \ge 0 \text{ and } (Qv)_s > 0. \tag{2.19}$$

(ii) *If $v_{\min} < 0$ and $v_{\max} > 0$, then there are indices r, s, such that the inequalities* (2.18) *and* (2.19) *are strict.*

Proof: By Remark 2.5.3, without loss of generality we need only consider the case that there is an $n \times n$ irreducible stochastic matrix A such that $Q = I - A$. Note that in this case the null space of Q is spanned by the vector $\mathbf{1}$.

Let $H_{\min} = \{j | v_j = v_{\min}\}$ and $H_{\max} = \{j | v_j = v_{\max}\}$. Since $v \notin \mathcal{N}(Q)$, we see that v is not a multiple of $\mathbf{1}$, so the sets $H_{\min}$ and $H_{\max}$ must be nonempty and disjoint subsets in $\{1, \dots, n\}$. Thus, as A is an irreducible matrix, there exist indices $r, h, s, k \in \{1, \dots, n\}$ such that $r \in H_{\min}, h \notin H_{\min}$ and $a_{r,h} > 0$, and $s \in H_{\max}, k \notin H_{\max}$ and $a_{s,k} > 0$. We now use the fact that A is stochastic as follows

$$(Qv)_r = [(I - A)v]_r = v_r - \sum_{j=1}^{n} a_{r,j} v_j \ <$$

$$v_r - \min_{1 \le m \le n} v_m \sum_{j=1}^{n} a_{r,j} = v_r - v_r = 0.$$

Thus (2.18) holds. Moreover if $v_r < 0$, then strict inequalities hold throughout (2.18). In a similar way we show that (2.19) holds and so if $x_s > 0$, then strict inequalities hold throughout (2.19). □

Lemma 2.5.6 helps to establish the following result, which appears in modified form in [98].

THEOREM 2.5.7 *Suppose that $Q \in \mathbb{R}^{n,n}$ is a singular and irreducible M-matrix. If $0 \neq w = \begin{bmatrix} w_1 & \dots & w_n \end{bmatrix}^t \in \mathcal{R}(Q)$, then there exist indices $1 \leq r, s \leq n$ such that*

$$\left(Q^{\#}w\right)_r < 0 \text{ and } w_r < 0 \tag{2.20}$$

and

$$\left(Q^{\#}w\right)_s > 0 \text{ and } w_s > 0. \tag{2.21}$$

<u>**Proof**</u>: Since $\mathcal{R}(Q)$ and $\mathcal{N}(Q)$ are complementary subspaces in $\mathbb{R}^n$, and since $0 \neq w \in \mathcal{R}(Q)$, there exists a vector $v \in \mathcal{R}(Q)$, with $v \neq 0$, such that $w = Qv$ and such that

$$v = Q^{\#}Qv = Q^{\#}w.$$

Now as v is a nonzero vector in $\mathcal{R}(Q)$, it must have both negative and positive entries (otherwise its inner product with a positive left null vector of $Q^{\#}$ fails to be 0). Hence, in the notation of Lemma 2.5.6, we have that $v_{\min} < 0 < v_{\max}$.

It now follows by Lemma 2.5.6 that there are indices r and s such that v_r, $(Qv)_r < 0$ and v_s, $(Qv)_s > 0$. Since $v = Q^{\#}w$ and $w = Qv$, we thus find that for those same indices r, s, we have w_r, $(Q^{\#}w)_r < 0$ and w_s, $(Q^{\#}w)_s > 0$. □

Thus we see that, in the language of Theorem 2.5.7, for each nonzero vector w in the range of Q, there are positive and negative entries whose signs are preserved under the mapping $w \mapsto Q^{\#}w$.

3

Group Inverses and Derivatives of Matrix Functions

In this chapter we develop one of the central tools of the book, namely, expressions for the partial derivatives of the Perron value and vector with respect to the matrix entries. With such tools at our disposal we will be able, among other things, to address the questions raised in sections 1.1 and 1.2. These will be discussed in Chapters 4 and 5, respectively.

In section 3.1, we present some background results on eigenvalues and eigenvectors as functions of matrix entries, then move on in section 3.2 to find expressions for the first and second derivatives of the Perron value of an irreducible nonnegative matrix, with respect to the entries in that matrix. As anticipated by our remarks in Chapter 1, those expressions will involve group inverses of certain singular M-matrices. Section 3.3 deals with the problem of determining whether the Perron value of an irreducible nonnegative matrix is a convex or concave function of a particular matrix entry; that leads naturally to the consideration of the sign patterns of group inverses of singular M-matrices. Finally in section 3.4 we derive expressions for the first and second derivatives of the Perron vectors of irreducible nonnegative matrices. Those derivatives are taken with respect to the entries of the nonnegative matrix in question, and depend on how the Perron vector in question is normalised.

3.1 Eigenvalues as functions

In this section we consider eigenvalues of matrices as functions of matrices, or, more specifically, of the matrix entries. We do not develop the theory in its greatest generality, but instead concentrate on nonnegative matrices, and, to a lesser extent, on essentially nonnegative matrices.

The latter is the set with which we begin.

Consider the set of the irreducible $n \times n$ essentially nonnegative matrices:

$$\tilde{\Phi}^{n,n} = \{A = [a_{i,j}] \in \mathbb{R}^{n,n} \mid a_{i,j} \geq 0, \ i \neq j, \text{ and } A \text{ is irreducible}\}.$$

For each $A \in \mathbb{R}^{n,n}$, let $f_A(\lambda) = \det(\lambda I - A)$ be characteristic polynomial of A. Then the *spectral abscissa* of A, $s(A)$, is given by

$$s(A) = \max\{\text{Re}(\lambda) \mid f_A(\lambda) = 0\}.$$

Evidently the spectral abscissa can be viewed as a function from $\mathbb{R}^{n,n}$ to $\mathbb{R}$.

The restriction of $s(\cdot)$ to $\tilde{\Phi}^{n,n}$, denoted by $r(\cdot)$, is commonly called the *Perron value*. Let $A \in \tilde{\Phi}^{n,n}$. Then for $\alpha \geq 0$ sufficiently large, $A + \alpha I$ is a nonnegative and irreducible matrix and so its Perron value $r(A + \alpha I) = s(A) + \alpha$ is a simple eigenvalue of $A + \alpha I$. It follows that $s(A)$ is a simple eigenvalue of A.

Since $s(A)$ is a simple eigenvalue of $A \in \tilde{\Phi}^{n,n}$, there exists a neighbourhood $\mathcal{N}_A$ of A in $\mathbb{R}^{n,n}$ such that each $B \in \mathcal{N}_A$ has a simple eigenvalue $\lambda(B)$ and such that for $B \in \mathcal{N}_A \cap \tilde{\Phi}^{n,n}$, $\lambda(B) = s(B)$. In these circumstances it is known, see, for example, Wilkinson's book [112, pp.66–67], or the paper of Andrew, Chu, and Lancaster [5], that $\lambda(\cdot)$ is an analytic function of each of the n^2 entries of the elements in $\mathcal{N}_A$. Denote by $E_{i,j} \in \mathbb{R}^{n,n}$ the matrix whose (i, j)–th entry is 1 and whose remaining entries are 0. Then by the analyticity property just mentioned, the first–order partial derivative of $\lambda(\cdot)$ with respect to the (i, j)-th entry is given by:

$$\lim_{t \to 0} \frac{\lambda(A + tE_{i,j}) - \lambda(A)}{t}.$$

In the cases that $i = j$, or $i \neq j$ and $a_{i,j} > 0$, then for sufficiently small $t > 0$, $A + tE_{i,j} \in \mathcal{N}_A \cap \tilde{\Phi}^{n,n}$ and so the partial derivative of $\lambda(\cdot)$ with respect to the (i, j)–th entry at A, coincides with the partial derivative of the Perron value $s(\cdot)$ with respect to the (i, j)-th entry at A. However, if $i \neq j$ and $a_{i,j} = 0$, then by the partial derivative of the Perron value with respect to the (i, j)-th entry at A, we understand this to mean the partial derivative from the right.

The situation with respect to the analyticity of an eigenvector corresponding to the Perron value, which is known as a *Perron vector*, is somewhat more complicated. First, even when an eigenvalue is simple, a corresponding eigenvector is only unique up to multiplication by a scalar. Thus, no continuity, and hence no differentiability, of an eigenvector can hold unless the eigenvector is normalised in some fixed fashion. However, normalising the eigenvector in some fixed way may still not be enough for the differentiability of its components. The following example illustrates this phenomenon.

EXAMPLE 3.1.1 Consider the parametric family in $\tilde{\Phi}^{2,2}$ given by:

$$A(h) = \begin{bmatrix} 1+h & 1+h \\ 1 & 1 \end{bmatrix}, \quad h > -1.$$

Here $r(h) = r(A(h)) = h+2$, while a corresponding Perron eigenvector is given by $x(h) = \begin{bmatrix} h+1 & 1 \end{bmatrix}^t$. The infinity norm of $x(h)$ is given by:

$$\|x(h)\|_\infty = \begin{cases} 1, & -1 < h \leq 0, \\ h+1, & h > 0, \end{cases}$$

If we now normalise $x(h)$ so that it is positive and has infinity norm 1, and if we let $w(h) = \begin{bmatrix} w_1(h) & w_2(h) \end{bmatrix}^t = x(h)/\|x(h)\|_\infty$, then, as one can check:

$$w_1(h) = \begin{cases} h+1, & -1 < h \leq 0 \\ 1, & h > 0 \end{cases} \quad \text{and } w_2(h) = \begin{cases} 1, & -1 < h \leq 0 \\ \dfrac{1}{1+h}, & h > 0. \end{cases}$$

Plotting $w_1(h)$ and $w_2(h)$ now reveals that neither function is differentiable at $h = 0$; Figure 3.1 illustrates.

In [5], it is shown that if $C(h)$ is a parametric family of matrices which has a simple eigenvalue $\lambda(h_0)$ at h_0, then there exists a neighbourhood of h_0 in which the family $C(h)$ has an eigenvalue function $\lambda(h)$ and right and left eigenvector functions that are both analytic functions in the neighbourhood. They further show that if $\hat{v}(h)$ is an eigenvector function which is analytic in the neighbourhood and we choose any analytic vector function $z(h)$ in the neighbourhood such that $z(h_0)^*\hat{v}(h_0) = 1$, then the components of the vector function $v(h) = \hat{v}(h)/(z(h)^*\hat{v}(h))$ are analytic functions in the neighbourhood. This result will be the starting point of our analysis in section 3.4. By choosing different analytic vector functions $z(h)$, we will be able to generate the derivatives of the Perron vector under different normalisations.

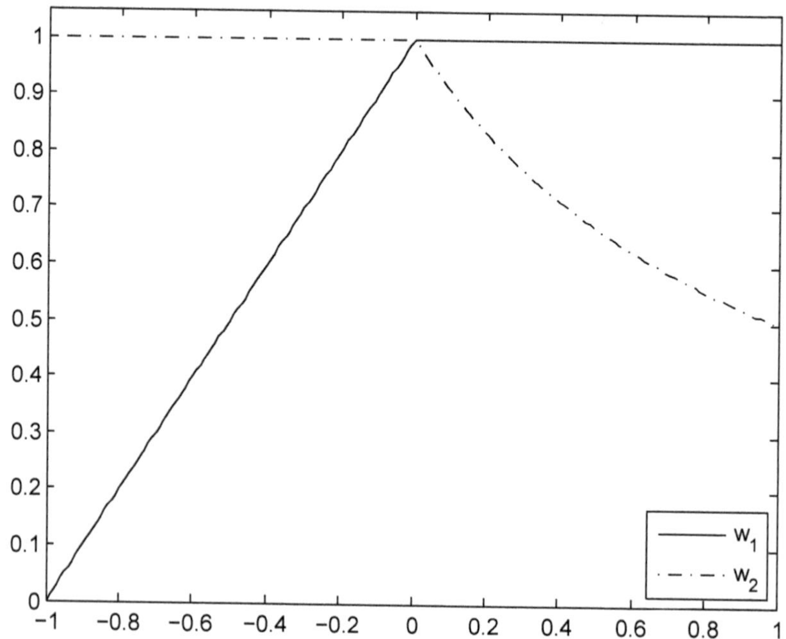

FIGURE 3.1
Graphs for $w_1(h)$ and $w_2(h), h \in [-1, 1]$

We now make a mild adjustment to our basic assumptions. Most of the applications that we will consider in this book involve nonnegative matrices rather than the less restrictive essentially nonnegative matrices. Consequently, we will generally assume that our matrices are nonnegative rather than essentially nonnegative, unless it is necessary to do otherwise. That said, we note that many of our results will continue to hold for the class $\tilde{\Phi}^{n,n}$. Henceforth, for a matrix $A \in \mathbb{R}^{n,n}$ we understand the inequality $A \geq 0$ to hold entrywise. (A similar notation applies for vectors $u, v \in \mathbb{R}^n$, where the inequalities $u \geq v, u > v$ are interpreted entrywise.) With this notation in place, we make the following definition:

$$\Phi^{n,n} = \{A \in \mathbb{R}^{n,n} \mid A \geq 0 \text{ and } A \text{ is irreducible}\}.$$

The spectral abscissa of $A \in \Phi^{n,n}$ now becomes the spectral radius of A and we will denote it by $r(A)$. We will continue to refer to $r(A)$ as the Perron value of A.

3.2 First and second derivatives of the Perron value

In this section we continue with the notation developed in the previous section and use the preliminaries obtained there to develop formulas for the first and second partial derivatives of the Perron value with respect to the matrix entries at a given matrix. For a function f of a matrix B, we will use the notation $\frac{\partial f(B)}{\partial_{i,j}}$ to denote the partial derivative of f with respect to the (i,j)-th entry of B. A similar notation is employed for higher order derivatives.

Suppose that $A \in \Phi^{n,n}$, let $B \in \mathcal{N}_A$, and consider the eigenvalue–eigenvector relation:

$$Bx(B) = \lambda(B)x(B),$$

where we assume that $x(B)$ is an eigenvector that is analytic in a suitable neighbourhood of B. If we vary $b_{i,j}$ and keep all other entries fixed, we see that $\frac{\partial B}{\partial_{i,j}} = E_{i,j}$. Let $y(B)$ be a left eigenvector of B corresponding to $\lambda(B)$, and normalise $y(B)$ so that $y(B)^t x(B) = 1$. Differentiating both sides of the eigenvalue–eigenvector equation with respect to the (i,j)–th entry, $i,j = 1,\dots,n$, we obtain that

$$\frac{\partial B}{\partial_{i,j}} x(B) + B\frac{\partial x(B)}{\partial_{i,j}} = \frac{\partial \lambda(B)}{\partial_{i,j}} x(B) + \lambda(B)\frac{\partial x(B)}{\partial_{i,j}}. \tag{3.1}$$

Substituting the relation $\frac{\partial B}{\partial_{i,j}} = E_{i,j}$ into (3.1), and then left–multiplying both sides of (3.1) by $y(B)^t$, we find that

$$y_i(B)x_j(B) = \frac{\partial \lambda(B)}{\partial_{i,j}}.$$

Finally, on substituting $B = A$, we have

$$\frac{\partial \lambda(A)}{\partial_{i,j}} = y_i(A)x_j(A).$$

As A is an irreducible nonnegative matrix, $\lambda(A)$ coincides with $r(A)$, so we find that

$$\frac{\partial r(A)}{\partial_{i,j}} = y_i(A)x_j(A). \tag{3.2}$$

In particular, we deduce that if we increase any entry in A, the Perron value of A will strictly increase.

Setting $Q = r(A)I - A$, we find from (vi) of section 2.2, that

$$QQ^{\#} = I - x(A)y(A)^t. \tag{3.3}$$

Consequently, we see that we can also express the derivative of the Perron value in terms of the product $QQ^{\#}$, i.e.,

$$\frac{\partial r(A)}{\partial_{i,j}} = (I - QQ^{\#})_{j,i}.$$

We note in passing that the representation above of the first derivative of the Perron value, or indeed of any simple eigenvalue of a matrix, in terms of right and left eigenvectors whose inner product is 1 can be found in many places, for example [109, p. 305], and Vahrenkamp's paper [110].

One of the main goals of this section is to produce an expression for the second partial derivative of $r(A)$, the Perron value function at A. The following result, due to Deutsch and Neumann in [30], shows how the group inverse of the M–matrix $r(A)I - A$ plays a central role in that expression.

THEOREM 3.2.1 *Let $A \in \Phi^{n,n}$ and put $Q = r(A)I - A$. Then for all $i, j, k, \ell = 1, \ldots, n$,*

$$\frac{\partial^2 r(A)}{\partial_{i,j}\partial_{k,\ell}} = (I - QQ^{\#})_{\ell,i} q^{\#}_{j,k} + (I - QQ^{\#})_{j,k} q^{\#}_{\ell,i}. \tag{3.4}$$

__Proof__: We follow the development and notation leading to (3.1). In particular, $x(B)$ is an analytic right eigenvector corresponding to $\lambda(B)$, while $y(B)$ is a corresponding left eigenvector, normalised so that $y(B)^t x(B) = 1$. Let $1 \leq i, j \leq n$ and consider the equality (3.1). It is clear that for any indices $1 \leq k, \ell \leq n$, $\frac{\partial^2 B}{\partial_{i,j}\partial_{k,\ell}} = 0$ and so if we differentiate both sides of (3.1) with respect to the (k, ℓ)-th entry we obtain

$$\begin{aligned} E_{i,j}\frac{\partial x(B)}{\partial_{k,\ell}} &+ E_{k,\ell}\frac{\partial x(B)}{\partial_{i,j}} + B\frac{\partial^2 x(B)}{\partial_{i,j}\partial_{k,\ell}} \\ &= \frac{\partial^2 \lambda(B)}{\partial_{i,j}\partial_{k,\ell}} x(B) + y(B)_i x(B)_j \frac{\partial x(B)}{\partial_{k,\ell}} \\ &+ y(B)_k x(B)_\ell \frac{\partial x(B)}{\partial_{i,j}} + \lambda(B)\frac{\partial^2 x(B)}{\partial_{i,j}\partial_{k,\ell}}. \end{aligned} \tag{3.5}$$

Premultiplying both sides of (3.5) by $y(B)^t$ and recalling that $y(B)^t x(B) = 1$, we find after a rearrangement that

$$\begin{aligned}\frac{\partial^2 \lambda(B)}{\partial_{i,j}\partial_{k,\ell}} &= y(B)_i \left[\left(\frac{\partial x(B)}{\partial_{k,\ell}} \right)_j - x(B)_j y(B)^t \frac{\partial x(B)}{\partial_{k,\ell}} \right] \\ &+ \; y(B)_k \left[\left(\frac{\partial x(B)}{\partial_{i,j}} \right)_\ell - x(B)_\ell y(B)^t \frac{\partial x(B)}{\partial_{i,j}} \right].\end{aligned} \tag{3.6}$$

Next, we seek an alternative expression for the differences in the square brackets above. Set $\hat{Q} = \lambda(B)I - B$. Then for any pair of indices $1 \le \mu, \nu \le n$, we obtain, from (3.1) after the substitution of $i = \mu$ and $j = \nu$, that

$$\hat{Q} \frac{\partial x(B)}{\partial_{\mu,\nu}} = E_{\mu,\nu} x(B) - y(B)_\mu x(B)_\nu x(B).$$

Premultiplying both sides of this equality by $\hat{Q}^\#$ and taking account of the facts that $\hat{Q}^\# x(B) = 0$ and $\hat{Q}^\# \hat{Q} = \hat{Q}\hat{Q}^\#$ leaves us with

$$\hat{Q}\hat{Q}^\# \frac{\partial x(B)}{\partial_{\mu,\nu}} = \hat{Q}^\# E_{\mu,\nu} x(B),$$

or, alternatively,

$$\frac{\partial x(B)}{\partial_{\mu,\nu}} - \left(I - \hat{Q}\hat{Q}^\# \right) \frac{\partial x(B)}{\partial_{\mu,\nu}} = \hat{Q}^\# E_{\mu,\nu} x(B).$$

Since $x(B)y(B)^t = I - \hat{Q}\hat{Q}^\#$, for any $1 \le \eta \le n$, the η-th components on both sides of the above display satisfy

$$\left(\frac{\partial x(B)}{\partial_{\mu,\nu}} \right)_\eta - x(B)_\eta y(B)^t \frac{\partial x(B)}{\partial_{\mu,\nu}} = \hat{q}^\#_{\eta,\mu} x(B)_\nu.$$

Substituting $\mu = k$, $\nu = \ell$, and $\eta = j$ followed by $\mu = i$, $\nu = j$, and $\eta = \ell$ in (3.6) this equality becomes:

$$\frac{\partial^2 \lambda(B)}{\partial_{i,j}\partial_{k,\ell}} = \left(I - \hat{Q}\hat{Q}^\# \right)_{\ell,i} \hat{q}^\#_{j,k} + \left(I - \hat{Q}\hat{Q}^\# \right)_{j,k} \hat{q}^\#_{\ell,i}.$$

To obtain (3.4), we now substitute $B = A$, so that $\lambda(B) = \lambda(A) = r(A)$, and $\hat{Q} = Q$. □

REMARK 3.2.2 Suppose that A is an irreducible nonnegative matrix of order n with right and left Perron vectors x, y, respectively, normalised so that $y^t x = 1$. Set $Q = r(A)I - A$, and recall that $QQ^\# = I - xy^t$. We find readily that for indices $i, j, k, \ell = 1, \dots, n$, we can recast (3.4) as follows:

$$\frac{\partial^2 r(A)}{\partial_{i,j} \partial_{k,\ell}} = x_\ell y_i q^\#_{j,k} + x_j y_k q^\#_{\ell,i}.$$

3.3 Concavity and convexity of the Perron value

From (3.2), we find that the Perron value $r(A)$ of an irreducible nonnegative matrix A is increasing in each entry of A. Is $r(A)$ concave up or concave down as a function of a particular entry of A? In order to address that question, we need to consider the sign of the second derivative of $r(A)$. Fortunately, equation (3.4) (or equivalently, Remark 3.2.2) provides us with a simple test for determining whether the Perron value as a function of a matrix entry at a matrix is concave up or concave down.

COROLLARY 3.3.1 ([30]) *Let $A \in \Phi^{n,n}$ and put $Q = r(A)I - A$. Then for all $i, j = 1, \dots, n$,*

$$\frac{\partial^2 r(A)}{\partial^2_{i,j}} = 2(I - QQ^\#)_{j,i} q^\#_{j,i}. \tag{3.7}$$

In particular, the Perron value is a concave up function of the (i,j)–th entry at A if $q^\#_{j,i} > 0$ and it is concave down if $q^\#_{j,i} < 0$.

Several authors, including Cohen [24], Elsner [35], and Friedland [43], have proved by means very different than ours here that, as a function of the diagonal entries, the Perron value is concave up. This same result will follow if we can show that the diagonal entries of $Q^\#$ are all positive. That fact has been shown in [94]; below we give two alternative proofs of the fact that $Q^\#$ has positive diagonal entries.

Suppose that we have an irreducible nonnegative matrix A of order n, and let $Q = r(A)I - A$. By appealing to Theorem 2.5.5, we find that $Q^\# \in \mathcal{P}_0^n$, so that all principal minors of $Q^\#$ of order less than n are positive; in particular we see that the diagonal entries of $Q^\#$ must be positive. For a more computational approach to the positivity of the diagonal entries of $Q^\#$, consider for concreteness the (n, n) entry of $Q^\#$.

According to Observation 2.3.4 (and substituting Q for B in the notation of that result) we have

$$q^{\#}_{n,n} = \frac{\overline{y}^t Q^{-1}_{1,1} \overline{x} x_n y_n}{y^t x},$$

where x and y are right and left Perron vectors for A and where $Q_{1,1}$ is the leading principal submatrix of Q of order $n-1$. As $Q_{1,1}$ is a nonsingular M–matrix, its inverse is nonnegative, and it now follows readily that $q^{\#}_{n,n} > 0$. As the choice of the index n was arbitrary, we find that every diagonal entry of $Q^{\#}$ is positive. Via either argument, we establish that the diagonal of $Q^{\#}$ is positive, and hence that the Perron value $r(A)$ is concave up as a function of any diagonal entry of A.

Our next result provides a contrasting conclusion for the Perron value as a function of the off-diagonal entries.

LEMMA 3.3.2 ([30]) *Let $A \in \Phi^{n,n}$. Then in each row and each column of A there exists an off-diagonal entry $a_{i,j}$ such that at A, the Perron value is concave down function as a function of $a_{i,j}$.*

Proof: Set $Q = r(A)I - A$. By Corollary 3.3.1, the sign of the second partial derivative of the Perron value with respect to the (i, j)–th entry coincides with the sign of the (j, i) entry of $Q^{\#}$. Let x and y be right and left Perron vectors of A, respectively, so that $Q^{\#}x = 0$ and $y^t Q^{\#} = 0^t$. As both x and y are positive, and as the diagonal entries of $Q^{\#}$ are all positive, $Q^{\#}$ must have a negative off-diagonal entry in each row and in each column. □

Next, we explore a quantitative relationship between the diagonal and off-diagonal entries of group inverses of singular and irreducible M-matrices. To motivate this exploration, we recall the following. It is a well-known property of an invertible row diagonally dominant M-matrix B that its inverse C, which is known to be a nonnegative matrix, is *elementwise column diagonally dominant*; that is

$$c_{j,j} \geq c_{i,j}, \quad \text{for all } i \neq j, \ j = 1, \ldots, n. \tag{3.8}$$

This property, due to Metzler [93], can be generalised to any nonsingular M-matrix B as follows. Any invertible M-matrix B admits a scaling of its columns by a positive diagonal matrix D so that the M-matrix $\tilde{B} = BD$ is row diagonally dominant, see [12, Condition M$_{37}$, p.137]. Setting $\tilde{C} = \tilde{B}^{-1}$ and applying Metzler's property to

$$\tilde{C} = (\tilde{c}_{i,j}) = \tilde{B}^{-1} = D^{-1}B^{-1} = D^{-1}C,$$

we obtain via (3.8) that

$$\frac{c_{j,j}}{d_{j,j}} = \tilde{c}_{j,j} \geq \tilde{c}_{i,j} = \frac{c_{i,j}}{d_{i,i}}, \quad \text{for all } i \neq j, \ j = 1, \ldots, n. \tag{3.9}$$

We now present a generalisation of (3.9) to singular and irreducible M-matrices. The result is due to Deutsch and Neumann [31].

THEOREM 3.3.3 *Let* $Q \in \mathbb{R}^{n,n}$ *be a singular and irreducible M–matrix. Let* $Q^{\#}$ *be its group inverse and suppose that* $x = \begin{bmatrix} x_1 & \ldots & x_n \end{bmatrix}^t$ *is a positive right null vector of* Q*. Then*

$$\frac{q^{\#}_{j,j}}{x_j} > \frac{q^{\#}_{i,j}}{x_i}, \quad \text{for all } i \neq j, \ j = 1, \ldots, n. \tag{3.10}$$

In particular, if $Q\mathbf{1} = 0$*, then*

$$q^{\#}_{j,j} > q^{\#}_{i,j}, \quad \text{for all } i \neq j, \ j = 1, \ldots, n.$$

Proof: Without loss of generality, we take $j = n$. Let y be a left null vector for Q, normalised so that $y^t x = 1$, let $\overline{x}$ be the vector formed from x by deleting its last entry, and let $Q_{1,1}$ denote the leading principal submatrix of Q of order $n-1$. As usual, for each $j = 1, \ldots, n$, we denote the j-th standard unit basis vector in $\mathbb{R}^n$ by e_j. From (2.13), we find that

$$\frac{q^{\#}_{n,n}}{x_n} - \frac{q^{\#}_{i,n}}{x_i} = \left(\frac{1}{x_n}e_n - \frac{1}{x_i}e_i\right)^t Q^{\#} e_n = \frac{y_n e_i^t (Q_{1,1})^{-1}\overline{x}}{x_i} > 0.$$

□

We saw above that for an irreducible nonnegative matrix A, $r(A)$ is concave up as a function of each diagonal entry of A. In [43], it is shown further that $r(A)$ is in fact a concave up function of the entire diagonal. That is, if $A \in \Phi^{n,n}$ and $D \in \mathbb{R}^{n,n}$ is any diagonal matrix, then for each $h \in [0, 1]$,

$$r(hA + (1-h)(A+D)) \leq hr(A) + (1-h)r(A+D). \tag{3.11}$$

Friedland shows also that equality holds in this inequality if and only if $D = \alpha I$, for some scalar α. We will reestablish these facts using a different approach than the one used in [43]. Thinking of the Perron value as a function of the vector $\begin{bmatrix} a_{1,1} & \ldots & a_{n,n} \end{bmatrix}^t \in \mathbb{R}^n$, we consider the Hessian of the transformation. From (3.4), we find that the Hessian H is given by

$$H \equiv \left[\frac{\partial^2 r(A)}{\partial_{i,i}\partial_{j,j}}\right] = (I - QQ^{\#})^t \circ Q^{\#} + Q^{\#t} \circ (I - QQ^{\#}), \tag{3.12}$$

where for square matrices M_1, M_2 of the same order, $M_1 \circ M_2$ denotes their entrywise (or Hadamard) product. The following lemma shows that H a positive semidefinite matrix.

LEMMA 3.3.4 ([30]) *Suppose that $A \in \mathbb{R}^{n,n}$ is an irreducible essentially nonnegative matrix, with right and left Perron vectors x and y respectively, normalised so that $y^t x = 1$. Define $Q = r(A)I - A$, and let H be given by (3.12). Then H is a positive semidefinite matrix whose null space is spanned by* **1**.

Proof: Let $X = \text{diag}(x), Y = \text{diag}(y)$, and set $G = YQX$. As in the proof of Theorem 2.5.5, we find that $G + G^t$ is a positive semidefinite M-matrix; further its null space is spanned by **1**. Setting $K = YQ^{\#}X$ and again appealing to the proof of Theorem 2.5.5, we deduce that for each $u \in \mathbb{R}^n$, there exists a vector $v \in \mathbb{R}^n$ such that $u^t K^t u = v^t G v$. Indeed, we may take $v = X^{-1}Q^{\#}Xu$. It now follows that since $G + G^t$ is positive semidefinite, so is $H = K + K^t$.

Suppose now that u is a null vector for H. Then necessarily the vector $v = X^{-1}Q^{\#}Xu$ is a null vector for $G + G^t$. Consequently, v must be a scalar multiple of **1**. Hence we must have $Q^{\#}Xu = ax$ for some scalar a, so that $QQ^{\#}Xu = (I - xy^t)Xu = 0$. As the null space of $I - xy^t$ is spanned by x, we find readily that u must be a scalar multiple of **1**. □

The preceding observations help to furnish a proof of Friedland's result in [43] using the group inverse as the basis of our approach.

THEOREM 3.3.5 *Let $A \in \Phi^{n,n}$ and let $D \in R^{n,n}$ be a diagonal matrix. Then for all $h \in [0, 1]$*

$$r(hA + (1 - h)(A + D)) \leq hr(A) + (1 - h)r(A + D). \qquad (3.13)$$

Equality holds in (3.13) *for some $h \in (0, 1)$ if and only if $D = \alpha I$, for some scalar $\alpha \in \mathbb{R}$.*

Proof: Let u be the vector in $\mathbb{R}^n$ such that $D = \text{diag}(u)$, and set

$$g(h) = r(hA + (1 - h)(A + D)) = r(A + (1 - h)D).$$

For each $h \in [0, 1]$, let

$$Q_h^{\#} = (r(hA + (1 - h)D)I - (hA + (1 - h)D))^{\#}.$$

Finally, for each $h \in [0, 1]$, we define H_h as

$$H_h = (I - Q_h Q_h^{\#})^t \circ Q_h^{\#} + Q_h^{\#t} \circ (I - Q_h Q_h^{\#}).$$

From the remarks leading up to (3.12), we find that for each $h \in [0,1]$, the Hessian of $g(h)$ is equal to H_h. Consequently it follows that for each such h,

$$\frac{d^2 g(h)}{dh^2} = u^t H_h u. \tag{3.14}$$

By Lemma 3.3.4, H_h is positive semidefinite, and so we find that for each $h \in [0,1]$, $u^t H_h u \geq 0$. Thus $g(h)$ is a concave up function of h on the interval $[0,1]$, from which (3.13) follows readily.

Next we consider the case of equality in (3.13). If D is a multiple of the identity matrix, say $D = \alpha I$, then for each $h \in [0,1]$ we have

$$r(hA + (1-h)(A+D)) = r(A) + (1-h)\alpha =$$
$$hr(A) + (1-h)(r(A)+\alpha) = hr(A) + (1-h)r(A+D),$$

so that equality holds in (3.13) for every $h \in [0,1]$.

On the other hand, if D is not a scalar multiple of the identity matrix, then the vector u is not a scalar multiple of $\mathbf{1}$. By Lemma 3.3.4, for each $h \in [0,1]$, the null space of H_h is spanned by the all ones vector. It now follows that for every $h \in [0,1]$,

$$\frac{d^2 g(h)}{dh^2} = u^t H_h u > 0.$$

Hence $g(h)$ is strictly concave up on $[0,1]$, and we readily deduce that strict inequality holds in (3.13) whenever $0 < h < 1$. □

As already noted in this section, the Perron value of an irreducible nonnegative matrix A is a concave up function of each diagonal entry; further the Perron value is concave down in at least one off-diagonal entry in every row and column. It is natural to wonder when the Perron value is a concave down function of every off-diagonal entry. From Corollary 3.3.1 we observe that this is equivalent to asking for all the off-diagonal entries of $(r(A)I - A)^\#$ to be negative. We investigate a weaker version of this condition by asking for the off-diagonal entries of $(r(A)I - A)^\#$ to be nonpositive. Since the diagonal entries of $(r(A)I - A)^\#$ are positive and since $(r(A)I - A)^\# x = 0$ for any positive Perron vector x for A, we can conclude from [12, Exercise 6.4.14] that $(r(A)I - A)^\#$ is an M-matrix if and only if each of its off-diagonal entries is nonpositive. The following two examples illustrate some of the possible behaviours of $(r(A)I - A)^\#$.

EXAMPLE 3.3.6 Consider the irreducible matrix

$$A = \begin{bmatrix} 0 & 1 & 0 \\ 0 & 0 & 1 \\ 1 & 0 & 0 \end{bmatrix}.$$

It is straightforward to verify that

$$(I - A)^{\#} = \begin{bmatrix} \frac{1}{3} & 0 & -\frac{1}{3} \\ -\frac{1}{3} & \frac{1}{3} & 0 \\ 0 & -\frac{1}{3} & \frac{1}{3} \end{bmatrix},$$

which evidently is an M-matrix.

EXAMPLE 3.3.7 Let

$$A = \begin{bmatrix} 0 & 1 & 0 & 0 \\ 0 & 0 & 1 & 0 \\ 0 & 0 & 0 & 1 \\ 1 & 0 & 0 & 0 \end{bmatrix}.$$

A computation shows that

$$Q^{\#} = \begin{bmatrix} \frac{3}{8} & \frac{1}{8} & -\frac{1}{8} & -\frac{3}{8} \\ -\frac{3}{8} & \frac{3}{8} & \frac{1}{8} & -\frac{1}{8} \\ -\frac{1}{8} & -\frac{3}{8} & \frac{3}{8} & \frac{1}{8} \\ \frac{1}{8} & -\frac{1}{8} & -\frac{3}{8} & \frac{3}{8} \end{bmatrix}$$

which is not an M–matrix.

We pose the following problem, which was first stated in [30].

QUESTION 3.3.8 *Characterise all nonnegative and irreducible matrices A for which $Q^{\#} = (r(A)I - A)^{\#}$ is an M-matrix.*

In Chapter 6 we will provide one answer to this question using mean first passage times for Markov chains. However, without appealing to mean first passage times, the authors of this book have studied several classes of matrices A in $\Phi^{n,n}$ for which the associated M-matrix $Q = r(A)I - A$ has a group inverse which is an M-matrix. For example, if x and y are positive n-vectors such that $y^t x = 1$, we find readily that for the rank one matrix $A = xy^t$, we have $(I - A)^{\#} = I - xy^t$, which is certainly an M-matrix. The following observation, which appears in the paper of Kirkland and Neumann [80], is inspired in part by the fact that the matrix xy^t has just one nonzero eigenvalue.

OBSERVATION 3.3.9 *Let $A \in \Phi^{n,n}$ be a diagonalisable matrix with Perron value equal to 1, and suppose that A has 2 distinct nonzero eigenvalues 1 and σ, the latter of multiplicity k. Let E_λ denote the eigenprojection of A corresponding to the eigenvalue λ, for $\lambda = 0, 1, \sigma$. Then*

$$(I-A)^{\#} = I + \frac{k}{k+1-\text{trace}(A)}A - \frac{2k+1-\text{trace}(A)}{k+1-\text{trace}(A)}E_1.$$

In particular, $(I-A)^{\#}$ is an M-matrix if and only if

$$a_{i,j} \leq \left[\frac{(2k+1)-\text{trace}(A)}{k}\right](E_1)_{i,j}, \quad \text{for all } i \neq j,\ i,j = 1, \ldots, n. \tag{3.15}$$

Proof: From the hypothesis, it follows that σ must be real and that A has the spectral resolution

$$A = E_1 + \sigma E_\sigma.$$

Moreover, it is known that

$$I = E_1 + E_\sigma + E_0.$$

Hence,

$$I - A = (1-\sigma)E_\sigma + E_0.$$

But then it follows from (2.2) and (2.3) that

$$\begin{aligned}(I-A)^{\#} &= \frac{1}{1-\sigma}E_\sigma + E_0 = I - E_1 + \frac{\sigma}{1-\sigma}E_\sigma \\ &= I - \frac{1}{\sigma-1}\left[A - (2-\sigma)E_1\right] \\ &= I + \frac{k}{k+1-\text{trace}(A)}A - \frac{2k+1-\text{trace}(A)}{k+1-\text{trace}(A)}E_1,\end{aligned}$$

where the last equality follows since $\text{trace}(A) = k\sigma + 1$. Thus the off-diagonal entries of $(I-A)^{\#}$ are nonpositive if and only if (3.15) holds. We deduce that condition (3.15) is equivalent to $(I-A)^{\#}$ being an M–matrix. □

EXAMPLE 3.3.10 Let $m \in \mathbb{N}$, and consider the stochastic matrix

$$A = \left[\begin{array}{c|c} 0 & \frac{1}{m}J \\ \hline \frac{1}{m}J & 0 \end{array}\right],$$

where J denotes an $m \times m$ all ones matrix. It is straightforward to determine that A is diagonalisable, and has three eigenvalues: $1, -1$ and 0, the last with multiplicity $2m-2$. Further, the eigenprojection matrix E_1 has every entry equal to $\frac{1}{2m}$. Referring to Observation 3.3.9, we find that for this example, (3.15) is equivalent to $a_{i,j} \le \frac{3}{2m}$ for all $i \ne j$. Thus Observation 3.3.9 implies that $(I-A)^{\#}$ is an M–matrix; indeed it is readily verified that

$$(I-A)^{\#} = \left[\begin{array}{c|c} I - \frac{3}{4m}J & -\frac{1}{4m}J \\ \hline -\frac{1}{4m}J & I - \frac{3}{4m}J \end{array}\right].$$

In Chapter 4 we will apply our results on the convexity and concavity of the Perron value as a function of the matrix entries to the Leslie model of population growth.

3.4 First and second derivatives of the Perron vector

As a continuation of our results on the derivatives of the Perron value, in this section we develop results on the first and second order derivatives of the Perron vector as a function of the matrix entries. To this end we will use the results of [5], as described in section 3.1. Since we will be working with a family of matrices in $\Phi^{n,n}$, it is not essential to develop our results in neighbourhoods. However, we will do so in order to be consistent with other parts of the book.

Let $\mathcal{J}$ be an interval in $\mathbb{R}$ and let $A(h)$, $h \in \mathcal{J}$, be a family of irreducible nonnegative matrices, namely, a family in $\Phi^{n,n}$. Let h_0 be an arbitrary, but fixed, value of h in $\mathcal{J}$. Then by [5, Theorem 2.1], there exists a neighbourhood of h_0 in which $r(h) = r(A(h))$ and $x(h) = x(A(h))$ are, respectively, an analytic eigenvalue function, and a corresponding analytic eigenvector function. Finally, let $z(h)$ be a real vector whose entries are analytic functions of h such that $z(h)^t x(h) = 1$ throughout $\mathcal{J}$. Then, from the relation

$$A(h)x(h) = r(h)x(h)$$

we have, in a similar fashion to (3.1), that upon differentiating both sides with respect to h,

$$\frac{dA(h)}{dh}x(h) + A\frac{dx(h)}{dh} = \frac{dr(h)}{dh}x(h) + r(h)\frac{dx(h)}{dh}.$$

On setting $Q(h) = r(h)I - A(h)$ and rearranging sides, we find that

$$Q(h)\frac{dx(h)}{dh} = \frac{dA(h)}{dh}x(h) - \frac{dr(h)}{dh}x(h), \quad \text{for all } h \in \mathcal{J}. \tag{3.16}$$

Multiplying (3.16) by $Q^{\#}(h)$, it now follows that for some scalar $\alpha(h)$ (which will be determined momentarily) we have

$$\begin{aligned}\frac{dx(h)}{dh} &= Q^{\#}(h)\frac{dA(h)}{dh}x(h) - \frac{dr(h)}{dh}Q^{\#}(h)x(h) + \alpha(h)x(h)\\ &= Q^{\#}(h)\frac{dA(h)}{dh}x(h) + \alpha(h)x(h).\end{aligned} \tag{3.17}$$

Premultiplying both sides of (3.17) by $z(h)$ and making use of the constraint that $z(h)^t x(h) = 1$, which implies that $z(h)^t \frac{dx(h)}{dh} = -\left(\frac{dz(h)}{dh}\right)^t x(h)$, we obtain

$$\alpha(h) = -z(h)^t Q^{\#}(h)\frac{dA(h)}{dh}x(h) - \left(\frac{dz(h)}{dh}\right)^t x(h).$$

If we now substitute this expression for $\alpha(h)$ in (3.17) we obtain the following representation for the derivative of the Perron vector:

$$\begin{aligned}\frac{dx(h)}{dh} &= Q^{\#}(h)\frac{dA(h)}{dh}x(h)\\ &\quad - \left(\left(\frac{dz(h)}{dh}\right)^t x(h)\right)x(h) - \left(z(h)^t Q^{\#}(h)\frac{dA(h)}{dh}x(h)\right)x(h).\end{aligned}$$

We now show how different choices of $z(h)$ lead to formulas for $\frac{dx(h)}{dh}$ subject to different normalisations of $x(h)$. We comment that for all the cases that we will consider here it will turn out that $\left(\frac{dz(h)}{dh}\right)^t x(h) = 0$.

Case 1: Let $p > 0$ be any number and normalise $x(h) = \begin{bmatrix} x_1(h) & \dots & x_n(h) \end{bmatrix}^t$ so that $\sum_{i=1}^n x_i^p(h) = 1$ (i.e., $\|x(h)\|_p = 1$). Next, letting $z(h) = \begin{bmatrix} x_1(h)^{p-1} & \dots & x_n(h)^{p-1} \end{bmatrix}^t$, we have $z^t(h)x(h) = 1$ and, as $\sum_{i=1}^n x_i^p(h) = 1$,

$$\left(\frac{dz(h)}{dh}\right)^t x(h) = \sum_i^n \left[(p-1)x_i^{p-2}(h)\frac{dx_i(h)}{dh}\right] x_i(h) =$$

$$\sum_{i=1}^n (p-1)x_i^{p-1}(h)\frac{dx_i(h)}{dh} = 0.$$

Let us consider two natural choices for p.

Subcase 1(i). Suppose that $p = 1$, so that the 1-norm of $x(h)$ is unity, a subcase which will be of interest to us when we consider applications to Markov chains. Then $z(h) = \mathbf{1}$ and so we find that

$$\frac{dx(h)}{dh} = Q^{\#}(h)\frac{dA(h)}{dh}x(h) - \left(\mathbf{1}^t Q^{\#}(h)\frac{dA(h)}{dh}x(h)\right)x(h). \tag{3.18}$$

In the special case that $A(h)$ has all column sums equal to 1 throughout $\mathcal{J}$, (3.18) simplifies as follows:

$$\frac{dx(h)}{dh} = Q^{\#}(h)\frac{dA(h)}{dh}x(h). \tag{3.19}$$

Subcase 1(ii). Suppose that $p = 2$, so that $x(h)$ is normalised to have Euclidean length 1. Let us choose $z(h) = x(h)$, $h \in \mathcal{J}$. Let $y(h)$ be a left Perron vector of $A(h)$ normalised so that $y(h)^t x(h) = 1$, in which case it also holds that $y(h)^t Q^{\#}(h) = 0$. Then on premultiplying both sides of (3.18) by $y(h)^t$ we obtain

$$y(h)^t\frac{dx(h)}{dh} = -x(h)^t Q^{\#}(h)\frac{dA(h)}{dh}x(h). \tag{3.20}$$

In the special case that the left and right Perron vectors of $A(h)$ coincide on $\mathcal{J}$ (for example if $A(h)$ is symmetric) then $y(h) = z(h)$, showing that $y(h)^t \frac{dx(h)}{dh} = 0$. A substitution of (3.20) into (3.18) yields that

$$\frac{dx(h)}{dh} = Q^{\#}(h)\frac{dA(h)}{dh}x(h).$$

Case 2: Here we assume that, in addition to $z(h)^t x(h) = 1$, $z(h)$ is a constant vector throughout $\mathcal{J}$. Then, again, $\left(\frac{dz(h)}{dh}\right)^t x(h) = 0$ and, once more, (3.18) reduces to

$$\frac{dx(h)}{dh} = Q^{\#}(h)\frac{dA(h)}{dh}x(h) - \left(z(h)^t Q^{\#}(h)\frac{dA(h)}{dh}x(h)\right)x(h). \tag{3.21}$$

On taking $z = \mathbf{1}$ we note that Subcase 1(i) is also a special case of Case 2.

Of special interest to us is to examine the effect upon the Perron vector when only one row, say the first row, of A is perturbed, and when only one entry in $z(h)$, say $z_1(h)$, is held to be a nonzero constant, say, $\delta > 0$. That is, we will let $z(h) = \delta e_1$. Since $z(h)^t x(h) = 1$, throughout $\mathcal{J}$, this choice of $z(h)$ forces $x_1(h)$ to be held constant at $\frac{1}{\delta}$ throughout the interval. Furthermore, as the last $n-1$ rows of $\frac{dA(h)}{dh}$ are all zero, we see that $\frac{dA(h)}{dh}x(h) = \gamma(h)e_1$, for some function $\gamma(h)$. Thus

$$z(h)^t Q^{\#}(h)\frac{dA(h)}{dh}x(h) = \gamma(h)\delta q^{\#}_{1,1}(h) = \frac{\gamma(h)}{x_1(h)}q^{\#}_{1,1}(h). \tag{3.22}$$

Now, since the first entry of $x(h)$ is held a constant, $\frac{dx_1(h)}{dh} = 0$ and so our interest focuses on the truncated derivative vector $\frac{d\bar{x}(h)}{dh} \equiv \begin{bmatrix} \frac{dx_2(h)}{dh} & \cdots & \frac{dx_n(h)}{dh} \end{bmatrix}^t$. From (3.21) and (3.22) we find that

$$\frac{d\bar{x}(h)}{dh} = \gamma(h)\left(\overline{Q^{\#}(h)e_1} - \frac{q^{\#}_{1,1}(h)}{x_1(h)}\bar{x}(h)\right), \tag{3.23}$$

where $\overline{Q^{\#}(h)e_1}$ denotes the vector formed from $Q^{\#}e_1$ by deleting its first entry. Representing (3.23) component by component, yields

$$\frac{dx_i(h)}{dh} = \gamma(h)\left(q^{\#}_{i,1}(h) - \frac{q^{\#}_{1,1}(h)}{x_1(h)}x_i(h)\right), \quad \text{for all } i = 2, \ldots, n. \tag{3.24}$$

Next, we provide an alternate expression for $\frac{d\bar{x}(h)}{dh}$. We begin by partitioning $Q(h) = r(h)I - A(h)$ as:

$$Q(h) = \left[\begin{array}{c|c} q_{1,1}(h) & Q_{1,2}(h) \\ \hline Q_{2,1}(h) & Q_{2,2}(h) \end{array}\right],$$

where $Q_{2,2}(h) \in \mathbb{R}^{n-1,n-1}$. Recalling that the last $n-1$ rows of $\frac{dA(h)}{dh}$ are zero, we obtain from (3.16) the following equality

$$Q_{2,2}(h)\frac{d\bar{x}(h)}{dh} = -\frac{dr(h)}{dh}\bar{x}(h). \tag{3.25}$$

Now $Q_{2,2}(h)$ is a principal submatrix of order $n-1$ of a singular and irreducible M–matrix, so it is invertible and has a nonnegative inverse. Whence

$$\frac{d\bar{x}(h)}{dh} = -\frac{dr(h)}{dh}Q_{2,2}^{-1}(h)\bar{x}(h) \text{ for all } h \in \mathcal{J}. \tag{3.26}$$

In particular, if each entry in the first row of $A(h)$ is a nondecreasing function of h, and at least one of those entries is strictly increasing in h, it follows that $\frac{d\bar{x}(h)}{dh} < 0$ for all $h \in \mathcal{J}$. Thus, we find that in this case, all entries of $\overline{x(h)}$ are decreasing in h.

We note that (3.23) is derived in [31], while (3.26) is a result of Elsner, Johnson, and Neumann [36]. We now combine (3.23) and (3.26) in one theorem summarising our results on the derivative of the Perron vector when one of its entries is held fixed and $A(h)$ changes only in a corresponding row.

THEOREM 3.4.1 *Suppose that $A(h) \in \Phi^{n,n}$, $h \in \mathcal{J}$, is a family of matrices such that all their entries outside row ℓ are a constant and in row ℓ they are analytic functions of h. Let $z(h) = \delta e_\ell$, for some $\delta > 0$, and let $x(h)$ be the right Perron vector of $A(h)$ normalised so that $z(h)^t x(h) = 1$. Let*

$$\bar{x}(h) = \begin{bmatrix} x_1(h) & \ldots & x_{\ell-1}(h) & x_{\ell+1}(h) & \ldots & x_n(h) \end{bmatrix}^t, \quad h \in \mathcal{J}.$$

Then there exists a function $\gamma(h)$, $h \in \mathcal{J}$, such that for all $h \in \mathcal{J}$,

$$\frac{d\bar{x}(h)}{dh} = \gamma(h)\left(\overline{Q^{\#}(h)e_\ell} - \frac{q^{\#}_{\ell,\ell}(h)}{x_\ell(h)}\bar{x}(h)\right), \tag{3.27}$$

where $\overline{Q^{\#}(h)e_\ell}$ denotes the vector formed from $Q^{\#}(h)e_\ell$ by deleting its ℓ–th entry. Moreover, in the case that each entry in the ℓ–th row of $A(h)$ is a nondecreasing function of h, and at least one of those entries is strictly increasing in h, we have

$$\frac{d\bar{x}(h)}{dh} < 0, \quad \text{for all } h \in \mathcal{J}. \tag{3.28}$$

We next move to finding the second order derivatives of the Perron vector under the assumption that the entries of $A(h)$ are a constant outside one row, say the ℓ–th, and where $e_\ell^t A(h)$ is a linear function of h for $h \in \mathcal{J}$. We will continue with the normalisation that the ℓ–th entry of the Perron vector is held constant throughout $\mathcal{J}$.

THEOREM 3.4.2 ([31]) *Suppose that $A(h) \in \Phi^{n,n}$, $h \in \mathcal{J}$, is a family of matrices such that all their entries outside row ℓ are a constant and in row ℓ they are linear functions of h. Let $x(h)$ denote the Perron vector of $A(h)$, normalised so that throughout $\mathcal{J}$, $x_\ell(h)$ is held a constant $\delta > 0$. Let $Q_{(\ell)}(h)$ denote the $(n-1) \times (n-1)$ principal submatrix of $Q(h)$ resulting from the deletion of its ℓ-th row and column and let $y(h)$ denote the left Perron vector of $A(h)$ normalised so that $y(h)^t x(h) = 1$. Finally, let*

$$\bar{x}(h) = \begin{bmatrix} x_1(h) & \ldots & x_{\ell-1}(h) & x_{\ell+1}(h) & \ldots & x_n(h) \end{bmatrix}^t, \quad h \in \mathcal{J}.$$

Then

$$\begin{aligned}\frac{d^2\bar{x}(h)}{dh^2} &= 2\left(\frac{dr(A(h))}{dh}\right)^2 Q_{(\ell)}^{-2}(h)\bar{x}(h) - \frac{d^2r(A(h))}{d^2h}Q_{(\ell)}^{-1}(h)\bar{x}(h)\\ &= 2\left[\sum_{j=1}^{n} x_j(h)y_\ell(h)\frac{da_{\ell,j}(h)}{dh}\right]^2 Q_{(\ell)}^{-2}(h)\bar{x}(h)\\ &- 2\left[\sum_{j,k=1}^{n} x_j(h)y_\ell(h)q^{\#}_{k,\ell}(h)\frac{da_{\ell,j}(h)}{dh}\frac{da_{\ell,k}(h)}{dh}\right] Q_{(\ell)}^{-1}(h)\bar{x}(h).\end{aligned} \tag{3.29}$$

Proof: Without loss of generality, we assume that $\ell = 1$ and follow the steps leading to (3.25). Observe that here $Q_{(1)}(h)$ corresponds to $Q_{2,2}(h)$ in that formula. But then, upon differentiating (3.25) with respect to h we have

$$\frac{dQ_{(1)}(h)}{dh}\frac{d\bar{x}(h)}{dh} + Q_{(1)}(h)\frac{d^2\bar{x}(h)}{dh^2} = -\frac{d^2r(h)}{dh^2}\bar{x}(h) - \frac{dr(h)}{dh}\frac{d\bar{x}(h)}{dh}. \tag{3.30}$$

However, as $Q(h) = r(h)I - A(h)$, and rows $2, \ldots, n$ of A do not depend on h, we see that $\frac{dQ_{(1)}(h)}{dh} = \left(\frac{dr(h)}{dh}\right) I$. Furthermore, we note from (3.26) that $\bar{x}(h) = -\frac{dr(h)}{dh}Q_{(1)}(h)^{-1}\bar{x}(h)$. Substituting these relations in (3.30) and rearranging yields

$$Q_{(1)}(h)\frac{d^2\bar{x}(h)}{dh^2} = 2\left(\frac{dr(h)}{dh}\right)^2 Q_{(1)}^{-1}(h)\bar{x}(h) - \frac{d^2r(h)}{dh^2}\bar{x}(h) \tag{3.31}$$

from which we obtain at once the first equality in (3.29) (recalling that $\ell = 1$).

To obtain the second equality in (3.29), we compute the second derivative of $r(h)$ explicitly. Using the chain rule and recalling the expressions for the partial derivatives of the Perron value with respect to the matrix entries given in (3.2) we have

$$\frac{dr(h)}{dh} = \sum_{j=1}^{n}\frac{\partial r(h)}{\partial_{1,j}}\frac{da_{1,j}(h)}{dh} = \sum_{j=1}^{n} x_j(h)y_1(h)\frac{da_{1,j}(h)}{dh}. \tag{3.32}$$

Thus, the second derivative of $r(h)$ can be computed as follows:

$$\frac{d^2r(h)}{dh^2} = \sum_{j=1}^{n} \frac{da_{1,j}(h)}{dh} \sum_{k=1}^{n} \frac{\partial^2 r(h)}{\partial_{1,j}\partial_{1,k}} \frac{da_{1,k}(h)}{dh}$$

$$= \sum_{j,k=1}^{n} \frac{\partial^2 r(h)}{\partial_{1,j}\partial_{1,k}} \frac{da_{1,j}(h)}{dh} \frac{da_{1,k}(h)}{dh}. \quad (3.33)$$

We now use Remark 3.2.2 to obtain our final representation for the second derivative of $r(h)$:

$$\frac{d^2r(h)}{dh^2} = 2 \sum_{j,k=1}^{n} x_j(h) y_k(h) q_{k,1}^{\#}(h) \frac{da_{1,j}(h)}{dh} \frac{da_{1,k}(h)}{dh}. \quad (3.34)$$

Finally, substituting (3.32) and (3.34) on the right hand side of the first equality of (3.29) yields the remaining equality. □

The second equality in (3.29) allows us to develop the following special case of Theorem 3.4.2.

COROLLARY 3.4.3 ([31]) *Suppose that $A(h) \in \Phi^{n,n}$, $h \in \mathcal{J}$, is a family of matrices such that all the entries outside row ℓ are a constant, and in row ℓ, they are linear functions of h. Let $x(h)$ be the Perron vector of $A(h)$, normalised so that throughout $\mathcal{J}$, $x_\ell(h)$ is held a constant $\delta > 0$. Let*

$$\bar{x}(h) = \begin{bmatrix} x_1(h) & \dots & x_{\ell-1}(h) & x_{\ell+1}(h) & \dots & x_n(h) \end{bmatrix}^t, \quad h \in \mathcal{J}.$$

For $h \in \mathcal{J}$, define $\eta(h)$ and $\zeta(h)$, respectively, as

$$\eta(h) = \sum_{k=1}^{n} \frac{da_{\ell,k}(h)}{dh} q_{k,\ell}^{\#}(h) \quad (3.35)$$

and

$$\zeta(h) = \sum_{j=1}^{n} x_j(h) \frac{da_{\ell,j}(h)}{dh}. \quad (3.36)$$

If h is a value in $\mathcal{J}$ for which

$$\eta(h)\zeta(h) \le 0, \quad (3.37)$$

then

$$\frac{d^2\bar{x}(h)}{dh^2} \ge 0.$$

Proof: Observe that under the definitions of $\eta(h)$ and $\zeta(h)$ given above, the final term in (3.29) is equal to

$$-2\eta(h)\zeta(h)y_\ell(h)Q_\ell^{-1}(h)\bar{x}(h)$$

which is nonnegative when (3.37) holds. □

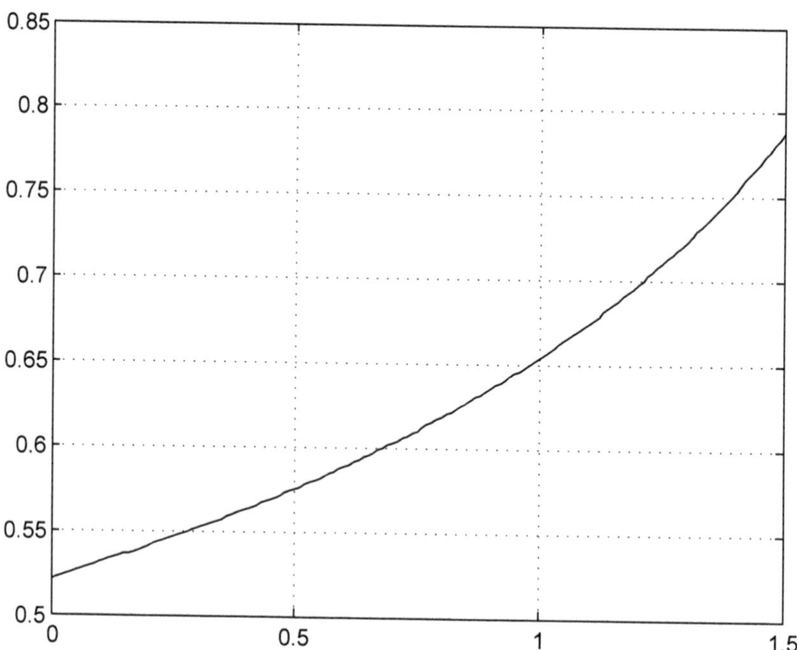

FIGURE 3.2
Graph of $x_1(h), h \in [0, \frac{3}{2}]$

EXAMPLE 3.4.4 Consider the matrix:

$$A(h) = \begin{bmatrix} \frac{1}{8} & 1 \\ 2-h & 1 \end{bmatrix}, \quad h \in \left[0, \frac{3}{2}\right].$$

Its Perron value is given by

$$r(A(h)) = \frac{\sqrt{561 - 256\,h}}{16} + \frac{9}{16},$$

while its Perron vector, normalised so that its 2nd component is equal to 1, is given by:

$$x(h) = \begin{bmatrix} \frac{\sqrt{561-256\,h}-7}{16(2-h)} \\ 1 \end{bmatrix}.$$

Observe that $\frac{da_{2,1}}{dh} = -1$ and $\frac{da_{2,2}}{dh} = 0$. Further, in the notation of Corollary 3.4.3, we have $q^{\#}_{1,2}(h) < 0$ (since $Q^{\#}(h)x(h) = 0$ and the diagonal entry $q^{\#}_{2,2}(h)$ must be positive). Hence, (3.37) holds, so that $x_1(h)$ must be concave up as a function of h, by Corollary 3.4.3. Figure 3.2 illustrates the convexity of $x_1(h)$ for h in the range $[0, \frac{3}{2}]$.

4

Perron Eigenpair in Demographic Applications

In this chapter, we apply the ideas and techniques developed in Chapter 3 to a class of nonnegative matrices arising in the modelling of population growth. We have already had a glimpse of this approach in section 1.1, where the Perron value and vector were seen to carry key information about the long-term behaviour of the Leslie population model. We begin by introducing the models in section 4.1, and then use the machinery of Chapter 3 to analyse the sensitivities of the Perron value and vector in sections 4.2 and 4.3. The related notion of elasticity is addressed in section 4.4.

4.1 Introduction to the size-classified population model

A standard model in mathematical demography considers a population that is closed to migration, where the individuals in the population are classified into various categories, known as *size classes.* For instance we might categorise the individuals in terms of stages of development (such as egg, larva, pupa, adult), or in terms of the size of a particular attribute (for example, for a population of tortoises, the length of the carapace) or in terms of the age of the individual (zero to five years, five to ten years, etc.). In each of these examples, observe that the individuals pass through the various categories in a consecutive manner as time goes on; for instance an individual can enter the pupa stage only having already passed from the egg to the larva stage, and once an individual enters the pupa stage, it cannot re-enter the egg stage at some later time. A single size-classified population model can be used to describe each of these examples, and we outline that model next.

Consider a population in which members are classified into n size classes, labelled $1, \ldots, n$. We fix a time unit, and model the population in discrete time at integer multiples of the selected time unit. For each $k \in \mathbb{N}$, the population is represented by the vector $v(k) \in \mathbb{R}^n$, where for each $i = 1, \ldots, n$ $v_i(k)$ is the number of members of the population that are in the i–th size class at time k. In order to reflect the fact that individuals move through the size classes in a consecutive manner, we assume that for each $i = 1, \ldots, n$, if an individual member of the population is in class i, then after one time unit has elapsed, the member either remains in class i, or it moves into class $i+1$, or it does not survive. (Note that here, the time unit must be short enough so that an individual cannot pass through two or more classes in one time unit.)

For each $i = 1, \ldots, n$, let the scalar f_i be the average number of surviving members of class 1 produced in one time unit by an individual in class i. Observe that for $i = 2, \ldots, n$ the scalar f_i arises from births to members in class i, while f_1 includes not only birth information for members in class 1, but also survival information from class 1 back into class 1. For each $i = 1, \ldots, n-1$, let the scalar p_i denote the proportion of individuals in class i that survive and move into class $i+1$ in one time unit; for each $i = 2, \ldots, n$, let the scalar b_i denote the proportion of individuals in class i that survive and remain in class i after one time unit. We note in passing that necessarily $p_i + b_i \leq 1, i = 2, \ldots, n-1$.

With these quantities in hand, we form the following matrix, which is known as the *population projection matrix*:

$$A = \begin{bmatrix} f_1 & f_2 & & \cdots & f_{n-1} & f_n \\ p_1 & b_2 & 0 & \cdots & 0 & 0 \\ 0 & p_2 & b_3 & 0 & 0 & 0 \\ \vdots & & \ddots & \ddots & \vdots & \vdots \\ & & & & & \\ 0 & & \cdots & 0 & p_{n-1} & b_n \end{bmatrix}. \tag{4.1}$$

Assuming that this regime of survivorship and fecundity remains constant over time, so that the f_is, p_is and b_is are unchanged over time, we find that for each $k \in \mathbb{N}, v(k+1) = Av(k)$. Consequently, $v(k) = A^k v(0), k \in \mathbb{N}$. Observe that in the special case that $b_i = 0, i = 2, \ldots, n$, our matrix A coincides with that arising in the Leslie model, which was introduced in section 1.1, for populations classified by age, and where the i–th age group consists of individuals of age between $i-1$ and i time units.

It is straightforward to determine that A is irreducible if and only if

$f_n > 0$ and $p_i > 0, i = 1, \ldots, n-1$, and henceforth we take that to be the case. Indeed in demographic applications, the positivity condition on the p_is is natural, otherwise there is a size class into which no member can survive. The condition that $f_n > 0$ can be ensured by, if necessary, restricting attention to the size classes up to and including the last class that is capable of reproducing.

Recall that for an $n \times n$ matrix M, the *directed graph associated with* M, $\mathcal{D}(M)$ is the directed graph on vertices labelled $1, \ldots, n$, such that for each $i, j \in \{1, \ldots, n\}, \mathcal{D}(M)$, contains the arc $i \to j$ if and only if $m_{i,j} \neq 0$ (see the book of Brualdi and Ryser [15, Chapter 3]). Considering the directed graph $\mathcal{D}(A)$ associated with the population projection matrix A of (4.1), we find that $\mathcal{D}(A)$ has a cycle of length 1 if and only if either $f_1 > 0$ or $b_i > 0$ for some $i = 2, \ldots, n$, while for any j between 2 and n, $\mathcal{D}(A)$ has a cycle of length j if and only if $f_j > 0$. It now follows that A is primitive if and only if either f_1 is positive, or some b_i is positive, or $gcd\{j | f_j > 0\} = 1$.

Suppose that A is primitive. Then from the Perron–Frobenius theorem, we find that the Perron value of A, $r(A)$, strictly dominates the moduli of the other eigenvalues of A. Observe that for each $k \in \mathbb{N}$, the vector $v(k)/\mathbf{1}^t v(k)$ represents the distribution of the population through classes $1, \ldots, n$, at time k. Thus when A is primitive, we find readily that as $k \to \infty, v(k)/\mathbf{1}^t v(k) \to x$, where x is the Perron vector of A normalised so that its entries sum to 1. The vector x is known as the *stable distribution vector* for the population. Note also that the total size of the population is given by $\mathbf{1}^t v(k)$ and that asymptotically, this quantity grows geometrically with ratio $r(A)$.

Thus the stable distribution vector carries information regarding the structure of the long-term behaviour of the population, and for this reason, it is a central quantity of interest for size-classified population models. Similarly, the Perron value $r(A)$ represents the asymptotic growth rate for the population, and this too is a key quantity for the population.

Evidently both the Perron value $r(A)$ and the stable distribution vector x are functions depending on the demographic parameters $f_i, i = 1, \ldots, n, p_i, i = 1, \ldots, n-1$, and $b_i, i = 2, \ldots, n$. Which of these parameters are highly influential on $r(A)$ and x, and which are not so influential? If one or more of the demographic parameters are altered, how will the resulting changes in $r(A)$ and x be manifested? These types of questions arise naturally if we seek to understand how changes in demographic parameters (due to a management strategy designed to protect a species, for instance) affect the key quantities associated with the population under consideration. In a different direction, if it can be determined that the Perron value $r(A)$, is quite sensitive to the value of a particular parameter, then that sensitivity can be used to inform the procedures used to estimate that parameter. Chapter 9 of Caswell's book [17], discusses these and other scenarios in which an understanding of the sensitivities of $r(A)$ and x to the demographic parameters brings valuable insight.

Our approach to these sensitivity questions will be in terms of derivatives. As we will see in the next sections, the structure of the matrix (4.1), coupled with the the techniques of Chapter 3, will yield a good deal of information about the derivatives of the Perron value and vector of the population projection matrix for the size-classified model.

4.2 First derivatives for the size-classified model

Consider the population matrix A given by (4.1), and suppose that it is irreducible, with Perron value $r(A)$. Set $\xi_1(A) = 1$ and

$$\xi_j(A) = \frac{p_1 p_2 \dots p_{j-1}}{(r(A) - b_2)(r(A) - b_3) \dots (r(A) - b_j)} \text{ for } j = 2, \dots, n.$$

A straightforward computation shows that the vector

$$\xi(A) = \begin{bmatrix} \xi_1(A) & \dots & \xi_n(A) \end{bmatrix}^t$$

serves as a right Perron vector for A. Similarly, defining $\eta_1(A) = 1, \eta_n(A) = \frac{f_n}{(r(A)-b_n)}$ and

$$\eta_j(A) = \frac{f_j}{(r(A) - b_j)} + \sum_{l=j+1}^{n} \frac{p_j \dots p_{l-1} f_l}{(r(A) - b_j) \dots (r(A) - b_l)}, j = 2, \dots, n-1,$$

it follows that the vector

$$\eta(A) = \begin{bmatrix} \eta_1(A) & \dots & \eta_n(A) \end{bmatrix}^t$$

is a left Perron vector for A. In view of (3.2), we find that for each $i, j = 1, \dots, n$, we have

$$\frac{\partial r(A)}{\partial_{i,j}} = \frac{\eta_i(A)\xi_j(A)}{\eta(A)^t \xi(A)}.$$

Note that $\eta(A)^t\xi(A)$ admits the explicit expression

$$1+\sum_{m=2}^{n}\frac{p_1\dots p_{m-1}f_m}{(r(A)-b_2)\dots(r(A)-b_m)}\left(\sum_{l=2}^{m}\frac{1}{(r(A)-b_l)}\right).$$

In the demographic setting, we have a particular interest in those pairs of indices (i,j) that identify entries in A that are in the top row, the main diagonal, and the first subdiagonal, for those positions correspond to the demographic parameters upon which the size-classified model is constructed. Making the appropriate substitutions yields

$$\frac{\partial r(A)}{\partial_{1,1}}=\frac{1}{1+\sum_{m=2}^{n}\frac{p_1\dots p_{m-1}f_m}{(r(A)-b_2)\dots(r(A)-b_m)}\left(\sum_{l=2}^{m}\frac{1}{(r(A)-b_l)}\right)}, \tag{4.2}$$

$$\frac{\partial r(A)}{\partial_{1,j}}=\frac{\left(\frac{p_1\dots p_{j-1}}{(r(A)-b_2)\dots(r(A)-b_j)}\right)}{1+\sum_{m=2}^{n}\frac{p_1\dots p_{m-1}f_m}{(r(A)-b_2)\dots(r(A)-b_m)}\left(\sum_{l=2}^{m}\frac{1}{(r(A)-b_l)}\right)}, \tag{4.3}$$
$$j=2,\dots,n,$$

$$\frac{\partial r(A)}{\partial_{2,1}}=\frac{\frac{f_2}{(r(A)-b_2)}+\sum_{l=3}^{n}\frac{p_2\dots p_{l-1}f_l}{(r(A)-b_2)\dots(r(A)-b_l)}}{1+\sum_{m=2}^{n}\frac{p_1\dots p_{m-1}f_m}{(r(A)-b_2)\dots(r(A)-b_m)}\left(\sum_{l=2}^{m}\frac{1}{(r(A)-b_l)}\right)}, \tag{4.4}$$

$$\frac{\partial r(A)}{\partial_{i+1,i}}=\frac{\left(\frac{1}{p_i}\sum_{l=i+1}^{n}\frac{p_1\dots p_{l-1}f_l}{(r(A)-b_2)\dots(r(A)-b_l)}\right)}{1+\sum_{m=2}^{n}\frac{p_1\dots p_{m-1}f_m}{(r(A)-b_2)\dots(r(A)-b_m)}\left(\sum_{l=2}^{m}\frac{1}{(r(A)-b_l)}\right)}, \tag{4.5}$$
$$i=1,\dots,n-1,$$

and

$$\frac{\partial r(A)}{\partial_{i,i}}=\frac{\left(\frac{1}{(r(A)-b_i)}\sum_{l=i}^{n}\frac{p_1\dots p_{l-1}f_l}{(r(A)-b_2)\dots(r(A)-b_l)}\right)}{1+\sum_{m=2}^{n}\frac{p_1\dots p_{m-1}f_m}{(r(A)-b_2)\dots(r(A)-b_m)}\left(\sum_{l=2}^{m}\frac{1}{(r(A)-b_l)}\right)}, \tag{4.6}$$
$$i=2,\dots,n.$$

Inspecting (4.2) and (4.3), we find that for each $j=1,\dots,n-1$, $\frac{\partial r(A)}{\partial_{1,j}}\geq\frac{\partial r(A)}{\partial_{1,j+1}}$ if and only if $r(A)\geq p_j+b_{j+1}$. Similarly, from (4.4), (4.5) and (4.6), It follows that for each $j=1,\dots,n-1$, $\frac{\partial r(A)}{\partial_{j+1,j}}\geq\frac{\partial r(A)}{\partial_{j+1,j+1}}$ if and only if $r(A)\geq p_j+b_{j+1}$. Thus we see that for each $j=1,\dots,n-1$, the quantity $p_j + b_{j+1}, j = 1, \ldots, n-1$, which coincides with the row sum of the $(j+1)$-st row of A, is critical in comparing the sensitivities of $r(A)$ to consecutive nonzero elements in a common row of A.

Referring to (4.5), we find that for each $i = 1, \ldots, n-2$, $\frac{\partial r(A)}{\partial_{i+1,i}} \geq \frac{\partial r(A)}{\partial_{i+2,i+1}}$ if and only if

$$\frac{p_{i+1}}{p_i} \geq \frac{\sum_{l=i+2}^{n} \frac{p_1 \ldots p_{l-1} f_l}{(r(A)-b_2)\ldots(r(A)-b_l)}}{\sum_{l=i+1}^{n} \frac{p_1 \ldots p_{l-1} f_l}{(r(A)-b_2)\ldots(r(A)-b_l)}}. \tag{4.7}$$

Observe that (4.7) holds trivially if $p_{i+1} > p_i$, which yields the conclusion that if the survivorship proportion from class $i+1$ into class $i+2$ exceeds the survivorship proportion from class i into class $i+1$, then $r(A)$ is more sensitive to changes in the latter than it is to changes in the former. From (4.6) it follows that for each $i = 2, \ldots, n-1$, $\frac{\partial r(A)}{\partial_{i,i}} \geq \frac{\partial r(A)}{\partial_{i+1,i+1}}$ if and only if

$$\frac{r(A) - b_{i+1}}{r(A) - b_i} \geq \frac{\sum_{l=i+1}^{n} \frac{p_1 \ldots p_{l-1} f_l}{(r(A)-b_2)\ldots(r(A)-b_l)}}{\sum_{l=i}^{n} \frac{p_1 \ldots p_{l-1} f_l}{(r(A)-b_2)\ldots(r(A)-b_l)}}. \tag{4.8}$$

If $b_i > b_{i+1}$, then (4.8) holds trivially, and we conclude that if survivorship proportion from class i into class i exceeds the survivorship proportion from class $i+1$ into class $i+1$, then $r(A)$ is more sensitive to changes in the former than it is to changes in the latter.

EXAMPLE 4.2.1 We consider the following population projection matrix for the desert tortoise that is based on data from Doak, Kareiva, and Klepetka's paper [33]:

$$A = \begin{bmatrix} 0 & 0 & 0 & 0 & 0 & 1.300 & 1.980 & 2.570 \\ 0.716 & 0.567 & 0 & 0 & 0 & 0 & 0 & 0 \\ 0 & 0.149 & 0.567 & 0 & 0 & 0 & 0 & 0 \\ 0 & 0 & 0.149 & 0.604 & 0 & 0 & 0 & 0 \\ 0 & 0 & 0 & 0.235 & 0.560 & 0 & 0 & 0 \\ 0 & 0 & 0 & 0 & 0.225 & 0.678 & 0 & 0 \\ 0 & 0 & 0 & 0 & 0 & 0.249 & 0.851 & 0 \\ 0 & 0 & 0 & 0 & 0 & 0 & 0.016 & 0.860 \end{bmatrix}. \tag{4.9}$$

The computed Perron value $r(A)$ for A is 0.9581, so that $A^k \to 0$ as $k \to \infty$; indeed [33] reports that the species is threatened. Computing $p_j + b_{j+1}$ for the various values of j, we find that $r(A) > p_j + b_{j+1}$ for $j = 2, 3, 4, 5, 7$. Consequently, $\frac{\partial r(A)}{\partial_{1,7}} > \frac{\partial r(A)}{\partial_{1,6}}, \frac{\partial r(A)}{\partial_{1,8}}$, so that among the positive entries in the first row of A, the Perron value $r(A)$ is most sensitive to the entry in the $(1,7)$ position. Evidently we also have $\frac{\partial r(A)}{\partial_{j+1,j}} > \frac{\partial r(A)}{\partial_{j+1,j+1}}$ for each $j = 2,3,4,5,7$. Finally, as $p_4 \geq p_3, p_6 \geq p_5, b_2 \geq b_3$, and $b_4 \geq b_5$, we readily find that $\frac{\partial r(A)}{\partial_{4,3}} \geq \frac{\partial r(A)}{\partial_{5,4}}$, $\frac{\partial r(A)}{\partial_{6,5}} \geq \frac{\partial r(A)}{\partial_{7,6}}$, $\frac{\partial r(A)}{\partial_{2,2}} \geq \frac{\partial r(A)}{\partial_{3,3}}$, and $\frac{\partial r(A)}{\partial_{4,4}} \geq \frac{\partial r(A)}{\partial_{5,5}}$.

Computing $\frac{\partial r(A)}{\partial_{i,j}}$ over all ordered pairs (i,j) that correspond to positive entries in A, we find that the maximum sensitivity (i.e., derivative) in that range of positions occurs for the pair $(8,7)$. We conclude then that in this example, $r(A)$ is most sensitive to the survival rate from the seventh size class into the eighth size class.

We now turn our attention to the derivatives of the stable distribution vector for A, and here we will make extensive use of the tools developed in Chapter 3. Many of the results in the remainder of this section can be found in the paper of Kirkland [72]. We remark that the issue of producing expressions for the derivatives of the stable distribution vector is also dealt with in [17, section 9.4]; there the problem is approached via the use of the eigendecomposition for A, under the hypothesis that A has distinct eigenvalues. As we will see below, the structure of the population projection matrix A leads to a corresponding structure for $(r(A)I - A)^{\#}$; the latter then facilitates some detailed analysis of the entries in the derivative of the stable distribution vector.

We begin with a preliminary result, which follows directly from (3.18).

PROPOSITION 4.2.2 *Let A be an $n \times n$ population projection matrix given by (4.1), and suppose that A is irreducible, with Perron value $r(A)$ and stable distribution vector x. Then for each $i, j = 1, \ldots, n$,*

$$\frac{\partial x}{\partial_{i,j}} = x_j (r(A)I - A)^{\#} e_i - (x_j \mathbf{1}^t (r(A)I - A)^{\#} e_i) x. \tag{4.10}$$

In order to compute $(r(A)I - A)^{\#}$, it will be convenient to apply the observations made in Remark 2.5.3. Let $X = \mathrm{diag}(x)$ and let T be the matrix given by

$$T = \frac{1}{r(A)} X^{-1} A X \equiv \begin{bmatrix} a_1 & a_2 & \cdots & \cdots & a_{n-1} & a_n \\ 1-c_1 & c_1 & 0 & \cdots & 0 & 0 \\ 0 & 1-c_2 & c_2 & 0 & 0 & 0 \\ \vdots & & \ddots & \ddots & & \vdots \\ 0 & \cdots & 0 & 1-c_{n-2} & c_{n-2} & 0 \\ 0 & \cdots & & 0 & 1-c_{n-1} & c_{n-1} \end{bmatrix}.$$

Here we note that $a_1 = \frac{f_1}{r(A)}$ and for each $i = 2, \ldots, n$,

$$a_i = \frac{p_1 \ldots p_{i-1} f_i}{r(A)(r(A) - b_2) \ldots (r(A) - b_i)},$$

while for $i = 1, \ldots, n-1, c_i = \frac{b_{i+1}}{r(A)}$. As is noted in Remark 2.5.3, T is stochastic, and further, setting $Q = I - T$, we have

$$(r(A)I - A)^{\#} = \frac{1}{r(A)} X Q^{\#} X^{-1}. \tag{4.11}$$

Maintaining the notation above, we have the following.

LEMMA 4.2.3 ([72]) *For each* $i, j = 1, \ldots, n$,

$$\frac{\partial x}{\partial_{i,j}} = \frac{x_j}{r(A)x_i} \left(X Q^{\#} e_i - (x^t Q^{\#} e_i) x \right).$$

In particular, for any index $l = 1, \ldots, n$,

$$\frac{\partial x_l}{\partial_{i,j}} = \frac{x_j x_l}{r(A)x_i} (e_l - x)^t Q^{\#} e_i.$$

Proof: Applying Proposition 4.2.2 and (4.11), we have

$$\frac{\partial x}{\partial_{i,j}} = \frac{x_j}{r(A)} X Q^{\#} X^{-1} e_i - \left(\frac{x_j}{r(A)} \mathbf{1}^t X Q^{\#} X^{-1} e_i \right) x.$$

Since $X^{-1} e_i = \frac{1}{x_i} e_i$ and $\mathbf{1}^t X = x^t$, it now follows that

$$\frac{\partial x}{\partial_{i,j}} = \frac{x_j}{r(A)x_i} \left(X Q^{\#} e_i - (x^t Q^{\#} e_i) x \right).$$

Next, fix an index l between 1 and n. Then

$$\frac{\partial x_l}{\partial_{i,j}} = e_l^t \frac{\partial x}{\partial_{i,j}} = \frac{x_j}{r(A)x_i} \left(e_l^t X Q^{\#} e_i - (x^t Q^{\#} e_i) e_l^t x \right).$$

Since $e_l^t X = x_l e_l^t$, the desired expression now follows. □

In order to produce explicit expressions for the derivatives of the stable distribution vector, we now turn our attention to deriving a formula for $Q^{\#}$. Let $s_i = \sum_{p=1}^{i} a_p, i = 1, \ldots, n-1$, and set $c_0 \equiv 0$ and $s_0 \equiv 0$. Next we define the vector w by

$$w = \frac{1}{\sum_{i=0}^{n-1}\left(\frac{1-s_i}{1-c_i}\right)} \begin{bmatrix} \frac{1-s_0}{1-c_0} \\ \frac{1-s_1}{1-c_1} \\ \vdots \\ \frac{1-s_{n-1}}{1-c_{n-1}} \end{bmatrix}; \tag{4.12}$$

we note that w is the left Perron vector for T, normalised so that $w^t\mathbf{1} = 1$. Let the trailing subvector of w of order $n-1$ be $\widehat{w}$, and note that from the eigen-equation $w^tT = w^t$, it follows that $\widehat{w}^t = w_1 \begin{bmatrix} a_2 & \ldots & a_n \end{bmatrix} (I - \widehat{T})^{-1}$, where $\widehat{T}$ denotes the trailing principal submatrix of T of order $n-1$.

A straightforward computation shows that

$$(I-\widehat{T})^{-1} \equiv R = \begin{bmatrix} \frac{1}{1-c_1} & 0 & 0 & \ldots & 0 & 0 \\ \frac{1}{1-c_1} & \frac{1}{1-c_2} & 0 & \ldots & 0 & 0 \\ \frac{1}{1-c_1} & \frac{1}{1-c_2} & \frac{1}{1-c_3} & 0 & \ldots & 0 \\ \vdots & & & \ddots & & \vdots \\ \frac{1}{1-c_1} & \frac{1}{1-c_2} & \frac{1}{1-c_3} & \ldots & \frac{1}{1-c_{n-2}} & 0 \\ \frac{1}{1-c_1} & \frac{1}{1-c_2} & \frac{1}{1-c_3} & \ldots & \frac{1}{1-c_{n-2}} & \frac{1}{1-c_{n-1}} \end{bmatrix}. \tag{4.13}$$

Applying a variant of the result in Proposition 2.5.1 (where we partition out the first row and column of T instead of the last row and column as in Proposition 2.5.1), it now follows that with the notation above,

$$Q^{\#} = \left[\begin{array}{c|c} w_1\widehat{w}^tR\mathbf{1} & -\widehat{w}^tR + (\widehat{w}^tR\mathbf{1})\widehat{w}^t \\ \hline -w_1R\mathbf{1} + w_1(\widehat{w}^tR\mathbf{1})\mathbf{1} & R - \mathbf{1}\widehat{w}^tR - R\mathbf{1}\widehat{w}^t + (\widehat{w}^tR\mathbf{1})\mathbf{1}\widehat{w}^t \end{array}\right]. \tag{4.14}$$

The structure of $Q^{\#}$ yields the following result.

PROPOSITION 4.2.4 ([72]) *Fix an index $i \in \{1, \ldots, n\}$. If $l < i$, then*

$$(e_l - x)^tQ^{\#}e_i =$$

$$\frac{-1}{1-c_{i-1}}\sum_{p=i}^{n} x_p + w_i\left[\sum_{p=l+1}^{n} x_p\left(\sum_{q=l}^{p-1}\frac{1}{1-c_q}\right) - \sum_{p=1}^{l-1} x_p\left(\sum_{q=p}^{l-1}\frac{1}{1-c_q}\right)\right],$$

while if $i \le l$, then

$$(e_l - x)^t Q^{\#} e_i =$$

$$\frac{1}{1-c_{i-1}} \sum_{p=1}^{i-1} x_p + w_i \left[\sum_{p=l+1}^{n} x_p \left(\sum_{q=l}^{p-1} \frac{1}{1-c_q} \right) - \sum_{p=1}^{l-1} x_p \left(\sum_{q=p}^{l-1} \frac{1}{1-c_q} \right) \right].$$

Proof: From (4.14) it follows that

$$Q^{\#} = \left[\begin{array}{c|c} 0 & 0^t \\ \hline -w_1 R\mathbf{1} & R - R\mathbf{1}\widehat{w}^t \end{array} \right] + \mathbf{1} \left[\begin{array}{c|c} w_1(\widehat{w}^t R\mathbf{1}) & -\widehat{w}^t R + (\widehat{w}^t R\mathbf{1})\widehat{w}^t \end{array} \right].$$

Since $e_l^t \mathbf{1} = x^t \mathbf{1} = 1$, we find that

$$(e_l - x)^t Q^{\#} = (e_l - x)^t \left[\begin{array}{c|c} 0 & 0^t \\ \hline -w_1 R\mathbf{1} & R - R\mathbf{1}\widehat{w}^t \end{array} \right].$$

Using the fact that $e_p^t R\mathbf{1} = \sum_{q=1}^{p} \frac{1}{1-c_q}$, we find that

$$\begin{aligned} (e_l - x)^t Q^{\#}\ e_1 &= w_1 \sum_{p=1}^{n} x_p \left(\sum_{q=1}^{p-1} \frac{1}{1-c_q} - \sum_{q=1}^{l-1} \frac{1}{1-c_q} \right) \\ &= w_1 \left(\sum_{p=l+1}^{n} x_p \left(\sum_{q=l}^{p-1} \frac{1}{1-c_q} \right) - \sum_{p=1}^{l-1} x_p \left(\sum_{q=p}^{l-1} \frac{1}{1-c_q} \right) \right). \end{aligned}$$

Similarly, for each $i = 2, \ldots, n$, we have

$$\begin{aligned} (e_l - x)^t Q^{\#} e_i = R_{l-1,i-1} - \frac{1}{1-c_{i-1}} \sum_{p=i}^{n} x_p + \\ w_i \left[\sum_{p=l+1}^{n} x_p \left(\sum_{q=l}^{p-1} \frac{1}{1-c_q} \right) - \sum_{p=1}^{l-1} x_p \left(\sum_{q=p}^{l-1} \frac{1}{1-c_q} \right) \right]. \end{aligned}$$

The expressions for $(e_l - x)^t Q^{\#} e_i$ now follow from the fact that $R_{l-1,i-1}$ is either 0 or $\frac{1}{1-c_{i-1}}$ according as $l < i$ or $i \le l$, respectively. □

Our next result gives formulas for the partial derivatives of x, and follows immediately from Lemma 4.2.3 and Proposition 4.2.4.

THEOREM 4.2.5 ([72]) *Let A be an irreducible population projection matrix given by (4.1), with Perron value $r(A)$ and stable distribution vector x. Fix indices $i, j, l \in \{1, \ldots, n\}$. If $l < i$, then*

$$\frac{\partial x_l}{\partial_{i,j}} = \frac{x_j x_l}{r(A) x_i} \left(\frac{-1}{1 - c_{i-1}} \sum_{p=i}^{n} x_p + \right.$$
$$\left. w_i \left[\sum_{p=l+1}^{n} x_p \left(\sum_{q=l}^{p-1} \frac{1}{1 - c_q} \right) - \sum_{p=1}^{l-1} x_p \left(\sum_{q=p}^{l-1} \frac{1}{1 - c_q} \right) \right] \right). \quad (4.15)$$

If $i \leq l$, then

$$\frac{\partial x_l}{\partial_{i,j}} = \frac{x_j x_l}{r(A) x_i} \left(\frac{1}{1 - c_{i-1}} \sum_{p=1}^{i-1} x_p + \right.$$
$$\left. w_i \left[\sum_{p=l+1}^{n} x_p \left(\sum_{q=l}^{p-1} \frac{1}{1 - c_q} \right) - \sum_{p=1}^{l-1} x_p \left(\sum_{q=p}^{l-1} \frac{1}{1 - c_q} \right) \right] \right). \quad (4.16)$$

Note that from Theorem 4.2.5, the index j influences the value of $\frac{\partial x_l}{\partial_{i,j}}$ only through the multiplicative term x_j. Thus, for any pair of indices j_1, j_2 between 1 and n we have $\frac{\partial x_l}{\partial_{i,j_2}} = \frac{x_{j_2}}{x_{j_1}} \frac{\partial x_l}{\partial_{i,j_1}}$. In particular, we see that if $x_{j_2} > x_{j_1}$, then each x_l will be more sensitive to changes in the (i, j_2) of A than to changes in the (i, j_1) entry.

A key question regarding the sensitivity of x_l with respect to the entry of A in position (i, j) is whether x_l is increasing or decreasing—i.e., whether its derivative with respect to the (i, j) entry is positive or negative. Our next result, which continues with the notation above, yields some insight into that question, and relies heavily on the explicit information about $Q^{\#}$ that has been developed so far.

COROLLARY 4.2.6 ([72]) *Suppose that $i \in \{1, \ldots, n\}$. Then $(e_l - x)^t Q^{\#} e_i$ is decreasing in l for $l = 1, \ldots i-1$, and for $l = i, \ldots, n$. Further, for each $i = 1, \ldots, n, (e_i - x)^t Q^{\#} e_i > 0$, and at least one of $(e_{i-1} - x)^t Q^{\#} e_i$ and $(e_n - x)^t Q^{\#} e_i$ is negative.*

In particular, if we have $\frac{\partial x_{l_0}}{\partial_{i,j}} < 0$ for some $l_0 \leq i-1$, then necessarily $\frac{\partial x_l}{\partial_{i,j}} < 0$ for all $l_0 \leq l \leq i-1$. If we have $\frac{\partial x_{l_1}}{\partial_{i,j}} < 0$ for some $i+1 \leq l_1 \leq n$, then necessarily $\frac{\partial x_l}{\partial_{i,j}} < 0$ for all $l_1 \leq l \leq n$. Further, $\frac{\partial x_i}{\partial_{i,j}}$ is positive for each $i = 1, \ldots, n$, while for each $i = 2, \ldots, n$, at least one of $\frac{\partial x_{i-1}}{\partial_{i,j}}$ and $\frac{\partial x_n}{\partial_{i,j}}$ is negative.

Proof: Evidently $\sum_{p=l+1}^{n} x_p(\sum_{q=l}^{p-1} \frac{1}{1-c_q})$ is decreasing with l while $\sum_{p=1}^{l-1} x_p(\sum_{q=p}^{l-1} \frac{1}{1-c_q})$ is increasing with l. From Proposition 4.2.4, we now find readily that $(e_l - x)^t Q^{\#} e_i > (e_{l+1} - x)^t Q^{\#} e_i$ for $l = 1, \ldots i-2$, and $(e_l - x)^t Q^{\#} e_i > (e_{l+1} - x)^t Q^{\#} e_i$ for $l = i, \ldots n-1$.

From (2.17) it follows that for each $i = 1, \ldots, n$, $q^{\#}_{i,i}$ is the unique maximum entry in the i−th column of $Q^{\#}$. Since the entries of x are positive and sum to 1, we find that $(e_i - x)^t Q^{\#} e_i > 0$. Since $0 = 0^t Q^{\#} e_i = \sum_{l=1}^{n} x_l (e_l - x)^t Q^{\#} e_i$, we see that for some $l, (e_l - x)^t Q^{\#} e_i < 0$. As $(e_l - x)^t Q^{\#} e_i$ is decreasing for $l = 1, \ldots, i-1$ and for $l = i, \ldots, n$, we conclude that at least one of $(e_{i-1} - x)^t Q^{\#} e_i$ and $(e_n - x)^t Q^{\#} e_i$ must be negative. The remaining conclusions now follow readily from Lemma 4.2.3. □

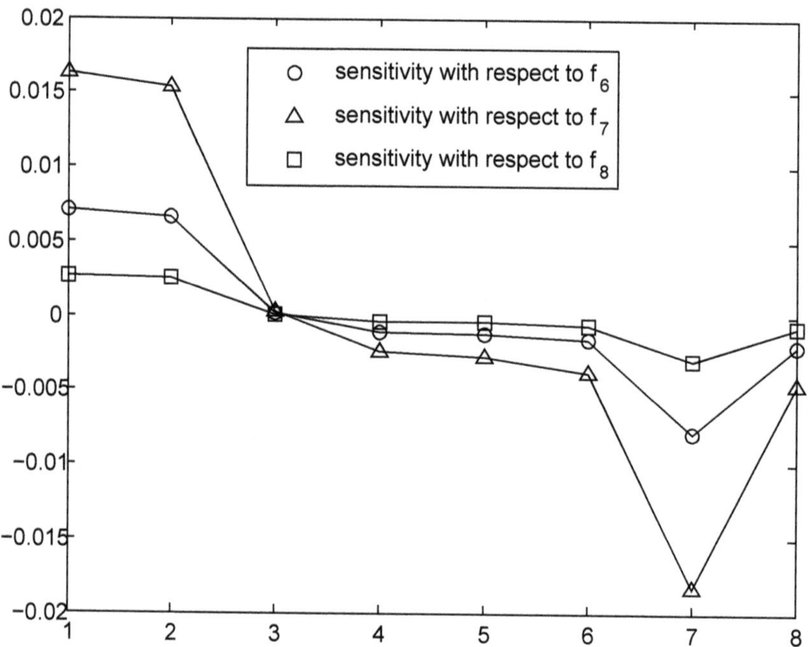

FIGURE 4.1
Sensitivities of the stable distribution vector in Example 4.2.7 with respect to f_6, f_7, f_8

EXAMPLE 4.2.7 In this example, we revisit the population projection matrix A for the desert tortoise given in (4.9). The corresponding

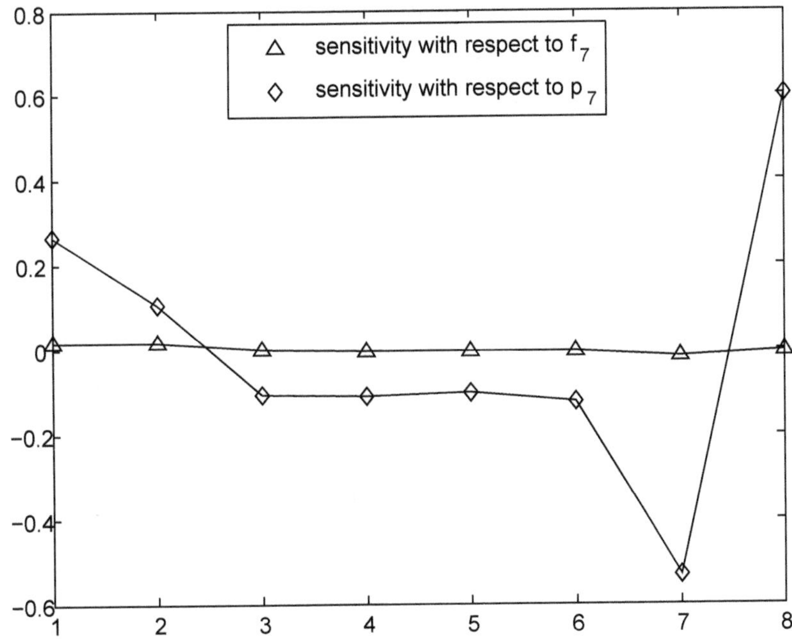

FIGURE 4.2
Sensitivities of the stable distribution vector in Example 4.2.7 with respect to f_7 and p_7

stable distribution vector is given by

$$x(A) = \begin{bmatrix} 0.2217 \\ 0.4058 \\ 0.1546 \\ 0.0651 \\ 0.0384 \\ 0.0309 \\ 0.0718 \\ 0.0117 \end{bmatrix}.$$

Figure 4.1 plots the sensitivities of the stable distribution vector with respect to f_6, f_7, and f_8—i.e., the entries of $\frac{\partial x(A)}{\partial_{1,j}}$, for $j = 6, 7, 8$. (In that figure and others, we have joined the data points by line segments in an effort to make the plots easier to follow.) From Figure 4.1, we observe that each of $\frac{\partial x(A)}{\partial_{1,6}}$, $\frac{\partial x(A)}{\partial_{1,7}}$, and $\frac{\partial x(A)}{\partial_{1,8}}$ exhibits the same sign pattern. Indeed the graphs of $\frac{\partial x(A)}{\partial_{1,j}}, j = 6, 7, 8$ all have the same shape, but the shape is more pronounced for $j = 7$, and less pronounced for $j = 6, 8$;

this is of course due to the fact that $x_7 > x_6 > x_8$. For each $j = 6, 7, 8$ we have $\frac{\partial x_l(A)}{\partial_{1,j}} > 0$ for $l = 1, 2, 3$ and $\frac{\partial x_l(A)}{\partial_{1,j}} < 0$ for $l = 4, \ldots, 8$, thus illustrating the conclusion of Corollary 4.2.6 that if $\frac{\partial x_l(A)}{\partial_{1,j}}$ is negative for some l, then $\frac{\partial x_{l+1}(A)}{\partial_{1,j}}$ is also necessarily negative. Note from Figure 4.1 that $\frac{\partial x_7(A)}{\partial_{1,j}} < \frac{\partial x_8(A)}{\partial_{1,j}}$, for $j = 6, 7, 8$, so that $\frac{\partial x_l(A)}{\partial_{1,j}}$ fails to be monotonically decreasing in l.

Figure 4.2 displays sensitivities of the stable distribution vector with respect to f_7 and p_7, that is, $\frac{\partial x(A)}{\partial_{1,7}}$ and $\frac{\partial x(A)}{\partial_{8,7}}$, on the same axes. Evidently the stable distribution vector is far more sensitive to changes in the survival rate into the final size class than it is to the birth rate corresponding to the seventh size class.

The class of Leslie matrices introduced in section 1.1 arises in an important and much-studied special case of our size-classified population model. Recall that a Leslie matrix has the general form

$$L = \begin{bmatrix} f_1 & f_2 & & \cdots & f_{n-1} & f_n \\ p_1 & 0 & 0 & \cdots & 0 & 0 \\ 0 & p_2 & 0 & 0 & 0 & 0 \\ \vdots & & \ddots & \ddots & \vdots & \vdots \\ & & & & & \\ 0 & & \cdots & 0 & p_{n-1} & 0 \end{bmatrix}. \tag{4.17}$$

Following the approach used earlier in this section, we find that an irreducible Leslie matrix L of the form (4.17) is diagonally similar to a scalar multiple of the stochastic Leslie matrix

$$\begin{bmatrix} a_1 & a_2 & \cdots & & a_{n-1} & a_n \\ 1 & 0 & 0 & \cdots & 0 & 0 \\ 0 & 1 & 0 & 0 & 0 & 0 \\ \vdots & \cdots & \ddots & \ddots & \vdots & \vdots \\ & & & & & \\ 0 & & \cdots & 0 & 1 & 0 \end{bmatrix}, \tag{4.18}$$

where $a_1 = \frac{f_1}{r(L)}$ and $a_i = \frac{p_1 \cdots p_{i-1} f_i}{r(L)^i}, i = 2, \ldots, n$. As above, we set $s_0 = 0$ and $s_j = \sum_{p=1}^{j} a_p, j = 1, \ldots, n-1$, and let w be given by (4.12). We note also that in this setting, the matrix R of (4.13) is just the lower triangular matrix of order $n-1$ with all entries on and below the diagonal equal to 1. We will maintain this notation throughout the remainder of this section.

The following result is immediate from Theorem 4.2.5.

COROLLARY 4.2.8 ([72]) *Suppose that L is an irreducible $n \times n$ Leslie matrix with Perron value $r(L)$ and stable distribution vector $x(L)$. If $i, j \in \{1, \ldots, n\}$, then, suppressing the explicit dependence of the entries of the stable distribution vector on L, we have*

$$\frac{\partial x_l}{\partial_{i,j}} = \frac{x_j x_l}{r(L)x_i}\left(-\sum_{p=i}^{n} x_p + w_i\left[\sum_{p=1}^{n} p x_p - l\right]\right)$$

for $l = 1, \ldots, i-1$, and

$$\frac{\partial x_l}{\partial_{i,j}} = \frac{x_j x_l}{r(L)x_i}\left(\sum_{p=1}^{i-1} x_p + w_i\left[\sum_{p=1}^{n} p x_p - l\right]\right)$$

for $l = i, \ldots, n$.

Observe that for the stable distribution vector $x(L)$ of a Leslie matrix L of the form (4.1), we have $x_l(L) \geq x_{l+1}(L)$ if and only if $r(L) \geq p_l$. In particular, if $r(L) > 1$, (so that, asymptotically, the total size of the population being modelled is growing with time) we have $x_1(L) \geq x_2(L) \geq \ldots \geq x_n(L)$. In developing the next two results, we we will assume that the entries of $x(L)$ are nonincreasing. Again we suppress the dependence of the entries of $x(L)$ on L.

COROLLARY 4.2.9 ([72]) *Let L be a Leslie matrix with stable population vector x, and suppose that $x_1 \geq x_2 \geq \ldots \geq x_n$. If $\frac{\partial x_l}{\partial_{i,j}} \geq 0$ and $l \neq i-1$, then $\frac{\partial x_l}{\partial_{i,j}} > \frac{\partial x_{l+1}}{\partial_{i,j}}$.*

Proof: Suppose that $1 \leq l \leq i-2$. Since $x_l \geq x_{l+1}$ and $\frac{\partial x_l}{\partial_{i,j}} \geq 0$, we find from Corollary 4.2.8 that

$$\frac{\partial x_l}{\partial_{i,j}} = \frac{x_j x_l}{r(L)x_i}\left(-\sum_{p=i}^{n} x_p + w_i\left[\sum_{p=1}^{n} p x_p - l\right]\right) \geq$$
$$\frac{x_j x_{l+1}}{r(L)x_i}\left(-\sum_{p=i}^{n} x_p + w_i\left[\sum_{p=1}^{n} p x_p - l\right]\right).$$

But

$$\frac{x_j x_{l+1}}{r(L)x_i}\left(-\sum_{p=i}^{n} x_p + w_i\left[\sum_{p=1}^{n} p x_p - l\right]\right) >$$

$$\frac{x_j x_{l+1}}{r(L)x_i}\left(-\sum_{p=i}^{n} x_p + w_i\left[\sum_{p=1}^{n} p x_p - (l+1)\right]\right) = \frac{\partial x_{l+1}}{\partial_{i,j}},$$

so we find that $\frac{\partial x_l}{\partial_{i,j}} > \frac{\partial x_{l+1}}{\partial_{i,j}}$. A similar argument applies if $i \le l \le n-1$. □

Next, we turn our attention to providing upper and lower bounds on the entries of the derivative of the stable distribution vector x with respect to the entries in a Leslie matrix L on the top row. As above, we restrict our attention to the situation that the entries in x are nonincreasing. Referring to (4.12), we see that in the Leslie matrix context, the vector w also has the property that its entries are nonincreasing and sum to 1.

THEOREM 4.2.10 *Suppose that L is an irreducible Leslie matrix with Perron value $r(L)$ and stable distribution vector $x(L)$. Suppose further that the entries of $x(L)$ are nonincreasing.*
a) For each $j = 1, \ldots, n$, we have

$$\frac{\partial x_l}{\partial_{1,j}} \ge -\frac{l-1}{r(L)max\{j,l\}}.$$

b) For each $j = 1, \ldots, n$, we have

$$\frac{\partial x_l}{\partial_{1,j}} \le \frac{n+1-2l}{2r(L)max\{j,l\}}$$

if $l \le (n+1)/2$, and

$$\frac{\partial x_l}{\partial_{1,j}} < 0$$

if $l > (n+1)/2$.

Proof: a) From Corollary 4.2.8, we find that

$$\frac{\partial x_l}{\partial_{1,j}} = \frac{x_j x_l w_1}{r(L)x_1}\left(\sum_{p=1}^{n} p x_p - l\right),$$

where w is given by (4.12). Note that $\sum_{p=1}^{n} p x_p - l \ge 1-l$, while $\frac{x_j x_l w_1}{x_1} \le$

$w_1 \min\{x_l \frac{x_j}{x_1}, x_j \frac{x_l}{x_1}\} \leq w_1 \min\{x_l, x_j\}$. Since $\mathbf{1}^t x = 1$ and $x_1 \geq \ldots \geq x_n$, it follows that $x_j \leq \frac{1}{j}, x_l \leq \frac{1}{l}$. Further, we have $w_1 \leq 1$, so we see that

$$\frac{x_j x_l w_1}{r(L) x_1} \left(\sum_{p=1}^{n} p x_p - l \right) \geq \frac{1-l}{r(L)} \min\left\{\frac{1}{j}, \frac{1}{l}\right\},$$

from which the conclusion follows readily.

b) As above, we have

$$\frac{\partial x_l}{\partial_{1,j}} = \frac{x_j x_l w_1}{r(L) x_1} \left(\sum_{p=1}^{n} p x_p - l \right).$$

Since the entries of x are nonincreasing and sum to 1, it follows that $\sum_{p=1}^{n} p x_p \leq \frac{n+1}{2}$. Hence we find that $\frac{x_j x_l w_1}{r(L) x_1} (\sum_{p=1}^{n} p x_p - l) \leq \frac{x_j x_l w_1}{r(L) x_1} (\frac{n+1}{2} - l)$.

Observing that the entries of w sum to 1, and that w_1 is the maximum entry in w, it follows that $\frac{1}{n} \leq w_1 \leq 1$. If $l \leq (n+1)/2$, then from the fact that $w_1 \leq 1$, we have that

$$\frac{x_j x_l w_1}{r(L) x_1} \left(\frac{n+1}{2} - l \right) \leq \frac{1}{r(L) max\{j, l\}} \left(\frac{n+1}{2} - l \right).$$

On the other hand, if $l > (n+1)/2$, we have

$$\frac{x_j x_l w_1}{r(L) x_1} \left(\frac{n+1}{2} - l \right) < 0.$$

□

We note that Theorem 4.2.10 above serves a correction to [72, Theorem 4.1]; the latter is in error in part b) for the case that $l > (n+1)/2$.

REMARK 4.2.11 Evidently our lower bound on $\frac{\partial x_l}{\partial_{1,j}}$ is nonpositive, nondecreasing in j and nonincreasing in l. For $l \leq \frac{n+1}{2}$, our upper bound on $\frac{\partial x_l}{\partial_{1,j}}$ is nonnegative, and nonincreasing in both j and l. For $l > \frac{n+1}{2}$, we find that x_l is a decreasing function of every entry in the first row of L.

Next, we turn our attention to $\frac{\partial x}{\partial_{i,i-1}}$. The following two examples from [72] show that some of the entries in that vector may not be bounded, even if the entries of x are nonincreasing.

EXAMPLE 4.2.12 Suppose that $i \geq 3$ and $1 \leq l \leq \frac{i-1}{2}$. Fix a small positive ϵ, and define the vector x via $x_p = \frac{1-\epsilon}{i-1}$ for $p = 1, \ldots, i-1, x_i = \epsilon - \epsilon^2$, and $x_p = \frac{\epsilon^2}{n-i}$ for $p = i+1, \ldots, n$. Note that the entries of x are nonincreasing and sum to 1.

Next we consider the Leslie matrix

$$L = \begin{bmatrix} 0 & 0 & \ldots & 0 & \frac{x_1}{x_n} \\ \frac{x_2}{x_1} & 0 & 0 & \ldots & 0 \\ 0 & \frac{x_3}{x_2} & 0 & \ldots & 0 \\ \vdots & & \ddots & \ddots & \vdots \\ 0 & \ldots & 0 & \frac{x_{n-1}}{x_n} & 0 \end{bmatrix};$$

it is straightforward to determine that $r(L) = 1$, and that x is the stable distribution vector for L. Note also that in this case, the vector w of (4.12) is given by $w = \frac{1}{n}\mathbf{1}$.

Evaluating the expression $\frac{x_{i-1}x_l}{x_i}\left(-\sum_{p=i}^n x_p + w_i\left[\sum_{p=1}^n px_p - l\right]\right)$ yields a value of $\frac{(1-\epsilon)^2}{(i-1)^2(\epsilon-\epsilon^2)}(-\epsilon + \frac{1}{2n}(i - 2l + i\epsilon + (n+i+1)\epsilon^2))$, which diverges to infinity as $\epsilon \to 0^+$. Consequently we see that if $3 \leq i$ and $1 \leq l \leq \frac{i-1}{2}$, there is a sequence of Leslie matrices with decreasingly ordered stable distribution vectors such that the corresponding sequence of derivatives $\frac{\partial x_l}{\partial_{i,i-1}}$ is unbounded from above.

EXAMPLE 4.2.13 Suppose that $2 \leq l \leq i-1$. Let ϵ be small and positive, and let x be given by $x_1 = 1-\epsilon, x_2 = \ldots, x_{i-1} = \frac{\epsilon-\epsilon^3}{i-2}, x_i = \epsilon^3 - \epsilon^4$, and $x_p = \frac{\epsilon^4}{n-i}, p = i+1\ldots, n$. Evidently the entries of x are nonincreasing and sum to 1; as in Example 4.2.12, the vector x is the stable distribution vector for the Leslie matrix

$$L = \begin{bmatrix} 0 & 0 & \ldots & 0 & \frac{x_1}{x_n} \\ \frac{x_2}{x_1} & 0 & 0 & \ldots & 0 \\ 0 & \frac{x_3}{x_2} & 0 & \ldots & 0 \\ \vdots & & \ddots & \ddots & \vdots \\ 0 & \ldots & 0 & \frac{x_{n-1}}{x_n} & 0 \end{bmatrix}.$$

Further, $r(L) = 1$, and the corresponding vector w of (4.12) is $\frac{1}{n}\mathbf{1}$.

Evaluating the expression

$$\frac{x_{i-1}x_l}{x_i}\left(-\sum_{p=i}^{n} x_p + w\left[\sum_{p=1}^{n} px_p - l\right]\right)$$

yields the value

$$\frac{(\epsilon-\epsilon^3)^2}{(i-2)^2(\epsilon^3-\epsilon^4)}(-\epsilon^3+\frac{1}{2n}(2-2l+(i-1)\epsilon+(i-1)\epsilon^3+(n-i+1)\epsilon^4)),$$

which diverges to $-\infty$ as $\epsilon \to 0^+$. Hence for $2 \leq l \leq i-1$, there is a sequence of Leslie matrices with decreasingly ordered stable distribution vectors such that the corresponding sequence of derivatives $\frac{\partial x_l}{\partial_{i,i-1}}$ is unbounded from below.

We remark that [72] provides an upper bound on $\frac{\partial x_l}{\partial_{i,i-1}}$ when $\frac{i}{2} \leq l \leq i-1$, as well as upper and lower bounds on $\frac{\partial x_l}{\partial_{i,i-1}}$ for $i \leq l \leq n$, and for $l=1, i=2$.

4.3 Second derivatives of the Perron value in the age-classified model

In section 1.1, we saw by example that for an irreducible Leslie matrix L, while the Perron value $r(L)$ is an increasing function of each entry of L, $r(L)$ may be concave up as a function of some entries, and concave down as a function of other entries. In this section we explore that phenomenon further by taking a close look at the signs of the quantities $\frac{\partial^2 r(L)}{\partial^2_{i,j}}$ corresponding to the demographically interesting pairs of indices (i,j), namely $(1,j), j=1,\ldots,n$ and $(k+1,k), k=1,\ldots,n-1$. In this section we follow the general approach of Kirkland and Neumann's paper [81], where the results in this section first appeared.

We begin with a discussion of the sign pattern of the entries in the first column of the group inverse associated with a stochastic Leslie matrix.

THEOREM 4.3.1 ([81]) *Let L be an irreducible stochastic Leslie ma-*

trix of the form

$$L = \begin{bmatrix} a_1 & a_2 & \dots & \dots & a_{n-1} & a_n \\ 1 & 0 & 0 & \dots & 0 & 0 \\ 0 & 1 & 0 & 0 & 0 & 0 \\ \vdots & \dots & \ddots & \ddots & \vdots & \vdots \\ \vdots & \dots & \dots & \ddots & \vdots & \vdots \\ 0 & \dots & \dots & \dots & 1 & 0 \end{bmatrix}, \tag{4.19}$$

and set $Q = I - L$. *Let* $s_0 = 0$, *and* $s_i = \sum_{p=1}^{i} a_p, i = 1, \dots, n-1$. *Then for each* $i = 1, \dots, n$, *we have*

$$q^{\#}_{i,1} = \frac{1}{\sum_{j=0}^{n-1}(1-s_j)} \left(\frac{\sum_{j=0}^{n-1}(1-s_j)j}{\sum_{j=0}^{n-1}(1-s_j)} - (i-1) \right). \tag{4.20}$$

In particular, $q^{\#}_{i,1}$ *is strictly decreasing with the index* i. *Further, there is an index* $k_0 < \frac{n+1}{2}$ *such that* $q^{\#}_{i,1} \geq 0$ *for* $i = 1, \dots, k_0$ *and* $q^{\#}_{i,1} \leq 0$ *for* $i = k_0 + 1, \dots, n$.

Proof: Referring to (4.14) we note that since L is a stochastic Leslie matrix, the matrix R of (4.13) is the lower triangular matrix with ones on and below the diagonal. We then find that the first column of $Q^{\#}$ can be written as

$$(w_1 \widehat{w}^t R\mathbf{1})\mathbf{1} - w_1 \begin{bmatrix} 0 \\ 1 \\ 2 \\ \vdots \\ n-1 \end{bmatrix},$$

where

$$w = \frac{1}{\sum_{i=0}^{n-1}(1-s_i)} \begin{bmatrix} 1-s_0 & 1-s_1 & \dots & 1-s_{n-1} \end{bmatrix}^t,$$

and where $\widehat{w}$ is the trailing subvector of w of order $n-1$. In particular, (4.20) now follows, and it is transparent that $q^{\#}_{i,1}$ is strictly decreasing with the index i.

To establish the second conclusion, it suffices to show that if $i \geq \frac{n+1}{2}$, then $q^{\#}_{i,1} \leq 0$. It is readily seen that $q^{\#}_{i,1} \leq 0$ if and only if

$$\sum_{j=1}^{n-1}(1-s_j)j \leq (i-1)\sum_{j=0}^{n-1}(1-s_j). \tag{4.21}$$

A little manipulation then establishes that (4.21) is equivalent to

$$\sum_{j=0}^{n-i-1}(1-s_{j+i})(j+1) \le \sum_{j=0}^{i-2}(1-s_{i-j-2})(j+1). \tag{4.22}$$

Now suppose that $i \ge \frac{n+1}{2}$. Then $n-i-1 \le i-2$, and that fact, in conjunction with the fact that $s_{j+1} \ge s_{i-j-2}, i = 0, \ldots, n-i-1$, establishes that (4.22) holds for any such i. □

Recall that in Example 1.1.1, we observed that for a certain family of 7×7 Leslie matrices, the corresponding Perron value is concave down as a function of the $(1,7)$ entry. Theorem 4.3.1 serves to explain that observation, since it follows from Theorem 4.3.1 that for an irreducible Leslie matrix of order 7, the second derivative of the Perron value with respect to the $(1,7)$ entry is negative. Example 1.1.1 also provides a family of 7×7 Leslie matrices whose Perron value is concave up as a function of the $(1,1)$ entry; that observation is explained by the discussion in section 3.3, where we saw that the Perron value is a concave up function of any diagonal entry.

EXAMPLE 4.3.2 ([81]) We now show that for any index k with $2 \le k < \frac{n+1}{2}$, there is a stochastic Leslie matrix L such that, in the language of Theorem 4.3.1, $q^{\#}_{k,1} > 0 > q^{\#}_{k+1,1}$. To construct the desired stochastic Leslie matrix, fix such a k, and consider the stochastic Leslie matrix L whose top row contains an α in the $(k-1)$-st position, and a $1-\alpha$ in the n–th position, where α is chosen so that

$$\frac{n(n-1-2k)}{(n-k)(n-k-1)-2} < \alpha < \frac{n(n+1-2k)}{(n-k)(n-k+1)}. \tag{4.23}$$

(We note in that from the hypothesis on k, we always have $0 < \frac{n(n-1-2k)}{(n-k)(n-k-1)-2} < \frac{n(n+1-2k)}{(n-k)(n-k+1)} < 1$.)

We find now that $\sum_{j=0}^{n-1}(1-s_j) = n - \alpha(n-k+1)$, while

$$\sum_{j=0}^{n-1}(1-s_j)j = \frac{n(n-1)}{2} - \alpha\left(\frac{n(n-1)}{2} - \frac{(k-2)(k-1)}{2}\right).$$

From (4.20) it follows that $q^{\#}_{k,1} > 0$ if and only if

$$\frac{\sum_{j=0}^{n-1}(1-s_j)j}{\sum_{j=0}^{n-1}(1-s_j)} > k-1,$$

and this last is readily seen to be equivalent to the condition

$$n(n-2k+1) - \alpha(n-k)(n-k+1) > 0.$$

Similarly, again from (4.20), we have $q^{\#}_{k+1,1} < 0$ if and only if

$$\frac{\sum_{j=0}^{n-1}(1-s_j)j}{\sum_{j=0}^{n-1}(1-s_j)} < k,$$

which is in turn equivalent to the condition

$$n(n-1-2k) < \alpha((n-k)(n-k-1)-2).$$

As α has been chosen so that (4.23) holds, we thus see that $q^{\#}_{k,1} > 0 > q^{\#}_{k+1,1}$. Thus, Theorem 4.3.1 establishes a constraint on the positions where a sign change can take place in the first column of $Q^{\#}$, while the above class of examples shows that the sign change can occur in any admissible position.

Next, we consider the superdiagonal entries of the group inverse associated with a stochastic Leslie matrix.

THEOREM 4.3.3 ([81]) *Let L be an irreducible stochastic Leslie matrix of the form (4.19), and set $Q = I - L$. For each $k = 1, \ldots, n-1$ we have*

$$q^{\#}_{k,k+1} = \tag{4.24}$$
$$-\frac{(k-1)(1-s_k)}{\sum_{j=0}^{n-1}(1-s_j)} - \frac{\sum_{j=k}^{n-1}(1-s_j)}{\sum_{j=0}^{n-1}(1-s_j)} + \frac{(1-s_k)}{\sum_{j=0}^{n-1}(1-s_j)}\frac{\sum_{j=0}^{n-1}(1-s_j)j}{\sum_{j=0}^{n-1}(1-s_j)}.$$

Any nonnegative entries on the superdiagonal of $Q^{\#}$ appear consecutively, starting from the $(1,2)$ position. If $k \geq \frac{n-1}{2}$ then $q^{\#}_{k,k+1} < 0$, and further, the entries $q^{\#}_{k,k+1}$ form a nondecreasing sequence for $k \geq \frac{n+1}{2}$.

Proof: From (4.14), we find that $q^{\#}_{1,2} = -\widehat{w}^t Re_1 + (\widehat{w}^t R\mathbf{1})(1-s_1)$, while for $k = 2, \ldots, n-1$,

$$q^{\#}_{k,k+1} = e_{k-1}Re_k - e_{k-1}R\mathbf{1}\widehat{w}^t e_k - \widehat{w}^t Re_k + (\widehat{w}^t R\mathbf{1})\widehat{w}^t e_k.$$

Substituting the formulas for R and $\widehat{w}$, (4.24) is readily established.

In order to show that any nonnegative entries on the superdiagonal of $Q^{\#}$ are consecutive, starting at the $(1,2)$ entry, it suffices to prove that if $q^{\#}_{k,k+1} \geq 0$ for some $k \geq 2$, then $q^{\#}_{k-1,k} \geq 0$. So, suppose that $q^{\#}_{k,k+1} \geq 0$ for some $k \geq 2$. We find from (4.24) that necessarily we must have $\frac{\sum_{j=0}^{n-1}(1-s_j)j}{\sum_{j=0}^{n-1}(1-s_j)} \geq k-1$. Next we consider $q^{\#}_{k-1,k} - q^{\#}_{k,k+1}$, and note that by substituting (4.24) for indices $k-1$ and k and simplifying, we find that

$$q^{\#}_{k-1,k} - q^{\#}_{k,k+1} = \frac{s_k - s_{k-1}}{\sum_{j=0}^{n-1}(1-s_j)} \left(\frac{\sum_{j=0}^{n-1}(1-s_j)j}{\sum_{j=0}^{n-1}(1-s_j)} - (k-1) \right), \quad (4.25)$$

which is necessarily nonnegative since $s_k \geq s_{k-1}$. Hence $q^{\#}_{k-1,k}$ is also nonnegative.

Next, suppose that $k \geq \frac{n-1}{2}$. Observe from (4.24) that in order to show that $q^{\#}_{k,k+1} \leq 0$, it suffices to prove that

$$-k\sum_{j=0}^{n-1}(1-s_j) + \sum_{j=0}^{n-1}(1-s_j)j \leq 0. \quad (4.26)$$

Observe that (4.26) can be rewritten as

$$-\sum_{j=0}^{k-1}(1-s_j)(k-j) + \sum_{j=k+1}^{n-1}(1-s_j)(j-k) \leq 0,$$

which is in turn equivalent to

$$-\sum_{j=1}^{k}(1-s_{k-j})j + \sum_{j=1}^{n-k-1}(1-s_{j+k})j \leq 0. \quad (4.27)$$

Since $k \geq n-k-1$ and $s_{k-j} \leq s_{j+k}$, for each $j = 1, \ldots n-k-1$, the inequality (4.27) now follows, which in turn is sufficient to show that $q^{\#}_{k,k+1} \leq 0$.

Finally, suppose that $k \geq \frac{n+1}{2}$. From (4.25), we find that $q^{\#}_{k-1,k} \leq q^{\#}_{k,k+1}$ if and only if

$$\sum_{j=0}^{n-1} j(1-s_j) \leq (k-1)\sum_{j=0}^{n-1}(1-s_j). \quad (4.28)$$

Observe that we may rewrite (4.28) in an equivalent form as

$$\sum_{j=1}^{n-k} j(1-s_{j+k-1}) \leq \sum_{j=1}^{k-1} j(1-s_{k-j-1}). \quad (4.29)$$

From the hypothesis that $k \geq \frac{n+1}{2}$, we find that $n-k \leq k-1$; we also have $1-s_{j+k-1} \leq 1-s_{k-j-1}$ for each $j=1,\ldots,n-k$. Consequently, (4.29) is readily seen to hold, and so we conclude that $q^{\#}_{k-1,k} \leq q^{\#}_{k,k+1}$ for $k \geq \frac{n+1}{2}$. □

The following result considers the Perron value for an irreducible Leslie matrix, and gives expressions for the second derivatives with respect to entries in the first row and subdiagonal of the matrix. The expressions are deduced directly from Corollary 3.3.1, Theorem 4.3.1, and Theorem 4.3.3.

THEOREM 4.3.4 ([81]) *Let L be an irreducible Leslie matrix of the form (4.17), with Perron value $r(L)$ and stable distribution vector x. Let $X = \text{diag}(x)$. Set $Q = I - \frac{1}{r(L)}X^{-1}LX$. For each $i = 1,\ldots,n$ we have*

$$\frac{\partial^2 r(L)}{\partial^2_{1,i}} = \frac{2}{r(L)}\left(\frac{1}{\sum_{j=0}^{n-1}(1-s_j)}\right)\left(\frac{p_1p_2\cdots p_{i-1}}{r(L)^{i-1}}\right)^2 q^{\#}_{i,1}. \tag{4.30}$$

For each $k = 1,\ldots,n-1$ we have

$$\frac{\partial^2 r(L)}{\partial^2_{k+1,k}} = \frac{2r(L)}{p_k^2}\left(\frac{1-s_k}{\sum_{j=0}^{n-1}(1-s_j)}\right) q^{\#}_{k,k+1}. \tag{4.31}$$

Further, there is an index $k_0 < \frac{n+1}{2}$ such that $\frac{\partial^2 r(L)}{\partial^2_{1,i}} \geq 0$ for $i = 1,\ldots,k_0$ and $\frac{\partial^2 r(L)}{\partial^2_{1,i}} \leq 0$ for $i = k_0+1,\ldots,n$. If, $\frac{\partial^2 r(L)}{\partial^2_{k+1,k}} \geq 0$ for some index k between 2 and n, then $\frac{\partial^2 r(L)}{\partial^2_{j+1,j}} \geq 0$ for each $j = 1,\ldots,k$.

Suppose that $r(L) \geq p_i, i = 1,\ldots,n-1$. If we have $\frac{\partial^2 r(L)}{\partial^2_{1,i}} \geq 0$ for some i, then necessarily $\frac{\partial^2 r(L)}{\partial^2_{1,i}} \geq \frac{\partial^2 r(L)}{\partial^2_{1,j}}$ for any index $j > i$. In particular,

$$\frac{\partial^2 r(L)}{\partial^2_{1,1}} = \max_{j=1,\ldots,n} \frac{\partial^2 r(L)}{\partial^2_{1,j}}.$$

Finally, if $p_j \geq p_i$ for indices i,j with $j > i$, and if $\frac{\partial^2 r(L)}{\partial^2_{i+1,i}} \geq 0$, then necessarily $\frac{\partial^2 r(L)}{\partial^2_{i+1,i}} \geq \frac{\partial^2 r(L)}{\partial^2_{j+1,j}}$.

We saw in section 4.2 that for a Leslie matrix L, if $r(L) > p_j, j = 1,\ldots,n-1$, then the first derivative of $r(L)$ with respect to the $(1,i)$ entry dominates the first derivative of $r(L)$ with respect to the $(1,j)$ entry whenever $j > i$. Theorem 4.3.4 provides some partial reinforcement of that inequality, as that result asserts that if the second derivative of $r(L)$ with respect to the $(1,i)$ entry is nonnegative, then it must dominate second derivative of $r(L)$ with respect to the $(1,j)$ entry whenever $j > i$. Similarly, in section 4.2 we saw that if $p_j \geq p_i$ and $j > i$, then first derivative of $r(L)$ with respect to the $(i+1,i)$ entry dominates the first derivative of $r(L)$ with respect to the $(j+1,j)$ entry; again Theorem 4.3.4 provides some partial reinforcement of that observation by establishing a corresponding inequality for second derivatives, under the hypothesis that $\frac{\partial^2 r(L)}{\partial^2_{i+1,i}} \geq 0$.

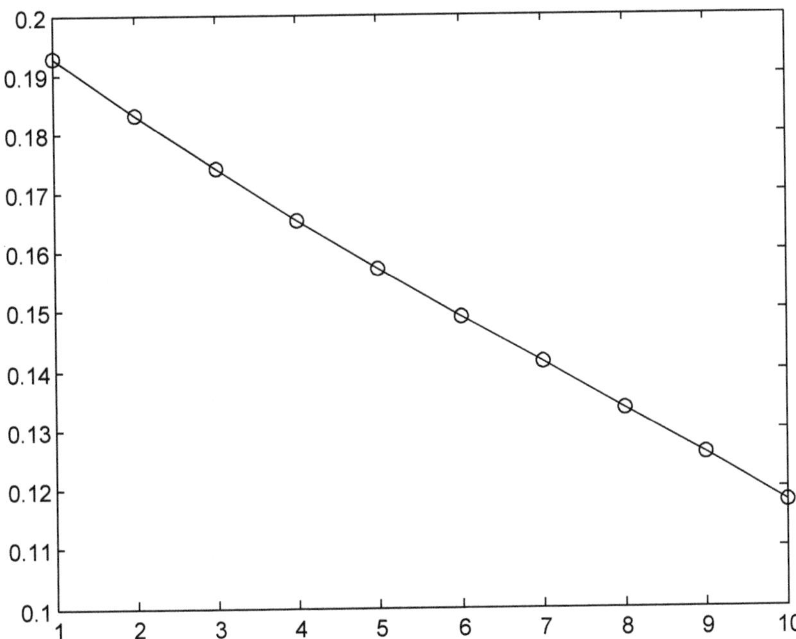

FIGURE 4.3
$\frac{\partial r(L)}{\partial_{1,k}}, k = 1, \ldots, 10$

EXAMPLE 4.3.5 In this example, we consider a Leslie matrix L arising from data reported in the book of Keyfitz and Flieger [67, p. 355] for the (female human) population of the United States in 1966. Here the population is classified into ten five-year age groups, yielding a 10×10

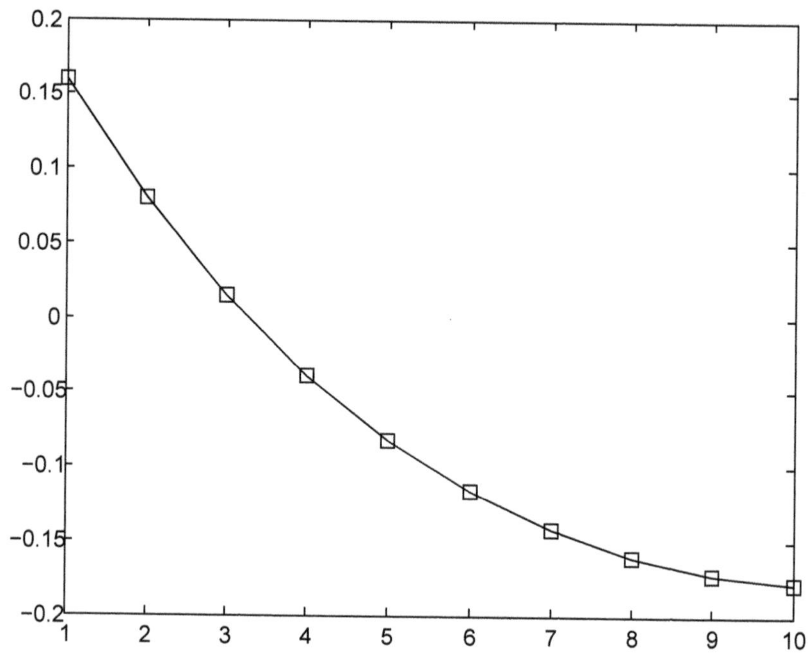

FIGURE 4.4
$\frac{\partial^2 r(L)}{\partial^2_{1,k}}, k = 1, \ldots, 10$

Leslie matrix L. The nonzero entries of L are (to four decimal places)

$$\begin{bmatrix} f_2 \\ \vdots \\ f_{10} \end{bmatrix} = \begin{bmatrix} 0.0010 \\ 0.0852 \\ 0.3057 \\ 0.4000 \\ 0.2806 \\ 0.1526 \\ 0.0642 \\ 0.0148 \\ 0.0009 \end{bmatrix} \text{ and } \begin{bmatrix} p_1 \\ \vdots \\ p_9 \end{bmatrix} = \begin{bmatrix} 0.9967 \\ 0.9984 \\ 0.9978 \\ 0.9967 \\ 0.9961 \\ 0.9947 \\ 0.9924 \\ 0.9887 \\ 0.9827 \end{bmatrix}.$$

The Perron value for L is 1.0498, while the stable distribution vector is

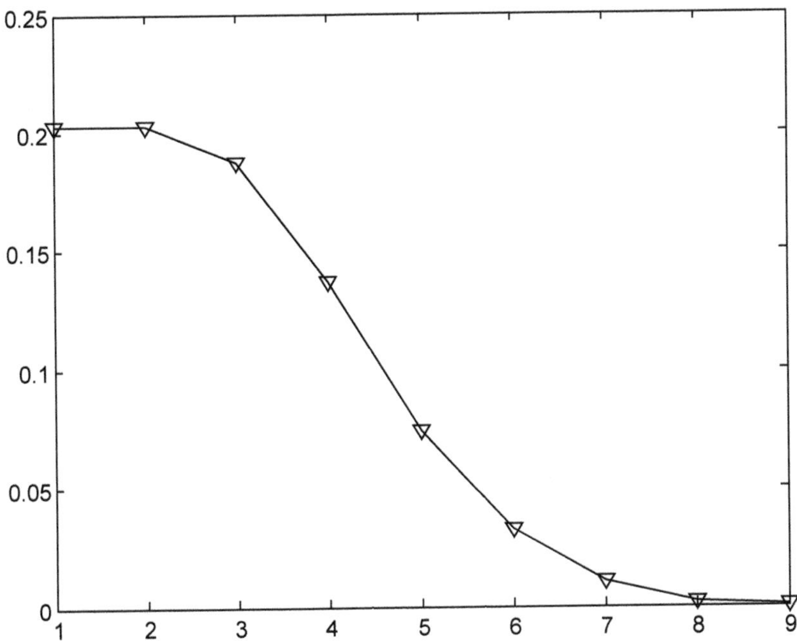

FIGURE 4.5
$\frac{\partial r(L)}{\partial_{k+1,k}}, k = 1, \ldots, 9$

computed as

$$x(L) = \begin{bmatrix} 0.1252 \\ 0.1189 \\ 0.1131 \\ 0.1075 \\ 0.1020 \\ 0.0968 \\ 0.0917 \\ 0.0867 \\ 0.0817 \\ 0.0765 \end{bmatrix}.$$

Using (3.2) and (3.7), we computed the first and second derivatives of $r(L)$ with respect to the entries in the first row of L, and with respect to the entries on the subdiagonal of L. Figures 4.3 and 4.4 depict $\frac{\partial r(L)}{\partial_{1,k}}$ and $\frac{\partial^2 r(L)}{\partial^2_{1,k}}$ for $k = 1 \ldots, 10$, respectively. As anticipated by our remarks in section 4.2, we see from Figure 4.3 that $\frac{\partial r(L)}{\partial_{1,k}}$ is decreasing

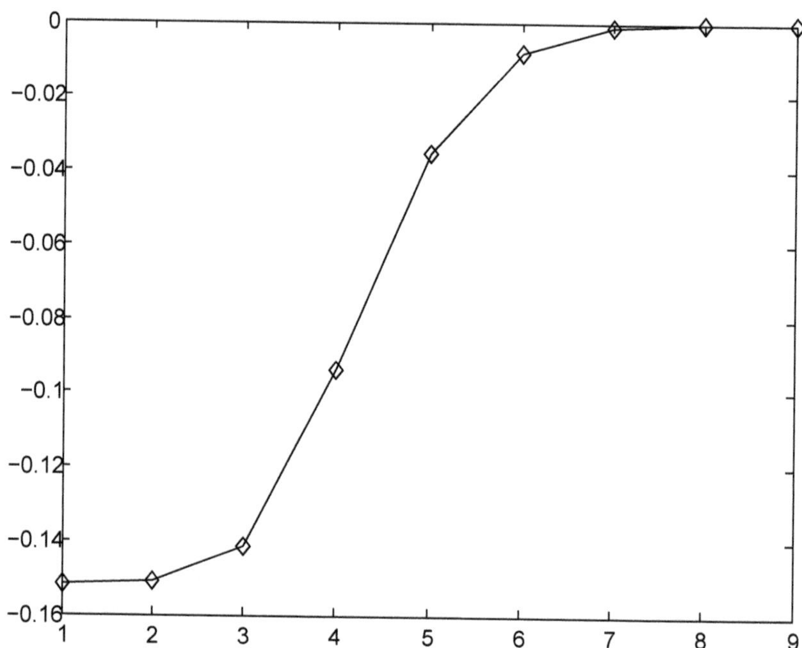

FIGURE 4.6
$\frac{\partial^2 r(L)}{\partial^2_{k+1,k}}, k = 1, \ldots, 9$

with the index k. This is of course due to the fact that $r(L) > p_k$, for each $k = 1, \ldots, 9$. Figure 4.4 illustrates the conclusion of Theorem 4.3.1, for we see that $r(L)$ is concave up as a function of $L_{1,k}$ for $k = 1, 2, 3$, and concave down as a function of $L_{1,k}$ for $k = 4, \ldots, 10$.

From Figures 4.5 and 4.6, we find that $r(L)$ is an increasing, concave down function of each subdiagonal entry of L. Note that $\frac{\partial r(L)}{\partial_{k+1,k}}$ is positive, but decreasing with the index k, while $\frac{\partial^2 r(L)}{\partial^2_{k+1,k}}$ is negative and increasing with the index k; thus we find that as the index k increases, there is a flattening of the sensitivity of $r(L)$ to changes in the $(k+1, k)$ entry of L.

4.4 Elasticity and its derivatives for the Perron value

In section 4.2, we considered, for a population projection matrix A, the sensitivities of the Perron value with respect to the matrix entries, i.e., expressions of the form $\frac{\partial r(A)}{\partial_{i,j}}$. Such derivatives thus measure the additive effect on $r(A)$ of a small additive perturbation of $a_{i,j}$. However, the nonzero entries of a population projection matrix may be of very different orders of magnitude, as is the case for the population projection matrix of (4.9). In that scenario, it may be more meaningful to consider a measure of the effect on the Perron value of a small *proportional* change in a matrix entry.

The *elasticity* $e_{i,j}$ of $r(A)$ with respect to $a_{i,j}$, which is defined in a paper of de Kroon, Plaisier, van Groenendael, and Caswell [29], provides just such a measure. For any irreducible nonnegative matrix A, the elasticity $e_{i,j}$ of $r(A)$ with respect to the (i,j) entry is defined as

$$e_{i,j} = \frac{\partial \ln r(A)}{\partial \ln a_{i,j}}.$$

Observe that $e_{i,j}$ admits the alternate representation

$$e_{i,j} = \frac{a_{i,j}}{r(A)} \frac{\partial r(A)}{\partial_{i,j}}. \tag{4.32}$$

Inspecting (4.32), we see that $e_{i,j}$ measures the proportional change in $r(A)$ arising from a small proportional change in $a_{i,j}$.

As is remarked in [29], the notion of elasticity is useful for population projection matrices, where the demographically relevant entries may be subject to different constraints on their size. For instance, for a population projection matrix of the form (4.1), the p_is and b_is, which measure survival rates, necessarily cannot exceed 1, while the f_is which reflect fecundity rates, are unbounded in general. So while the sensitivities of the Perron value (i.e, the partial derivatives with respect to matrix entries) provide some valuable information, they may neglect the constraints placed on the sizes of those matrix entries. Consequently, elasticities are proposed to alleviate that concern.

The following example contrasts the elasticities with the sensitivities for a particular population projection matrix. A similar comparison (in pictorial form) is made in [17, section 9.2.2].

EXAMPLE 4.4.1 We return to the population projection matrix A for the desert tortoise given in (4.9). Computing right and left Perron vectors for A, we produce the following matrix of sensitivities:

$$\left[\ \frac{\partial r(A)}{\partial_{i,j}}\ \right]_{i,j=1,\ldots,8} =$$

$$\begin{bmatrix}
0.0433 & 0.0793 & 0.0302 & 0.0127 & 0.0075 & \underline{0.0060} & \underline{0.0140} & \underline{0.0023} \\
\underline{0.0580} & \underline{0.1062} & 0.0405 & 0.0170 & 0.0100 & 0.0081 & 0.0188 & 0.0031 \\
0.1522 & \underline{0.2786} & \underline{0.1062} & 0.0447 & 0.0264 & 0.0212 & 0.0493 & 0.0080 \\
0.3994 & 0.7313 & \underline{0.2786} & \underline{0.1173} & 0.0692 & 0.0556 & 0.1294 & 0.0211 \\
0.6018 & 1.1018 & 0.4198 & \underline{0.1767} & \underline{0.1043} & 0.0838 & 0.1949 & 0.0318 \\
1.0646 & 1.9492 & 0.7427 & 0.3126 & \underline{0.1845} & \underline{0.1482} & 0.3448 & 0.0563 \\
0.9712 & 1.7782 & 0.6775 & 0.2851 & 0.1683 & \underline{0.1352} & \underline{0.3145} & 0.0513 \\
1.1357 & 2.0794 & 0.7923 & 0.3334 & 0.1968 & 0.1581 & \underline{0.3678} & \underline{0.0600}
\end{bmatrix}.$$

Here the underlined entries are those that correspond to positions in which A has a nonzero entry. As noted in Example 4.2.1, the maximum sensitivity corresponding to the nonzero entries in A is found in the $(8, 7)$ position.

A different picture emerges when we consider the matrix of elasticities for A, which is given by

$$\left[\ e_{i,j}\ \right]_{i,j=1,\ldots,8} =$$

$$\begin{bmatrix}
0 & 0 & 0 & 0 & 0 & 0.0082 & 0.0290 & 0.0061 \\
0.0433 & 0.0628 & 0 & 0 & 0 & 0 & 0 & 0 \\
0 & 0.0433 & 0.0628 & 0 & 0 & 0 & 0 & 0 \\
0 & 0 & 0.0433 & 0.0739 & 0 & 0 & 0 & 0 \\
0 & 0 & 0 & 0 & 0.0433 & 0.1049 & 0 & 0 \\
0 & 0 & 0 & 0 & 0 & 0.0351 & 0.2794 & 0 \\
0 & 0 & 0 & 0 & 0 & 0 & 0.0061 & 0.0539
\end{bmatrix}.$$

Here we find that for the elasticities, the $(8, 7)$ position yields the minimum of the positive elasticity values, while the maximum elasticity is found in the $(7, 7)$ position. We thus conclude that the proportional change in the Perron value is most responsive to proportional changes in the survival rate from the seventh size class back into itself.

EXAMPLE 4.4.2 Here we consider the Leslie matrix L of Example 4.3.5 for the 1966 U.S. human population classified into five-year age groups. Computing the derivatives of the Perron value of L with respect to the entries in the top row of L, we have

$$\begin{bmatrix} \frac{\partial r(L)}{\partial_{1,1}} \\ \\ \vdots \\ \\ \frac{\partial r(L)}{\partial_{1,10}} \end{bmatrix} = \begin{bmatrix} 0.1928 \\ 0.1830 \\ 0.1741 \\ 0.1655 \\ 0.1571 \\ 0.1491 \\ 0.1413 \\ 0.1335 \\ 0.1258 \\ 0.1177 \end{bmatrix}.$$

Observe that the quantities $\frac{\partial r(L)}{\partial_{1,j}}$ are monotonically decreasing with the index j, indicating that $r(L)$ has diminishing sensitivity (as a function of the age class) to small additive perturbations in the fecundity rates.

The vector of elasticities with respect to the first row of L is given by

$$\begin{bmatrix} e_{1,1} \\ \\ \vdots \\ \\ e_{1,10} \end{bmatrix} = \begin{bmatrix} 0 \\ 0.0002 \\ 0.0141 \\ 0.0482 \\ 0.0599 \\ 0.0398 \\ 0.0205 \\ 0.0082 \\ 0.0018 \\ 0.0001 \end{bmatrix},$$

and here we see a different pattern. The elasticities $e_{1,j}$ are unimodal with respect to the index j, i.e., they are increasing for $j = 1, \ldots, 5$ and decreasing for $j = 5, \ldots, 10$. In particular, we find that, with respect to small proportional changes in fecundity rates, the largest proportional change in $r(L)$ arises from the fifth age class.

The following result, which appears in [29], establishes a dependence between the various elasticities.

THEOREM 4.4.3 *Let A be an irreducible nonnegative matrix of order n. Then*

$$\sum_{i=1}^{n}\sum_{j=1}^{n} e_{i,j} = 1.$$

Proof: Let x and y denote right and left Perron vectors for A, respectively, normalised so that $y^tx = 1$. From (3.2) and (4.32), we have $e_{i,j} = \frac{a_{i,j}x_jy_i}{r(A)}, i,j = 1,\ldots,n$. Summing this last relation, and using the fact that $\sum_{j=1}^n a_{i,j}x_j = r(A)x_i, i = 1,\ldots,n$ yields

$$\sum_{i=1}^n\sum_{j=1}^n e_{i,j} = \frac{1}{r(A)}\sum_{i=1}^n y_i \sum_{j=1}^n a_{i,j}x_j = \frac{1}{r(A)}\sum_{i=1}^n y_i r(A)x_i = y^tx = 1,$$

as desired. □

From Theorem 4.4.3, we have the additional interpretation that $e_{i,j}$ can be thought of as measuring the contribution of the entry $a_{i,j}$ to $r(A)$.

In view of the interest in elasticities, it is natural to wonder how $e_{i,j}$ behaves as a function of $a_{i,j}$ We investigate this question by considering the partial derivative of $e_{i,j}$ with respect to $a_{i,j}$. From (4.32) we find that for any $n \times n$ irreducible nonnegative matrix A,

$$\frac{\partial e_{i,j}}{\partial a_{i,j}} = \frac{a_{i,j}}{r(A)}\frac{\partial^2 r(A)}{\partial^2_{i,j}} - \frac{a_{i,j}}{r(A)^2}\left(\frac{\partial r(A)}{\partial_{i,j}}\right)^2 + \frac{1}{r(A)}\frac{\partial r(A)}{\partial_{i,j}}, i,j = 1,\ldots,n. \tag{4.33}$$

Using results from Chapter 3, the following lemma provides an alternate expression for $\frac{\partial e_{i,j}}{\partial a_{i,j}}$. In the remainder of this section, we will closely follow the matrix-theoretic approach in the paper of Kirkland, Neumann, Ormes, and Xu [77].

LEMMA 4.4.4 *Suppose that $A \in \Phi^{n,n}$. Denote the Perron value of A by $r(A)$, and let x, y denote right and left Perron vectors for A, respectively, normalised so that $y^tx = 1$. Let $Q = r(A)I - A$. Then for each $i, j = 1,\ldots,n$ we have*

$$\frac{\partial e_{i,j}}{\partial a_{i,j}} = \frac{y_ix_j}{r(A)}\left(2a_{i,j}q^{\#}_{j,i} - a_{i,j}\frac{y_ix_j}{r(A)} + 1\right). \tag{4.34}$$

In particular, $e_{i,j}$ is a nondecreasing function of $a_{i,j}$ if and only if

$$2a_{i,j}q^{\#}_{j,i} - a_{i,j}\frac{y_ix_j}{r(A)} + 1 \geq 0. \tag{4.35}$$

Proof: From (3.2) we find that $\frac{\partial r(A)}{\partial_{i,j}} = y_ix_j$, while from (3.7), we have $\frac{\partial^2 r(A)}{\partial^2_{i,j}} = 2y_ix_jq^{\#}_{j,i}$. Substituting those expressions into (4.33) and simplifying now yields (4.34). □

The goal in this section is to establish that $\frac{\partial e_{i,j}}{\partial a_{i,j}} \geq 0$ for all i, j, and we do so by first proving a sequence of technical results. Given an irreducible nonnegative matrix A as in Lemma 4.4.4, let $\tilde{A}$ be the stochastic matrix $\tilde{A} = \frac{1}{r(A)} X^{-1}AX$, where $X = \mathrm{diag}(x)$. Observe that a right Perron vector for $\tilde{A}$ is $\tilde{x} \equiv \mathbf{1}$, while $\tilde{y} \equiv Xy$ serves as a left Perron vector, and is normalised so that $\tilde{y}^t\tilde{x} = 1$. Letting $\tilde{Q} = I - \tilde{A}$, we find from Remark 2.5.3 that $\tilde{Q}^{\#} = r(A)X^{-1}Q^{\#}X$. Making the appropriate substitutions, we find readily that

$$2a_{i,j}q^{\#}_{j,i} - a_{i,j}\frac{y_i x_j}{r(A)} + 1 = 2\tilde{a}_{i,j}\tilde{q}^{\#}_{j,i} - \tilde{a}_{i,j}\frac{\tilde{y}_i\tilde{x}_j}{r(\tilde{A})} + 1$$

for all $i, j = 1, \ldots, n$. From these considerations, we see that any discussion of the sign of the left-hand side of (4.35) for the matrix A can be reformulated into an equivalent discussion of the sign of the left-hand side of (4.35) for the stochastic matrix $\tilde{A}$. We note that this reformulation in terms of the stochastic matrix $\tilde{A}$ is slightly simpler, since we have $r(\tilde{A}) = 1$ and $\tilde{x} = \mathbf{1}$. For this reason, several of the results that follow will be stated in the context of stochastic matrices only.

We first consider the special case that $i = j$, and note that, in the terminology above, the sign of $\frac{\partial e_{i,i}}{\partial a_{i,i}}$ coincides with the sign of $2\tilde{a}_{i,i}\tilde{q}^{\#}_{i,i} - \tilde{a}_{i,i}\tilde{y}_i\tilde{x}_j + 1$. Applying (2.17), we find that $\tilde{q}^{\#}_{i,i} > 0$, so that

$$2\tilde{a}_{i,i}\tilde{q}^{\#}_{i,i} - \tilde{a}_{i,i}\tilde{y}_i\tilde{x}_j + 1 \geq 1 - \tilde{a}_{i,i}\tilde{y}_i\tilde{x}_j > 0.$$

Consequently we have $\frac{\partial e_{i,i}}{\partial a_{i,i}} > 0, i = 1, \ldots, n$. Thus, in order to establish the nonnegativity of $\frac{\partial e_{i,j}}{\partial a_{i,j}}$ for $i, j = 1, \ldots, n$, we need only focus on the cases where $i \neq j$.

We begin with the following technical lemma. Recall that a square nonnegative matrix is *substochastic* if each of its row sums is bounded above by 1.

LEMMA 4.4.5 ([77]) *Let B be a substochastic matrix of order $n \geq 2$ with spectral radius less than 1. Fix an index j between 1 and n, and let $\alpha_l = e_j^t B^l \mathbf{1}$ for each $l \in \mathbb{N}$. Then*

$$\sum_{l=1}^{\infty} \alpha_l^2 + 2\sum_{l=1}^{\infty}\sum_{m=l+1}^{\infty} \alpha_l\alpha_m \leq \sum_{l=1}^{\infty}\alpha_l + 2\sum_{l=1}^{\infty}(l-1)\alpha_l. \qquad (4.36)$$

Suppose that for each vertex k in $\mathcal{D}(B)$, there is a walk from vertex j to *vertex k in $\mathcal{D}(B)$. In that case, if equality holds in (4.36), then there is a $p \in \mathbb{N}$ such that B is permutationally similar to a matrix of the form*

$$\left[\begin{array}{c|c|c|c|c|c} 0 & X_1 & 0 & 0 & \cdots & 0 \\ \hline 0 & 0 & X_2 & 0 & \cdots & 0 \\ \hline \vdots & & \ddots & \ddots & & \vdots \\ \hline 0 & 0 & \cdots & 0 & X_{p-1} & 0 \\ \hline 0 & 0 & 0 & \cdots & 0 & X_p \\ \hline 0 & 0 & 0 & \cdots & 0 & 0 \end{array}\right], \tag{4.37}$$

where $X_i\mathbf{1} = \mathbf{1}$, for $i = 1, \ldots, p$, and where X_1 has only one row, which corresponds to index j.

<u>**Proof**</u>: For each $l \in \mathbb{N}$, we have $0 \le \alpha_l \le 1$, so that $\sum_{l=1}^{\infty} \alpha_l^2 \le \sum_{l=1}^{\infty} \alpha_l$. Further,

$$\sum_{l=1}^{\infty} \sum_{m=l+1}^{\infty} \alpha_l \alpha_m = \sum_{m=2}^{\infty} \sum_{l=1}^{m-1} \alpha_l \alpha_m \le \sum_{m=2}^{\infty} (m-1)\alpha_m,$$

and so (4.36) follows readily.

Next, we suppose that equality holds in (4.36). From the hypothesis, every vertex in $\mathcal{D}(B)$ can be reached from vertex j by some directed walk, and without loss of generality we assume that $j = 1$. Then in particular for each $l \in \mathbb{N}$ we must have $\alpha_l = \alpha_l^2$, so that each α_l is either 0 or 1. By hypothesis, the spectral radius of B is less than 1, so necessarily $B^l \to 0$ as $l \to \infty$. Hence $\alpha_q < 1$ for some index q, from which we conclude that $\alpha_q = 0$ for some q. We thus find that in $\mathcal{D}(B)$, there is no walk of length q starting from vertex 1, and hence no walk of length longer than q starting from vertex 1. In particular, since every vertex in $\mathcal{D}(B)$ can be reached from 1 by some walk, we conclude that $\mathcal{D}(B)$ has no cycles. It follows that for some $p \in \mathbb{N}$, we have $\alpha_l = 1$ if $l \le p$ and $\alpha_l = 0$ if $l \ge p + 1$.

We claim that we can partition the vertices in $\mathcal{D}(B)$ that are distinct from 1 into sets $S_1, \cdots, S_p$ such that
i) for each i, S_i is the set of vertices v such that the distance from 1 to v in $\mathcal{D}(B)$ is i, and
ii) B is permutationally similar to a matrix of the form (4.37), where the diagonal blocks in the partitioning correspond to the index sets $\{1\}, S_1, S_2, \ldots, S_p$.

In order to establish the claim, we proceed by induction on p. Suppose first that $p = 1$; since every vertex distinct from 1 can be reached from 1 by some walk, and since $\mathcal{D}(B)$ contains no walks of length two or more starting at 1, it follows that for each $i \neq 1, 1 \to i$ is an arc in $\mathcal{D}(B)$. Hence, $S_1 = \{2, \ldots, n\}$. Since $\mathcal{D}(B)$ contains no cycles, it cannot contain an arc of the form $i \to 1$. It follows that the last $n-1$ rows and columns of B may be simultaneously permuted to the form

$$\left[\begin{array}{c|c} 0 & X_1 \\ \hline 0 & C \end{array}\right],$$

where X_1 is a positive row vector in $\mathbb{R}^{n-1}$ such that $X_1\mathbf{1} = 1$. Since $\alpha_2 = 0$, we see that $X_1 C\mathbf{1} = 0$; in view of the fact that X_1 is a positive vector, C must be the zero matrix. Hence B can be permuted to the form (4.37). Next, suppose that the claim holds for some $p \geq 1$, and that we have that $\alpha_l = 1$, if $l \leq p+1$, and that $\alpha_l = 0$, if $l \geq p+2$. Let S_1 be set of vertices i such that $1 \to i$ in $\mathcal{D}(B)$. Note that for each $l \geq 2$, $\alpha_l = \sum_{i=1}^{n} b_{j,i}\alpha_{i,l-1}$, where $\alpha_{i,l-1} = e_i^t B^{l-1}\mathbf{1}$. It follows that $\alpha_{i,l} = 1$, for $1 \leq l \leq p$ and that $\alpha_{i,l} = 0$, for $l \geq p+1$. Hence, for each vertex $i \in S_1$, the induction hypothesis applies to those vertices reachable from i by some walk in $\mathcal{D}(B)$, yielding a corresponding partitioning of the vertex set. But a vertex at distance d from i is necessarily at distance $d+1$ from j, and it now follows that the rows and columns of B can be simultaneously permuted to the form (4.37). This completes the proof of the induction step, and the proof of the claim.

From our claim above, it follows that B can be written in the form (4.37). Finally, since $\alpha_l = 1$ for $1 \leq l \leq p$, we see that that $X_i\mathbf{1} = \mathbf{1}$, for $i = 1, \ldots, p$. □

Lemma 4.4.5 helps to establish the following result.

PROPOSITION 4.4.6 ([77]) *Let B be a substochastic matrix of order $n \geq 2$ with spectral radius less than 1, and fix an index j. Then*

$$e_j^t(I-B)^{-1}\mathbf{1} + \left[e_j^t(I-B)^{-1}\mathbf{1}\right]^2 \leq 2e_j^t(I-B)^{-2}\mathbf{1}. \tag{4.38}$$

Suppose further that for each vertex k of $\mathcal{D}(B)$, there is a walk from vertex j to vertex k. In that case, if equality holds in (4.38), *then there is* *a* $p \in \mathbb{N}$ *such that* B *is permutationally similar to a matrix of the form*

$$\left[\begin{array}{c|c|c|c|c|c} 0 & X_1 & 0 & 0 & \cdots & 0 \\ \hline 0 & 0 & X_2 & 0 & \cdots & 0 \\ \hline \vdots & & \ddots & \ddots & & \vdots \\ \hline 0 & 0 & \cdots & 0 & X_{p-1} & 0 \\ \hline 0 & 0 & 0 & \cdots & 0 & X_p \\ \hline 0 & 0 & 0 & \cdots & 0 & 0 \end{array}\right],$$

where $X_i\mathbf{1} = \mathbf{1}$, *for* $i = 1, \ldots, p$, *and where* X_1 *has only one row, which corresponds to index* j.

Proof: Let $\alpha_l = e_j^t B^l \mathbf{1}$ for each $l \in \mathbb{N}$. Observe that since the spectral radius of B is less than 1, we have $(I-B)^{-1} = \sum_{l=0}^{\infty} B^l$ and $(I-B)^{-2} = \sum_{l=1}^{\infty} lB^{l-1}$. Thus we see that $2e_j^t(I-B)^{-2}\mathbf{1} = 2 + 2\sum_{l=2}^{\infty} l\alpha_{l-1}$, while $e_j^t(I-B)^{-1}\mathbf{1} = 1 + \sum_{l=1}^{\infty} \alpha_l$. Hence the inequality (4.38) is equivalent to the inequality

$$2 + 2\sum_{l=2}^{\infty} l\alpha_{l-1} \geq 1 + \sum_{l=1}^{\infty} \alpha_l + \left(1 + \sum_{l=1}^{\infty} \alpha_l\right)^2. \tag{4.39}$$

But (4.39) readily simplifies to

$$\sum_{l=1}^{\infty} \alpha_l + 2\sum_{l=1}^{\infty}(l-1)\alpha_l \geq \sum_{l=1}^{\infty} \alpha_l^2 + 2\sum_{l=1}^{\infty}\sum_{m=l+1}^{\infty} \alpha_l\alpha_m,$$

and so the conclusions, including the discussion of the equality case in (4.38), now follow from Lemma 4.4.5. □

We next use Proposition 4.4.6 in establishing the following result.

THEOREM 4.4.7 ([77]) *Suppose that* $n \geq 3$ *and that* $A \in \Phi^{n,n}$, *with* A *written as*

$$A = \left[\begin{array}{c|ccc} m_0 & m_1 & \cdots & m_{n-1} \\ \hline u & & B & \end{array}\right].$$

Then for each $1 \leq j \leq n-1$,

$$\frac{\partial e_{1,j+1}}{\partial a_{1,j+1}} \geq 0. \tag{4.40}$$

Further, we have

$$\frac{\partial e_{1,j+1}}{\partial a_{1,j+1}} = 0 \tag{4.41}$$

if and only if A is permutationally similar to a matrix $\overline{A}$ of the form

$$\overline{A} = \left[\begin{array}{c|c|c|c|c|c} 0 & 1 & 0 & 0 & \cdots & 0 \\ \hline 0 & 0 & X_1 & 0 & \cdots & 0 \\ \hline 0 & 0 & 0 & X_2 & \cdots & 0 \\ \hline \vdots & & \ddots & \ddots & & \vdots \\ \hline 0 & 0 & \cdots & 0 & X_{p-1} & 0 \\ \hline 0 & 0 & 0 & \cdots & 0 & X_p \\ \hline \mathbf{1} & 0 & 0 & \cdots & 0 & 0 \end{array}\right], \tag{4.42}$$

where the first row of A corresponds to the first row of $\overline{A}$, and where X_1 has just one row, which corresponds to row $j+1$ of A.

Proof: Let $m = \left[\begin{array}{ccc} m_1 & \ldots & m_{n-1} \end{array}\right]^t \in \mathbb{R}^{1,n-1}$ and let w^t be the left Perron vector for A, normalised so that $w^t\mathbf{1} = 1$. It follows from the eigen–equation $w^tA = w^t$ that $w_1 = 1/[1+m^t(I-B)^{-1}\mathbf{1}]$. Set $Q = I-A$; then for $j = 1, \ldots, n-1$, it follows from (2.17) (in conjunction with a permutation similarity) that

$$q^{\#}_{1,j+1} = w_1^2 m^t(I-B)^{-2}\mathbf{1} - w_1 e_j^t(I-B)^{-1}\mathbf{1}.$$

Define the quantity f as follows:

$$\begin{aligned} f = {} & 2m_j m^t(I-B)^{-2}\mathbf{1} \\ & - 2m_j e_j^t(I-B)^{-1}\mathbf{1}\left[1+m^t(I-B)^{-1}\mathbf{1}\right] \\ & - m_j\left[1+m^t(I-B)^{-1}\mathbf{1}\right] + \left[1+m^t(I-B)^{-1}\mathbf{1}\right]^2. \end{aligned}$$

We now find that from (4.35) that $\frac{\partial e_{1,j+1}}{\partial a_{1,j+1}} \geq 0$ if and only if $f \geq 0$.

Suppose that $i \neq 0, j$. Then, considering f as a function of m_i, we have

$$\begin{aligned} \frac{\partial f}{\partial m_i} = {} & m_j\left[2e_i^t(I-B)^{-2}\mathbf{1} - e_i^t(I-B)^{-1}\mathbf{1}\right] \\ & + 2e_i^t(I-B)^{-1}\mathbf{1}\left[1+m^t(I-B)^{-1}\mathbf{1} - m_j e_j^t(I-B)^{-1}\mathbf{1}\right]. \end{aligned}$$

Since $(I-B)^{-2} \geq (I-B)^{-1}$ and $m^t \geq m_j e_j^t \geq 0^t$, we find that $\frac{\partial f}{\partial m_i} \geq 0$ for any such i. Thus, in order to show that f is nonnegative, it suffices to consider the case that $m_i = 0$ for $i \neq 0, j$, i.e., that for some $t \in [0,1]$, $m_j = t$, and $m_0 = 1-t$. But in that case, f simplifies to the following expression:

$$1 - t + t^2\left[2e_j^t(I-B)^{-2}\mathbf{1} - e_j^t(I-B)^{-1}\mathbf{1} - (e_j^t(I-B)^{-1}\mathbf{1})^2\right].$$

Appealing to Proposition 4.4.6 we see that $f \geq 0$.

Next suppose that $f = 0$. Then from the above, necessarily we have $m_j = 1$ and

$$2e_j^t(I - B)^{-2}\mathbf{1} = e_j^t(I - B)^{-1}\mathbf{1} + \left[e_j^t(I - B)^{-1}\mathbf{1})\right]^2.$$

Since A is irreducible, the spectral radius of B is less than 1. Further, since each vertex in $\mathcal{D}(A)$ can be reached by a walk starting from vertex 1, and since each walk starting from vertex 1 passes immediately through vertex $j+1$, it follows that each vertex distinct from 1 and $j+1$ is reachable by a walk from vertex $j+1$, and that this walk is contained in $\mathcal{D}(B)$. Consequently, the hypothesis of Proposition 4.4.6 applies to the matrix B. Hence, from Proposition 4.4.6 we find that there is a p such that B can be written as

$$\left[\begin{array}{c|c|c|c|c|c} 0 & X_1 & 0 & 0 & \cdots & 0 \\ \hline 0 & 0 & X_2 & 0 & \cdots & 0 \\ \hline \vdots & & \ddots & \ddots & & \vdots \\ \hline 0 & 0 & \cdots & 0 & X_{p-1} & 0 \\ \hline 0 & 0 & 0 & \cdots & 0 & X_p \\ \hline 0 & 0 & 0 & \cdots & 0 & 0 \end{array}\right],$$

where $X_i\mathbf{1} = \mathbf{1}$, for $i = 1, \ldots, p$, and where the X_1 has only one row, which corresponds to index $j+1$. Using the fact that A is both irreducible and stochastic, we find that A is permutationally similar to a matrix $\overline{A}$ having the form in (4.42).

Finally, suppose that A is permutationally similar to the matrix $\overline{A}$ of (4.42), where X_1 has just one row, which corresponds to index $j+1$. Let B denote the trailing principal submatrix of $\overline{A}$ of order $n-1$. Then $(I - B)^{-1}$ can be written as

$$\left[\begin{array}{c|c|c|c|c|c} 1 & X_1 & X_1X_2 & X_1X_2X_3 & \cdots & X_1 \ldots X_p \\ \hline 0 & I & X_2 & X_2X_3 & \cdots & X_2 \ldots X_p \\ \hline \vdots & & \ddots & \ddots & & \vdots \\ \hline 0 & \cdots & 0 & I & X_{p-1} & X_{p-1}X_p \\ \hline 0 & 0 & 0 & \cdots & I & X_p \\ \hline 0 & 0 & 0 & \cdots & 0 & I \end{array}\right].$$

Since $X_i\mathbf{1} = \mathbf{1}, i = 1, \ldots, p$, we find readily that

$$(I - B)^{-1}\mathbf{1} = \begin{bmatrix} (p+1)\mathbf{1} \\ p\mathbf{1} \\ \vdots \\ 2\mathbf{1} \\ \mathbf{1} \end{bmatrix}.$$

We then deduce that $2e_1^t(I - B)^{-2}\mathbf{1} = (p+1)(p+2)$ while $e_1^t(I - B)^{-1}\mathbf{1} + \left(e_1^t(I - B)^{-1}\mathbf{1}\right)^2 = (p+1) + (p+1)^2$, which yields that

$$2e_1^t(I - B)^{-2}\mathbf{1} = e_1^t(I - B)^{-1}\mathbf{1} + \left(e_1^t(I - B)^{-1}\mathbf{1}\right)^2.$$

From our arguments above, we conclude that necessarily $\frac{\partial e_{1,j+1}}{\partial a_{1,j+1}} = 0$. □

The following, which is our main result in this section, is immediate from Theorem 4.4.7.

COROLLARY 4.4.8 ([77]) *Suppose that $n \geq 3$, and that $A \in \Phi^{n,n}$, with Perron value $r(A)$ and right Perron vector x. Then for each $1 \leq i, j \leq n$,*

$$\frac{\partial e_{i,j}}{\partial a_{i,j}} \geq 0. \tag{4.43}$$

Further,

$$\frac{\partial e_{i,j}}{\partial a_{i,j}} = 0 \tag{4.44}$$

if and only if $i \neq j$ and A is permutationally similar to the matrix $r(A)X\tilde{A}X^{-1}$, where $X = \text{diag}(x)$, and where the stochastic matrix $\tilde{A}$ has the following form:

$$\tilde{A} = \left[\begin{array}{c|c|c|c|c|c} 0 & 1 & 0 & 0 & \cdots & 0 \\ \hline 0 & 0 & \tilde{A}_1 & 0 & \cdots & 0 \\ \hline 0 & 0 & 0 & \tilde{A}_2 & \cdots & 0 \\ \hline \vdots & & \ddots & \ddots & & \vdots \\ \hline 0 & 0 & \cdots & 0 & \tilde{A}_{p-1} & 0 \\ \hline 0 & 0 & 0 & \cdots & 0 & \tilde{A}_p \\ \hline \mathbf{1} & 0 & 0 & \cdots & 0 & 0 \end{array}\right], \tag{4.45}$$

where the i–th row of A corresponds to the first row of $\tilde{A}$, and where $\tilde{A}_1$ has just one row, which corresponds to row j of A.

Corollary 4.4.8 is another example of the utility of the group generalised inverse in the analysis of the Perron value in general, and of Corollary 3.3.1 in particular: close analysis of the entries in the group inverse yields insight into the behaviour of the elasticity $e_{i,j}$ as a function of (i,j) entry. We note that a proof of (4.43) using the machinery of symbolic dynamics is also presented in [77].

REMARK 4.4.9 While Corollary 4.4.8 shows that $\frac{\partial e_{i,j}}{\partial a_{i,j}}$ is bounded below by zero for any irreducible nonnegative matrix A, it is not difficult to construct examples to show that these same quantities cannot be bounded from above in general.

For instance, suppose that $t > 0$, and consider the $n \times n$ matrix $A = \frac{t}{n}J$ (recall that J denotes an all ones matrix). Evidently $r(A) = t$, and it is readily determined that for the matrix $Q = tI - A$, we have $Q^{\#} = \frac{1}{t}\left(I - \frac{1}{n}J\right)$. Referring to (4.34) and making the appropriate substitutions, we find that for $i \neq j$, we have

$$\frac{\partial e_{i,j}}{\partial a_{i,j}} = \frac{1}{tn}\left(\frac{2t}{n}\left(\frac{-1}{tn}\right) - \frac{t}{n}\frac{1}{tn} + 1\right),$$

while

$$\frac{\partial e_{i,i}}{\partial a_{i,i}} = \frac{1}{tn}\left(\frac{2t}{n}\left(\frac{n-1}{tn}\right) - \frac{t}{n}\frac{1}{tn} + 1\right), i = 1, \ldots, n.$$

Evidently both expressions diverge to infinity as $t \to 0+$, and consequently the derivatives $\frac{\partial e_{i,j}}{\partial a_{i,j}}$ are not bounded from above as A ranges over the set of irreducible nonnegative matrices of order n.

As another example of this phenomenon, fix $t > 0$ and consider the following 3×3 matrix:

$$B = \begin{bmatrix} 0 & 1 & 0 \\ 0 & 0 & t \\ \frac{1}{t} & 0 & 0 \end{bmatrix}.$$

Then $r(B) = 1$, and the right and left Perron vectors for B can be taken as

$$x = \begin{bmatrix} \frac{1}{3} \\ \frac{1}{3} \\ \frac{1}{3t} \end{bmatrix}, y = \begin{bmatrix} 1 \\ 1 \\ t \end{bmatrix},$$

respectively. Observe that $y^t x = 1$. From (4.34) we find that $\frac{\partial e_{1,3}}{\partial b_{1,3}} = \frac{1}{3t}$, which is evidently unbounded from above as $t \to 0^+$.

We close this section with an example of a stochastic Leslie matrix with a particularly simple structure.

EXAMPLE 4.4.10 Suppose that $t \in (0,1)$, and consider the stochastic Leslie matrix A of order n given by

$$A = \begin{bmatrix} 0 & \dots & 0 & t & 1-t \\ 1 & 0 & 0 & \dots & 0 \\ 0 & 1 & 0 & \dots & 0 \\ \vdots & & \ddots & \ddots & \vdots \\ & & & & \\ 0 & \dots & 0 & 1 & 0 \end{bmatrix}.$$

We note that necessarily A is primitive, so that in particular from Theorem 4.4.7, $\frac{\partial e_{i,j}}{\partial a_{i,j}} > 0, i,j = 1,\dots,n$. However, in view of the special structure enjoyed by A, we might suspect that for some choices of (i,j), $\frac{\partial e_{i,j}}{\partial a_{i,j}}$ may be small.

Note that in the language of section 4.3, we have $s_j = 0, j = 0,\dots,n-2$ and $s_{n-1} = t$. Using Theorems 4.3.1 and 4.3.3 in conjunction with (4.34), we find, after some computations, that

$$\frac{\partial e_{1,n-1}}{\partial a_{1,n-1}} = \frac{n^2(1-t)}{(n-t)^3}, \frac{\partial e_{1,n}}{\partial a_{1,n}} = \frac{(n-1)^2 t}{(n-t)^3},$$
$$\frac{\partial e_{k+1,k}}{\partial a_{k+1,k}} = \frac{t(1-t)}{(n-t)^3}, k = 1,\dots,n-2, \text{and } \frac{\partial e_{n,n-1}}{\partial a_{n,n-1}} = \frac{(n-1)^2 t(1-t)}{(n-t)^3}.$$

In particular we see that for each $k = 1,\dots,n-1$, we have $\frac{\partial e_{k+1,k}}{\partial a_{k+1,k}} < \frac{\partial e_{1,j}}{\partial a_{1,j}}$ for $j = n-1, n$. Indeed it is not difficult to determine that $\frac{\partial e_{i,j}}{\partial a_{i,j}}$ is minimised (over pairs of indices (i,j) corresponding to nonzero entries in A) by the pairs $(k+1,k), k = 1,\dots,n-2$; interestingly enough, the collection of minimising indices does not depend on the values of $t \in (0,1)$. In view of the structure of the matrix A, and the characterisation of the equality (4.41), it is perhaps not so surprising that the pairs $(k+1,k), k = 1,\dots,n-2$ minimise $\frac{\partial e_{i,j}}{\partial a_{i,j}}$.

5

The Group Inverse in Markov Chains

We have already made extensive use of stochastic matrices in developing the results in Chapters 2,3, and 4. In this chapter, we specifically focus on stochastic matrices and their associated group inverses in the theory of discrete-time, time homogeneous Markov chains on a finite state space. We begin with a brief introduction to Markov chains in section 5.1, and then consider the special case of periodic stochastic matrices in section 5.2. This is followed in section 5.3 by an analysis of the conditioning properties of the stationary distribution under perturbation of the underlying stochastic matrix, and the use of the group inverse in deriving bounds on the subdominant eigenvalues of a stochastic matrix in section 5.4. The chapter concludes with a few illustrative examples in section 5.5.

5.1 Introduction to Markov chains

Recall that a square matrix T is called *stochastic* if it is entrywise nonnegative, and in addition $T\mathbf{1} = \mathbf{1}$. Such matrices are at the heart of a class of stochastic processes known as *Markov chains*, and in this section we briefly outline some of the key notions regarding stochastic matrices and Markov chains that will be of use in developing the results in this chapter and Chapter 6. For more detailed treatments of this material we refer the reader to the books of Kemeny and Snell [66], and Seneta [107].

Suppose that $n \in \mathbb{N}$, and consider a sequence of random variables defined on the set $\{1, \dots, n\}$, say $u_k, k \in \mathbb{N}$. The indices $1, \dots, n$ are known as the *states* of the Markov chain, and the set $\{1, \dots, n\}$ is the *state space*. Let $Pr\{A\}$ and $Pr\{A|B\}$ denote the probability of the event A, and the conditional probability of the event A given event B, respectively. We say that the sequence u_k has the *Markov property* if, for each $k \in \mathbb{N}$

$$Pr\{u_{k+1}|u_1, u_2, \dots, u_k\} = Pr\{u_{k+1}|u_k\}.$$

A sequence of random variables that enjoys the Markov property is known as a *Markov chain.* A Markov chain is said to be *time homogeneous* if there is a collection of fixed probabilities $t_{i,j}, i, j = 1, \ldots, n$ such that for each $k \in \mathbb{N}$ and $i, j = 1, \ldots, n$ we have

$$Pr\{u_{k+1} = j | u_k = i\} = t_{i,j}.$$

In this setting we refer to the quantity $t_{i,j}$ as the *transition probability from state i to state j* for the Markov chain. For a time homogeneous Markov chain, we may construct the corresponding *transition matrix* T as $T = [t_{i,j}]_{i,j=1,\ldots,n}$. We can also represent the iterates u_k of the Markov chain as vectors

$$v_k \equiv \begin{bmatrix} Pr\{u_k = 1\} \\ Pr\{u_k = 2\} \\ \vdots \\ Pr\{u_k = n\} \end{bmatrix}.$$

Observe that $v_k^t \mathbf{1} = 1$ for each $k \in \mathbb{N}$. It is straightforward to verify that for each $k \in \mathbb{N}$,

$$v_{k+1}^t = v_k^t T.$$

This last relation can be reformulated equivalently as

$$v_{k+1}^t = v_1^t T^k, k \in \mathbb{N}. \tag{5.1}$$

Evidently we may view the iterates of a time homogeneous Markov chain on the state space $\{1, \ldots, n\}$ as realisation of the power method, whereby powers of the stochastic matrix T are applied to an initial vector v_1^t.

Suppose now that our transition matrix T is primitive. Applying the Perron–Frobenius theorem to T (see [107, section 1.1]), we find that since $T\mathbf{1} = \mathbf{1}$, the eigenvalue 1 must coincide with the spectral radius of T, as the corresponding eigenvector $\mathbf{1}$ is positive. Consequently, there is a unique positive vector w such that $w^tT = w^t$ and $w^t\mathbf{1} = 1$; the vector w is known as the *stationary distribution vector*, or *stationary vector*, for T. (Evidently w is a left Perron vector for T.) From the hypothesis that T is primitive we find, again from the Perron–Frobenius theorem, that if $\lambda \neq 1$ is an eigenvalue of T, then $|\lambda| < 1$. In view of (5.1), we find that as $\lim_{k\to\infty} v_k = w$ – that is, the iterates of a Markov chain with primitive transition matrix T must converge to the corresponding stationary distribution w. Thus we see that the entries in w carry information about the long–term behaviour of the iterates of the corresponding Markov chain.

In the case that T is irreducible but not primitive, a collection of slightly weaker conclusions holds. We still have a unique stationary distribution vector $w > 0$ with $w^tT = w^t$ and $w^t\mathbf{1} = 1$; however, there is an integer parameter $d \geq 2$, known as the *period* of T, such that T has exactly d eigenvalues of modulus 1, namely $e^{\frac{2\pi i j}{d}}, j = 0, \ldots, d-1$. Indeed, d is given by the greatest common divisor of the cycle lengths in $\mathcal{D}(T)$ (see [15, section 3.4]). In this case, we have that as $k \to \infty$,

$$\frac{1}{d}\left(v_k + v_{k+1} + \ldots + v_{k+d-1}\right) \to w.$$

We use the notation $\mathcal{IS}_n$ to denote the set of all irreducible stochastic matrices of order n.

Finally, we note that if our stochastic matrix T is reducible, but has 1 as an algebraically simple eigenvalue, then we find that there is a unique nonnegative vector w such that $w^t\mathbf{1} = 1$ and $w^tT = w^t$; as above, we refer to this vector as the stationary distribution vector for T. We denote the set of stochastic matrices of order n having 1 as an algebraically simple eigenvalue by $\mathcal{AS}_n$.

Suppose that we have a Markov chain $u_k, k \in \mathbb{N}$, with irreducible transition matrix T of order n. For each $i, j = 1, \ldots, n$, we may define the *first passage time from i to j* as $\min\{k | k \geq 1, u_{k+1} = j, u_1 = i\}$. In other words, the first passage time from i to j is the random variable that takes on the value k in the case that the Markov chain enters state j for the first time after k steps, given that the Markov chain started in state i. The *mean first passage time from i to j*, denoted $m_{i,j}$ is the expected value of the first passage time from i to j. A standard result (relying on so-called absorbing chain techniques) shows that if $i \neq j$, then $m_{i,j}$ may be computed as follows: let $T_{(j)}$ denote the principal submatrix of T formed by deleting its j–th row and column; then

$$m_{i,j} = \begin{cases} e_i^t(I - T_{(j)})^{-1}\mathbf{1}, & \text{if } 1 \leq i \leq j-1 \\ e_{i-1}^t(I - T_{(j)})^{-1}\mathbf{1}, & \text{if } j+1 \leq i \leq n. \end{cases} \tag{5.2}$$

It is also known that for $i = 1, \ldots, n$, the *mean first return time to i*, $m_{i,i}$, is given by

$$m_{i,i} = \frac{1}{w_i}. \tag{5.3}$$

We refer the reader to [107, Theorem 4.5 and pp. 128–130], for derivations of (5.2) and (5.3). The *mean first passage matrix* for the Markov chain is the $n \times n$ matrix $M = [m_{i,j}]_{i,j=1,\ldots,n}$.

The final key idea that we present in this section begins with a partitioning of a stochastic matrix. Suppose that $T \in \mathcal{IS}_n$ and is partitioned as follows:

$$T = \left[\begin{array}{c|c} T_{1,1} & T_{1,2} \\ \hline T_{2,1} & T_{2,2} \end{array}\right], \tag{5.4}$$

where $T_{1,1}$ is of order $1 \le k \le n-1$ and $T_{2,2}$ is of order $1 \le n-k \le n-1$. Let $S = \{1, \ldots, k\}$; then the *stochastic complement of* $T_{1,1}$ *in* T is the matrix $\mathcal{P}(T)_S$ given by

$$\mathcal{P}(T)_S = T_{1,1} + T_{1,2}(I - T_{2,2})^{-1}T_{2,1}.$$

It is straightforward to see that $\mathcal{P}(T)_S$ is a $k \times k$ nonnegative matrix. Further, from the fact that $T\mathbf{1} = \mathbf{1}$, we have $(I - T_{2,2})^{-1}T_{2,1}\mathbf{1}_k = \mathbf{1}_{n-k}$ (where the subscripts on the all ones vectors denote their orders). We find in turn that $T_{1,1}\mathbf{1} + T_{1,2}(I - T_{2,2})^{-1}T_{2,1}\mathbf{1} = \mathbf{1}$. Hence $\mathcal{P}(T)_S$ is also a stochastic matrix. For each $i, j = 1, \ldots, k$, the (i, j) entry of $\mathcal{P}(T)_S$ is the probability that the original Markov chain with transition matrix T re-enters S for the first time with a transition into state j, given that the original Markov chain started in state i.

It is not difficult to determine that since T is irreducible, so is $\mathcal{P}(T)_S$; see the article of Johnson and Xenophontos [65] for a proof. Further, it is shown by Meyer in [96] that if the stationary distribution vector w for T is partitioned conformally with T as

$$w = \left[\begin{array}{c} w_1 \\ \hline w_2 \end{array}\right],$$

then w_1 is in fact a scalar multiple of the stationary distribution vector for $\mathcal{P}(T)_S$. Indeed, letting $\overline{S} = \{k+1, \ldots, n\}$, and denoting the stationary distribution vectors of $\mathcal{P}(T)_S$ and $\mathcal{P}(T)_{\overline{S}}$ by u_1 and u_2, respectively, the stationary distribution vector for T is given by

$$w = \left[\begin{array}{c} a_1 u_1 \\ \hline a_2 u_2 \end{array}\right],$$

where the vector $a = \left[\begin{array}{cc} a_1 & a_2 \end{array}\right]^t$ is the stationary distribution of the 2×2 *coupling matrix*, given by

$$\left[\begin{array}{cc} u_1^t T_{1,1}\mathbf{1} & u_1^t T_{1,2}\mathbf{1} \\ u_2^t T_{2,1}\mathbf{1} & u_2^t T_{2,2}\mathbf{1} \end{array}\right].$$

A straightforward computation reveals that this coupling matrix is stochastic; further it is readily determined that since T is irreducible, so is the coupling matrix.

Evidently given any nonempty proper subset α of $\{1, \dots, n\}$, we may define the corresponding stochastic complement $\mathcal{P}(T)_\alpha$ as follows. Let $\beta = \{1, \dots, n\} \setminus \alpha$, and denote by $T[\alpha, \beta]$ the submatrix of T on rows indexed by α and columns indexed by β; define $T[\beta, \alpha], T[\alpha, \alpha]$, and $T[\beta, \beta]$ analogously. Then the stochastic complement of $T[\alpha, \alpha]$ in T is

$$\mathcal{P}(T)_\alpha = T[\alpha, \alpha] + T[\alpha, \beta](I - T[\beta, \beta])^{-1} T[\beta, \alpha].$$

5.2 Group inverse in the periodic case

Suppose that we have an irreducible stochastic matrix T, and that T is periodic with period d. Then there is a permutation matrix P so that

$$PTP^t = \left[\begin{array}{c|c|c|c|c|c} 0 & T_1 & 0 & 0 & \dots & 0 \\ \hline 0 & 0 & T_2 & 0 & \dots & 0 \\ \hline \vdots & & \ddots & \ddots & & \vdots \\ \hline 0 & \dots & & 0 & T_{d-2} & 0 \\ \hline 0 & 0 & \dots & & 0 & T_{d-1} \\ \hline T_d & 0 & 0 & \dots & 0 & 0 \end{array}\right], \tag{5.5}$$

where the diagonal blocks on the right-hand side of (5.5) are square, though not necessarily of the same size (see [15, section 3.4], for example). We say that the stochastic matrix on the right hand side of (5.5) is in *periodic normal form.*

Setting $S_i = T_i T_{i+1} \dots T_d T_1 \dots T_{i-1}$, it follows that S_i is primitive and stochastic for each $i = 1, \dots, d$. One of the standard approaches for analysing the properties of an irreducible period stochastic matrix is to first write it in periodic normal form, then use the corresponding properties of the S_is to infer the desired information about the original periodic matrix. For example, it is straightforward to determine that if S_i has stationary distribution vector $w_i, i = 1, \dots, d$, then the vector

$$w = \frac{1}{d} \left[\begin{array}{c} w_1 \\ \hline w_2 \\ \hline \vdots \\ \hline w_d \end{array}\right]$$

serves as the stationary distribution vector for the right-hand side of

(5.5). It is readily established that $w_j^t = w_1^t T_1 \dots T_{j-1}$ for $j = 2, \dots, d$.

Our goal in this section is to develop an analogous approach to finding $(I-T)^\#$ for an irreducible stochastic and periodic matrix T in periodic normal form. Specifically, we will give a formula for $(I-T)^\#$ as a block partitioned matrix, where the partitioning coincides with that of the periodic normal form (5.5) for T. Our development will follow the line of reasoning in Kirkland's paper [68].

The following lemma will be useful in the sequel.

LEMMA 5.2.1 *Suppose that A and B are nonnegative matrices of orders $n \times k$ and $k \times n$, respectively. Suppose further that A and B both have all row sums equal to 1, and that both AB and BA are primitive. Let u denote the left Perron vector of AB, normalised so that $u^t\mathbf{1} = 1$. Then $(I-AB)^\# = I - \mathbf{1}u^t + A(I-BA)^\# B$. Further, $(I-AB)^\# A = A(I-BA)^\#$.*

Proof: Set $G = I - \mathbf{1}u^t + A(I-BA)^\# B$. In order to show that $G = (I-AB)^\#$, it suffices to show $G\mathbf{1} = 0, u^tG = 0^t$, and that $G(I-AB) = (I-AB)G = I - \mathbf{1}u^t$. Note that

$$G\mathbf{1} = (I-\mathbf{1}u^t+A(I-BA)^\# B)\mathbf{1} = A(I-BA)^\# B\mathbf{1} = A(I-BA)^\#\mathbf{1} = 0;$$

also

$$u^tG = u^t(I-\mathbf{1}u^t + A(I-BA)^\# B) = u^tA(I-BA)^\# B = 0^t,$$

the last since u^tA is the stationary distribution vector for BA.

Next we observe that

$$ABG = AB - \mathbf{1}u^t + ABA(I-BA)^\# B = -\mathbf{1}u^t + A(I+BA(I-BA)^\#)B.$$

Since $I + BA(I-BA)^\# = (I-BA)^\# + \mathbf{1}u^tA$, we thus find that

$$ABG = -\mathbf{1}u^t + A(I-BA)^\# B + A\mathbf{1}u^tAB = A(I-BA)^\# B = G - I + \mathbf{1}u^t.$$

It now follows that $(I-AB)G = I - \mathbf{1}u^t$; a similar argument shows that $G(I-AB) = I - \mathbf{1}u^t$ as well. We thus conclude that $(I-AB)^\# = I - \mathbf{1}u^t + A(I-BA)^\# B$.

From the expression above we thus have that

$$(I-AB)^\# A = A - \mathbf{1}u^tA + A(I-BA)^\# BA = -\mathbf{1}u^tA + A(I+(I-BA)^\# BA).$$

Again using the fact that $I + BA(I - BA)^{\#} = (I - BA)^{\#} + \mathbf{1}u^tA$, we thus find that

$$(I - AB)^{\#}A = -\mathbf{1}u^tA + A((I - BA)^{\#} + \mathbf{1}u^tA) = A(I - BA)^{\#},$$

as desired. □

We note in passing that our expression for $(I-AB)^{\#}$ in Lemma 5.2.1 is analogous to the standard formula $(I-XY)^{-1} = I + X(I-YX)^{-1}Y$ which applies in the case that both of the relevant inverses exist (see [57, section 0.7.4]).

THEOREM 5.2.2 ([68]) *Let T be an $n \times n$ irreducible stochastic matrix that is periodic with period $d \geq 2$. Suppose that T is in periodic normal form, with*

$$T = \left[\begin{array}{c|c|c|c|c|c} 0 & T_1 & 0 & 0 & \dots & 0 \\ \hline 0 & 0 & T_2 & 0 & \dots & 0 \\ \hline \vdots & & \ddots & \ddots & & \vdots \\ \hline 0 & \dots & & 0 & T_{d-2} & 0 \\ \hline 0 & 0 & \dots & & 0 & T_{d-1} \\ \hline T_d & 0 & 0 & \dots & 0 & 0 \end{array}\right]. \tag{5.6}$$

Write $S_i = T_iT_{i+1}\dots T_dT_1\dots T_{i-1}, i = 1,\dots,d$, let w_1 denote the stationary distribution vector for S_1, and for each $j = 2,\dots,d$, let w_j be given by $w_j^t = w_1^tT_1T_2\dots T_{j-1}$.

Let G be the $n \times n$ matrix such that when it is partitioned conformally with T as a $d \times d$ block matrix, the blocks $G_{i,j}, i, j = 1,\dots,d$ are given as follows:

$$G_{i,j} = \begin{cases} (I - S_i)^{\#}T_iT_{i+1}\dots T_{j-1} + \left(\frac{d-1}{2d} - \frac{j-i}{d}\right)\mathbf{1}w_j^t & \text{if } i < j, \\ (I - S_i)^{\#} + \left(\frac{d-1}{2d}\right)\mathbf{1}w_i^t & \text{if } i = j, \\ (I - S_i)^{\#}T_i\dots T_dT_1\dots T_{j-1} + \left(\frac{d-1}{2d} - \frac{d+j-i}{d}\right)\mathbf{1}w_j^t & \text{if } i > j. \end{cases} \tag{5.7}$$

Then $(I - T)^{\#} = G$.

<u>Proof</u>: We begin by recalling that the stationary distribution vector for T is given by

$$w = \frac{1}{d}\left[\begin{array}{c} w_1 \\ \hline w_2 \\ \hline \vdots \\ \hline w_d \end{array}\right].$$

Consider the matrix G whose blocks $G_{i,j}, i, j = 1, \ldots, d$, are given by (5.7). For each $i = 1, \ldots, d$ we have

$$\sum_{j=1}^{d} G_{i,j}\mathbf{1} = \left(\sum_{j<i} \left(\frac{d-1}{2d} - \frac{d+j-i}{d} \right) + \sum_{j \geq i} \left(\frac{d-1}{2d} - \frac{j-i}{d} \right) \right) \mathbf{1} = 0.$$

Hence $G\mathbf{1} = 0$. Similarly, note that for each $j = 1, \ldots, d$, the j–th block of w^tG is given by $\left(\sum_{i \leq j} \left(\frac{d-1}{2d} - \frac{j-i}{d} \right) + \sum_{i>j} \left(\frac{d-1}{2d} - \frac{d+j-i}{d} \right) \right) w_j^t$. Since $\sum_{i \leq j} \left(\frac{d-1}{2d} - \frac{j-i}{d} \right) + \sum_{i>j} \left(\frac{d-1}{2d} - \frac{d+j-i}{d} \right) = 0$, we see that $w^tG = 0^t$.

The proof will be complete provided that we can show that

$$(I - T)G = G(I - T) = I - \mathbf{1}w^t.$$

First consider the product $(I - T)G$, and fix an index i between 1 and d. Observe that the (i, i) block of $(I - T)G$ is given by

$$G_{i,i} - T_i G_{i+1,i} = (I - S_i)^{\#} + \left(\frac{d-1}{2d} \right) \mathbf{1}w_i^t$$
$$-T_i \left((I - S_{i+1})^{\#} T_{i+1} \ldots T_d T_1 \ldots T_{i-1} + \left(\frac{d-1}{2d} - \frac{d-1}{d} \right) \mathbf{1}w_i^t \right).$$

Applying Lemma 5.2.1, we find that

$$T_i((I - S_{i+1})^{\#} T_{i+1} \ldots T_d T_1 \ldots T_{i-1} =$$
$$(I - S_i)^{\#} T_i \ldots T_d T_1 \ldots T_{i-1} = (I - S_i)^{\#} S_i.$$

Thus we find that

$$G_{i,i} - T_i G_{i+1,i} = (I - S_i)(I - S_i)^{\#} + \frac{d-1}{d} \mathbf{1}w_i^t = I - \frac{1}{d} \mathbf{1}w_i^t.$$

Next, suppose that $i > j$, and consider the (i, j) block of the product $(I - T)G$. That block is given by

$$G_{i,j} - T_i G_{i+1,j} =$$
$$(I - S_i)^{\#} T_i \ldots T_d T_1 \ldots T_{j-1} + \left(\frac{d-1}{2d} - \frac{d+j-i}{d} \right) \mathbf{1}w_j^t$$
$$-T_i \left((I - S_{i+1})^{\#} T_{i+1} \ldots T_d T_1 \ldots T_{j-1} + \left(\frac{d-1}{2d} - \frac{d+j-i-1}{d} \right) \mathbf{1}w_j^t \right).$$

From Lemma 5.2.1, we find that

$$(I - S_i)^{\#} T_i \ldots T_d T_1 \ldots T_{j-1} = T_i (I - S_{i+1})^{\#} T_{i+1} \ldots T_d T_1 \ldots T_{j-1},$$

and it readily follows that $G_{i,j} - T_i G_{i+1,j} = \frac{-1}{d}\mathbf{1}w_j^t$. A similar argument applies if $i < j$, and we deduce that $(I - T)G = I - \mathbf{1}w^t$. Proceeding analogously to the above, we find that $G(I-T) = I - \mathbf{1}w^t$; it now follows that $G = (I - T)^{\#}$. □

EXAMPLE 5.2.3 Here we revisit the matrix of Example 2.3.3. Let T be the $n \times n$ matrix given by

$$T = \begin{bmatrix} 0 & 1 & 0 & 0 & \dots & 0 \\ 0 & 0 & 1 & 0 & \dots & 0 \\ \vdots & & \ddots & \ddots & & \vdots \\ & & & & & \\ 0 & 0 & \dots & & 0 & 1 \\ 1 & 0 & 0 & \dots & & 0 \end{bmatrix}.$$

Evidently T is irreducible, stochastic, and periodic with period n. In the language of Theorem 5.2.2, each S_i is the 1×1 matrix $\begin{bmatrix} 1 \end{bmatrix}$, so that each $(I - S_i)^{\#}$ is the 1×1 zero matrix. Similarly, each w_j is the vector $\begin{bmatrix} 1 \end{bmatrix}$. It now follows from (5.7) that

$$(I-T)^{\#} = \begin{bmatrix} \frac{n-1}{2n} & \frac{n-3}{2n} & \frac{n-5}{2n} & \dots & -\frac{n-3}{2n} & -\frac{n-1}{2n} \\ -\frac{n-1}{2n} & \frac{n-1}{2n} & \frac{n-3}{2n} & \frac{n-5}{2n} & \dots & -\frac{n-3}{2n} \\ -\frac{n-3}{2n} & -\frac{n-1}{2n} & \frac{n-1}{2n} & \frac{n-3}{2n} & \dots & -\frac{n-5}{2n} \\ \vdots & & & \ddots & & \vdots \\ & & & & & \\ \frac{n-3}{2n} & \frac{n-5}{2n} & \dots & -\frac{n-3}{2n} & -\frac{n-1}{2n} & \frac{n-1}{2n} \end{bmatrix}.$$

Suppose that we have a connected undirected graph $\mathcal{G}$ on vertices $1, \dots, n$, with $(0,1)$ adjacency matrix A. That is, $a_{i,j} = 1$ if $i \neq j$ and vertices i and j are adjacent, and $a_{i,j} = 0$ otherwise. Letting $D = \mathrm{diag}(A\mathbf{1})$, the Markov chain with transition matrix $D^{-1}A$ is known as the *simple random walk on* $\mathcal{G}$. Here we think of the vertices of $\mathcal{G}$ as the states of the Markov chain, with a particle, or "random walker," that moves from one vertex of $\mathcal{G}$ to another at each time step, as follows: if the particle is on vertex i at time k, then at time $k+1$ it moves to one of the vertices adjacent to vertex i, following any of the edges incident with vertex i with equal probability. Our next example deals with the simple random walk on the graph depicted in Figure 5.1.

EXAMPLE 5.2.4 Observe that for a periodic stochastic matrix T, Theorem 5.2.2 expresses $(I - T)^{\#}$ in terms of the lower–order matrices $(I - S_i)^{\#}, i = 1, \dots, d$. Further, note that by Lemma 5.2.1, we can

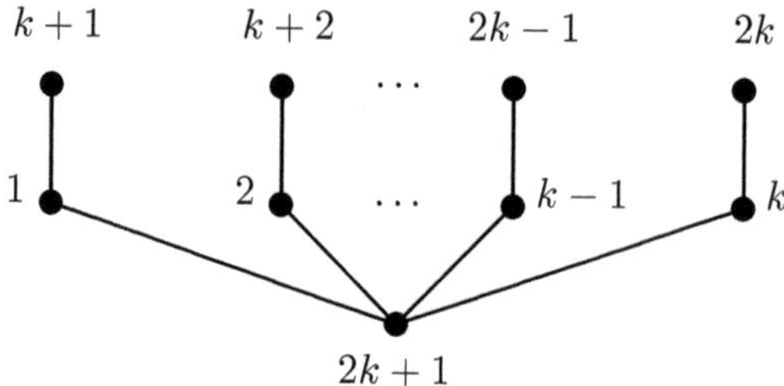

FIGURE 5.1
The graph for Example 5.2.4

also express $(I - S_{i+1})^{\#}$ in terms of S_i and $T_1, \ldots, T_d$. Consequently, by using the two results in tandem, we can recursively construct all of the blocks of $(I - T)^{\#}$ from $(I - S_1)^{\#}, w_1$, and the blocks $T_1, \ldots, T_d$. In this example we briefly illustrate this approach to computing $(I-T)^{\#}$.

Fix an integer $k \geq 2$, let T_1 be the $k \times (k+1)$ matrix given by $T_1 = \left[\begin{array}{c|c} \frac{1}{2}I & \frac{1}{2}\mathbf{1} \end{array} \right]$, and let T_2 be the $(k+1) \times k$ matrix given by $T_2 = \left[\begin{array}{c} I \\ \hline \frac{1}{k}\mathbf{1}^t \end{array} \right]$. We note that the stochastic matrix $T = \left[\begin{array}{c|c} 0 & T_1 \\ \hline T_2 & 0 \end{array} \right]$ is the transition matrix for the simple random walk on the graph shown in Figure 5.1.

We have $T_1T_2 = \frac{1}{2}I + \frac{1}{2k}J$ (recall that J denotes an all ones matrix of the appropriate order); hence $(I - T_1T_2)^{\#} = 2(I - \frac{1}{k}J)$, while $w_1 = \frac{1}{k}\mathbf{1}$. Since $w_2^t = w_1^tT_1$, we find that

$$w_2 = \left[\begin{array}{c} \frac{1}{2k}\mathbf{1} \\ \hline \frac{1}{2} \end{array} \right].$$

Using Lemma 5.2.1, we have $(I - T_2T_1)^{\#} = I - \mathbf{1}w_2^t + T_2(I - T_1T_2)^{\#}T_1$. It now follows that

$$(I - T_2T_1)^{\#} = \left[\begin{array}{c|c} 2I - \frac{3}{2k}J & -\frac{1}{2}\mathbf{1} \\ \hline -\frac{1}{2k}\mathbf{1}^t & \frac{1}{2} \end{array} \right].$$

Next, using these expressions for $(I - T_1T_2)^{\#}$ and $(I - T_2T_1)^{\#}$, along with those for T_1, T_2, w_1, and w_2, we can now compute the blocks of $(I - T)^{\#}$ by using Theorem 5.2.2. The resulting computation yields the following:

$$(I-T)^{\#} = \left[\begin{array}{c|c|c} 2I - \frac{7}{4k}J & I - \frac{9}{8k}J & -\frac{1}{8}\mathbf{1} \\ \hline 2I - \frac{9}{4k}J & 2I - \frac{11}{8k}J & -\frac{3}{8}\mathbf{1} \\ \hline -\frac{1}{4k}\mathbf{1}^t & -\frac{3}{8k}\mathbf{1}^t & \frac{5}{8} \end{array}\right].$$

Recall that Question 3.3.8 asks for a characterisation of the irreducible nonnegative matrices A such that $(r(A)I - A)^{\#}$ is an M-matrix, where $r(A)$ denotes the Perron value of A. Our next result provides an answer to Question 3.3.8 in the case that the matrix A is stochastic and periodic.

THEOREM 5.2.5 ([68]) *Let T be an irreducible stochastic matrix that is periodic with period $d \geq 2$. Suppose further that T is written in periodic normal form as*

$$T = \left[\begin{array}{c|c|c|c|c|c} 0 & T_1 & 0 & 0 & \dots & 0 \\ \hline 0 & 0 & T_2 & 0 & \dots & 0 \\ \hline \vdots & & \ddots & \ddots & & \vdots \\ \hline 0 & \dots & & 0 & T_{d-2} & 0 \\ \hline 0 & 0 & \dots & & 0 & T_{d-1} \\ \hline T_d & 0 & 0 & \dots & 0 & 0 \end{array}\right]. \tag{5.8}$$

Denote the stationary distribution vector for T by w, and partition w conformally with T in (5.8) as

$$w = \frac{1}{d}\left[\begin{array}{c} w_1 \\ \hline \vdots \\ \hline w_d \end{array}\right].$$

Then we have the following:
i) if $d \geq 4$, then $(I-T)^{\#}$ is not an M-matrix;
ii) if $d = 3$, then $(I-T)^{\#}$ is an M-matrix if and only if $T_1 = \mathbf{1}w_2^t, T_2 = \mathbf{1}w_3^t$ and $T_3 = \mathbf{1}w_1^t$;
iii) if $d = 2$, then $(I-T)^{\#}$ is an M-matrix if and only if $(I - T_1T_2)^{\#}T_1 \leq \frac{1}{4}\mathbf{1}w_2^t$ and $(I - T_2T_1)^{\#}T_2 \leq \frac{1}{4}\mathbf{1}w_1^t$.

Proof: Let $G = (I-T)^{\#}$, and partition G conformally with T as a $d \times d$ block matrix. From (5.7) of Theorem 5.2.2, it follows that $G_{1,2}\mathbf{1} = \frac{d-3}{2}\mathbf{1}$. In particular, we find that if $d \geq 4$, then G must have some positive off-diagonal entries, and so it cannot be an M-matrix. This establishes i).

ii) Let $d = 3$, and suppose that G is an M-matrix. In that case, since $G_{1,2}\mathbf{1} = 0$, it must be the case that $G_{1,2}$ is a zero matrix. Referring to (5.7), we find that if that is the case, then necessarily $(I - T_1T_2T_3)^{\#}T_1$ is a zero matrix. But then each column of T_1 must be a null vector of $(I - T_1T_2T_3)^{\#}$ — i.e., each column of T_1 is a scalar multiple of $\mathbf{1}$. By considering $G_{2,3}$ and $G_{3,1}$ we similarly deduce that each column of T_2 is a multiple of $\mathbf{1}$, as is each column of T_3. It now follows readily that $T_1 = \mathbf{1}w_2^t, T_2 = \mathbf{1}w_3^t$ and $T_3 = \mathbf{1}w_1^t$. Conversely, if $T_1 = \mathbf{1}w_2^t, T_2 = \mathbf{1}w_3^t$ and $T_3 = \mathbf{1}w_1^t$, a direct application of (5.7) yields that G is an M-matrix.

iii) Let $d = 2$, and suppose that G is an M-matrix. Then referring to (5.7), we see that since $G_{1,2} \le 0$ and $G_{2,1} \le 0$, it must certainly be the case that $(I-T_1T_2)^{\#}T_1 \le \frac{1}{4}\mathbf{1}w_2^t$ and $(I-T_2T_1)^{\#}T_2 \le \frac{1}{4}\mathbf{1}w_1^t$. Conversely if $(I - T_1T_2)^{\#}T_1 \le \frac{1}{4}\mathbf{1}w_2^t$ and $(I - T_2T_1)^{\#}T_2 \le \frac{1}{4}\mathbf{1}w_1^t$, then both $G_{1,2}$ and $G_{2,1}$ are nonpositive matrices. Further, in that case we also find that $(I - T_1T_2)^{\#}T_1T_2 \le \frac{1}{4}\mathbf{1}w_2^tT_2$. This last inequality is equivalent to $(I - T_1T_2)^{\#} + \frac{1}{4}\mathbf{1}w_1^t \le I - \frac{1}{2}\mathbf{1}w_1^t$, so that necessarily the off-diagonal entries of $G_{1,1}$ are nonpositive. A similar analysis shows that $G_{2,2}$ has nonpositive off-diagonal entries, it now follows that G is an M-matrix. □

Using Remark 2.5.3 and Theorem 5.2.5, the following result is readily established.

COROLLARY 5.2.6 ([68]) *Let A be an irreducible nonnegative matrix that is periodic with period $d \ge 2$. Suppose that A is written in periodic normal form as*

$$A = \left[\begin{array}{c|c|c|c|c|c} 0 & A_1 & 0 & 0 & \cdots & 0 \\ \hline 0 & 0 & A_2 & 0 & \cdots & 0 \\ \hline \vdots & & \ddots & \ddots & & \vdots \\ \hline 0 & \cdots & & 0 & A_{d-2} & 0 \\ \hline 0 & 0 & \cdots & & 0 & A_{d-1} \\ \hline A_d & 0 & 0 & \cdots & 0 & 0 \end{array}\right]. \tag{5.9}$$

Denote the Perron value of A by r, and let x, y denote right and left Perron vectors for A, respectively, normalised so that $y^tx = 1$. Partition x and y conformally with A as

$$\left[\begin{array}{c} x_1 \\ \hline \vdots \\ \hline x_d \end{array}\right] \text{ and } \frac{1}{d}\left[\begin{array}{c} y_1 \\ \hline \vdots \\ \hline y_d \end{array}\right],$$

respectively. Then we have the following:
i) if $d \geq 4$, *then* $(rI - A)^{\#}$ *is not an M-matrix;*
ii) if $d = 3$, *then* $(rI - A)^{\#}$ *is an M-matrix if and only if* $A_1 = rx_1y_2^t, A_2 = rx_2y_3^t$, *and* $A_3 = rx_3y_1^t$;
iii) if $d = 2$, *then* $(rI - A)^{\#}$ *is an M-matrix if and only if* $r(r^2I - A_1A_2)^{\#}A_1 \leq \frac{1}{4}x_1y_2^t$ *and* $r(r^2I - A_2A_1)^{\#}A_2 \leq \frac{1}{4}x_2y_1^t$.

EXAMPLE 5.2.7 In this example, we consider a highly structure class of periodic stochastic matrices, and determine when the associated group inverse is an M-matrix. Fix an integer $k \geq 2$ and a scalar $\alpha \in (0,1)$. Let T_1 and T_2 be the $k \times k$ matrices given by $T_1 = \alpha I + \frac{1-\alpha}{k}J$ and $\frac{1}{k}J$, respectively. Consider the periodic stochastic matrix given by

$$T = \left[\begin{array}{c|c} 0 & T_1 \\ \hline T_2 & 0 \end{array}\right].$$

Note that in the language of Theorem 5.2.5, we are in the case that $d = 2$, with $w_1 = w_2 = \frac{1}{k}\mathbf{1}$.

Applying Theorem 5.2.5, we find that $(I - T)^{\#}$ is an M-matrix if and only if both of the following hold: a) $(I - T_1T_2)^{\#}T_1 \leq \frac{1}{4k}J$; and b) $(I - T_2T_1)^{\#}T_2 \leq \frac{1}{4k}J$. Since each column of T_2 is a null vector for $(I - T_2T_1)^{\#}$, condition b) is certainly satisfied. Since $T_1T_2 = \frac{1}{k}J$, we find that $(I - T_1T_2)^{\#} = I - \frac{1}{k}J$, so that condition a) is equivalent to

$$\left(I - \frac{1}{k}J\right)\left(\alpha I + \frac{1-\alpha}{k}J\right) \leq \frac{1}{4k}J. \tag{5.10}$$

Equation (5.10) is readily seen to be equivalent to $\alpha \leq \frac{1}{4(k-1)}$. We thus deduce that the matrix $(I - T)^{\#}$ is an M-matrix if and only if $\alpha \leq \frac{1}{4(k-1)}$.

5.3 Perturbation and conditioning of the stationary distribution vector

Suppose that we have an irreducible stochastic matrix T with corresponding stationary distribution vector w. If the entries of T are perturbed to yield another stochastic matrix, what is the effect on the corresponding stationary distribution? (Note that a foreshadowing of that type of question appears in Example 1.2.1.) This question is relevant in trying to develop an understanding of the sensitivity of the stationary distribution vector in much the same way as we developed tools for understanding Perron vector sensitivities in Chapter 3. Further, if the entries in our matrix T have been estimated from data, and so have some uncertainty attached to them, we may wish glean some understanding of how those uncertainties affect the entries in the stationary distribution. The following key lemma due to Meyer in [95] shows how the group inverse can be used to address this question.

LEMMA 5.3.1 *Let T and $\tilde{T}$ be matrices in $\mathcal{AS}_n$. Denote the stationary distribution vectors for T and $\tilde{T}$ by w and $\tilde{w}$, respectively. Set $\tilde{T} = T + E$ and $Q = I - T$. Then the matrix $I - EQ^{\#}$ is invertible, and we have $\tilde{w}^t = w^t(I - EQ^{\#})^{-1}$.*

Proof: Since $I - \tilde{T} = Q - E$, we have, upon post–multiplying both sides by $Q^{\#}$, that $I - EQ^{\#} = \mathbf{1}w^t + (I - \tilde{T})Q^{\#}$. Let U be an orthogonal matrix whose first column is equal to $\frac{1}{\sqrt{n}}\mathbf{1}$. We find readily that $U^t\mathbf{1} = \sqrt{n}e_1$ and that w^tU can be written as $w^tU = \frac{1}{\sqrt{n}}\left[\begin{array}{c|c} 1 & z^t \end{array}\right]$ for some vector $z \in \mathbb{R}^{n-1}$. Further, there are nonsingular matrices B_1, B_2 and accompanying vectors v_1, v_2 such that

$$U^t(I - \tilde{T})U = \left[\begin{array}{c|c} 0 & v_1^t \\ \hline 0 & B_1 \end{array}\right] \text{ and } U^tQ^{\#}U = \left[\begin{array}{c|c} 0 & v_2^t \\ \hline 0 & B_2 \end{array}\right]$$

(here the nonsingularity of B_1 and B_2 follows from the fact that 1 is a simple eigenvalue for both T and $\tilde{T}$). Hence we find that

$$U^t(\mathbf{1}w^t + (I - \tilde{T})Q^{\#})U = \left[\begin{array}{c|c} 1 & z^t + v_1^tB_2 \\ \hline 0 & B_1B_2 \end{array}\right],$$

which is evidently nonsingular. Hence $I - EQ^{\#}$ is invertible.

Since $\tilde{w}^t(T + E) = \tilde{w}^t$, we have $\tilde{w}^tE = \tilde{w}^tQ$. Multiplying on the right by $Q^{\#}$ we obtain $\tilde{w}^tEQ^{\#} = \tilde{w}^tQQ^{\#} = \tilde{w}^t(I - \mathbf{1}w^t) = \tilde{w}^t - w^t$. Hence we have $w^t = \tilde{w}^t(I - EQ^{\#})$; the desired expression for $\tilde{w}^t$ now follows. □

The following result from Golub and Meyer's paper [45] can be established in essentially the same manner as (3.19). However, we provide an independent proof here for completeness.

COROLLARY 5.3.2 *Suppose that $T \in \mathcal{AS}_n$ and denote the stationary distribution vector of T by w. Let E be an $n \times n$ matrix with row sums 0 such that $t_{i,j} > 0$ whenever the (ij) entry of E is nonzero. For all*

s such that $T + sE \in \mathcal{AS}_n$, let $w(s)$ denote the corresponding stationary distribution vector. Letting $Q = I - T$, we have

$$\left.\frac{dw(s)}{ds}\right|_{s=0} = w^t E Q^{\#}.$$

Proof: We begin by observing that since T has 1 as an algebraically simple eigenvalue, there is an $\epsilon > 0$ such that for all s with $|s| < \epsilon$, the stochastic matrix $T + sE$ has 1 as an algebraically simple eigenvalue. Thus $w(s)$ is well-defined whenever $|s| < \epsilon$.

From Lemma 5.3.1, we have, for all sufficiently small s, that $w(s)^t = w(0)^t(I - sEQ^{\#})^{-1}$. When s is sufficiently small we can write

$$(I - sEQ^{\#})^{-1} = I + sEQ^{\#} + \sum_{j=2}^{\infty}(sEQ^{\#})^j,$$

and so in that case we find that

$$\frac{w(s)^t - w(0)^t}{s} = w^t EQ^{\#} + \sum_{j=2}^{\infty} s^{j-1} w^t (EQ^{\#})^j;$$

the formula for $\left.\frac{dw(s)}{ds}\right|_{s=0}$ now follows. □

Suppose that we have $T \in \mathcal{AS}_n$ with stationary distribution vector w. If we are given a fixed matrix E with zero row sums, Corollary 5.3.2 then gives us a way of measuring the sensitivity of the stationary distribution when T is perturbed by a small scalar multiple of E. In a related vein, we can also discuss the *conditioning* of the stationary distribution under perturbation of T. If $\tilde{T} = T + E$ is another stochastic matrix with 1 as an algebraically simple eigenvalue and stationary distribution vector $\tilde{w}$, we seek to measure the size of $w - \tilde{w}$ in terms of the size of the (not necessarily small) perturbing matrix E. Again, the group inverse is useful in addressing that question. From the eigen-equation $\tilde{w}^t\tilde{T} = \tilde{w}^t$ we deduce that $\tilde{w}^t(I - T) = \tilde{w}^t E$. Setting $Q = I - T$ and post-multiplying by $Q^{\#}$ now yields

$$\tilde{w}^t - w^t = \tilde{w}^t(I - \mathbf{1}w^t) = \tilde{w}^t QQ^{\#} = \tilde{w}^t EQ^{\#}. \qquad (5.11)$$

In particular, as we will see below, (5.11) allows us to produce inequalities of the form

$$||\tilde{w} - w||_p \leq ||E||_q f(Q^{\#}) \qquad (5.12)$$

for suitable vector and matrix norms $||\cdot||_p$ and $||\cdot||_q$, respectively, and a corresponding suitable function f. There are a number of such inequalities in the literature for various choices of p, q and f; the paper of Cho and Meyer [23] surveys eight such inequalities, while the results of [23] and the companion paper of Kirkland [69] provide comparisons between these various inequalities. We remark here that because w and $\tilde{w}$ are normalised to have 1-norm equal to 1, the most natural choices for p in (5.12) are 1 and ∞. It will transpire that for our results in this section, we will employ the ∞-norm for E; recall that for any $B \in \mathbb{R}^{n,n}$,

$$||B||_\infty = \max_{i=1,\ldots,n} \sum_{j=1}^{n} |b_{i,j}|.$$

We note here for convenience that the 1-norm for B is given by

$$||B||_1 = \max_{j=1,\ldots,n} \sum_{i=1}^{n} |b_{i,j}|,$$

so that $||B^t||_1 = ||B||_\infty$. We refer the reader to Chapter 5 of [57] for further material on matrix and vector norms.

If we have a function f such that (5.12) holds for any pair of stochastic matrices T and $\tilde{T} = T + E$ in $\mathcal{AS}_n$,, then in the literature on Markov chains, the function f is known as a *condition number* for the Markov chain (see [23] and the references therein). This notion differs somewhat from the definition of a *structured condition number* considered by Higham and Higham in [54], which we discuss now.

Suppose that $T \in \mathcal{AS}_n$ with stationary distribution vector w. Fix a vector norm $||\cdot||_p$ and matrix norm $||\cdot||_q$; then a corresponding structured condition number for the stationary distribution vector is given by

$$C^{(p,q)}_{\mathcal{AS}_n}(T) = \lim_{\epsilon \to 0} \sup \left\{ \frac{||\tilde{w} - w||_p}{\epsilon} \middle| T + E \in \mathcal{AS}_n, \right.$$
$$\left. \tilde{w}^t(T + E) = \tilde{w}^t, \tilde{w}^t\mathbf{1} = 1, ||E||_q \le \epsilon \right\}. \quad (5.13)$$

Observe that for a condition number f, the inequality (5.12) provides an upper bound on the norm of $\tilde{w} - w$ without a restriction on the norm of the perturbing matrix E; however, the bound arising from (5.12) may or may not be tight as $\tilde{T}$ ranges over $\mathcal{AS}_n$. On the other hand, the structured condition number $C^{(p,q)}_{\mathcal{AS}_n}$ yields sharp information on changes in the stationary distribution vector, but under the restriction that the norm of the perturbing matrix is small. The following example illustrates this distinction.

EXAMPLE 5.3.3 Consider the 2×2 stochastic matrix

$$T = \begin{bmatrix} 0 & 1 \\ 1 & 0 \end{bmatrix},$$

which has stationary distribution vector $w = \begin{bmatrix} \frac{1}{2} & \frac{1}{2} \end{bmatrix}^t$. From Corollary 5.3.14 below, it will follow that $C^{(\infty,\infty)}_{\mathcal{AS}_n}(T) = \frac{1}{8}$.

On the other hand, note that for any $x \in (0,1)$, if we consider the perturbing matrix $E = xe_1(e_1 - e_2)^t$, then the stationary distribution vector for $T + E$ is given by $\tilde{w} = \begin{bmatrix} \frac{1}{2-x} & \frac{1-x}{2-x} \end{bmatrix}^t$. We thus find that $||w - \tilde{w}||_\infty = \frac{x}{2(2-x)}$, and $||E||_\infty = 2x$. So in that instance we have $||w - \tilde{w}||_\infty = \frac{||E||_\infty}{4(2-x)} > \frac{1}{8}||E||_\infty$. In particular, if x is close to 1, then $||w - \tilde{w}||_\infty$ is roughly twice the size of $C^{(\infty,\infty)}_{\mathcal{AS}_n}(T)||E||_\infty$. This highlights the fact that the structured condition number provides information on changes in the stationary distribution vector only as the norm of the perturbing matrix E tends to zero.

The following result rests on a key claim that appears at the beginning of the proof; the lemma itself will be useful in the development of our results below on condition numbers for Markov chains. Our proof follows the approach of [107].

LEMMA 5.3.4 *Let v be a vector in $\mathbb{R}^n$ such that $v^t\mathbf{1} = 0$.*
a) Suppose that B is an $n \times n$ matrix with complex entries. Then $||B^tv||_1 \leq ||v||_1 \frac{1}{2} \max_{i,j=1,\ldots,n} \sum_{k=1}^n |b_{i,k} - b_{j,k}|$.
b) Suppose that $z \in \mathbb{C}^n$. Then $|v^tz| \leq ||v||_1 \max_{i,j=1,\ldots,n} \frac{|z_i - z_j|}{2}$.

Proof: Throughout we suppose that $v \neq 0$, otherwise the conclusions are obvious. We begin with a claim that there are scalars $a(i,j), 1 \leq i < j \leq n$ such that $\sum_{1 \leq i < j \leq n} |a(i,j)| = ||v||_1$ and in addition, v can be written as $v = \sum_{1 \leq i < j \leq n} a(i,j) \left(\frac{e_i - e_j}{2}\right)$.

We will establish the claim by induction on n, and note that if $n = 2$, then $v = v_1(e_1 - e_2)$, so that we may take $a(1,2) = 2v_1$. Suppose now that the claim holds for vectors of order $n - 1 \geq 2$, and that $v \in \mathbb{R}^n$ with $v^t\mathbf{1} = 0$. Without loss of generality, we may assume that $|v_1| = \min\{|v_j||v_j \neq 0\}$ and that $v_1 > 0$. Since some entry of v is negative, we may also assume without loss of generality that $v_2 < 0$.

Observe that we can write v as

$$v = 2v_1 \begin{bmatrix} \frac{1}{2} \\ -\frac{1}{2} \\ 0 \\ \vdots \\ 0 \end{bmatrix} + \begin{bmatrix} 0 \\ v_2 + v_1 \\ v_3 \\ \vdots \\ v_n \end{bmatrix}.$$

Set $a(1,2) = 2v_1$, and note that we can apply the induction hypothesis to the vector

$$\hat{v} = \begin{bmatrix} v_2 + v_1 \\ v_3 \\ \vdots \\ v_n \end{bmatrix}.$$

Thus, there are scalars $\hat{a}(i,j), 1 \leq i < j \leq n-1$ such that $\sum_{1 \leq i < j \leq n-1} |\hat{a}(i,j)| = ||\hat{v}||_1$ and $\hat{v} = \sum_{1 \leq i < j \leq n-1} \hat{a}(i,j) \left(\frac{e_i - e_j}{2}\right)$. It now follows that v can be written as $v = a(1,2)\frac{e_1 - e_2}{2} + \sum_{1 \leq i < j \leq n-1} \hat{a}(i,j) \left(\frac{e_{i+1} - e_{j+1}}{2}\right)$. Further, we have

$$|a(1,2)| + \sum_{1 \leq i < j \leq n-1} |\hat{a}(i,j)| = 2v_1 + ||\hat{v}||_1 = 2v_1 + |v_2 + v_1| + \sum_{j=3}^{n} |v_j| =$$

$$2v_1 + |v_2| - v_1 + \sum_{j=3}^{n} |v_j| = ||v||_1$$

(note that since $v_2 < 0$ and $-v_2 \geq v_1, |v_2 + v_1| = |v_2| - v_1$). This completes the proof of the induction step, and the claim.

a) Applying the claim above, we find that

$$B^t v = \frac{1}{2} \sum_{1 \leq i < j \leq n} a(i,j)(B^t e_i - B^t e_j).$$

Hence,

$$||B^t v||_1 \leq \frac{1}{2} \sum_{1 \leq i < j \leq n} |a(i,j)| ||B^t e_i - B^t e_j||_1 \leq$$

$$\frac{1}{2} \left(\sum_{1 \leq i < j \leq n} |a(i,j)| \right) \max_{i,j=1,\ldots,n} \sum_{k=1}^{n} |b_{i,k} - b_{j,k}|.$$

Since $\sum_{1 \leq i < j \leq n} |a(i,j)| = ||v||_1$, the conclusion follows.

b) Now we consider $|v^t z|$. Using the claim above we have $|v^t z| \leq \sum_{1\leq i<j\leq n} |a(i,j)| \left(\frac{|z_i - z_j|}{2}\right) \leq \max_{i,j=1,\ldots,n} \frac{|z_i - z_j|}{2} \sum_{1\leq i<j\leq n} |a(i,j)| = ||v||_1 \max_{i,j=1,\ldots,n} \frac{|z_i - z_j|}{2}$. □

We note that part b) of Lemma 5.3.4 appears in a paper of Alpin and Gabassov [4].

Next, we define the following quantities, which will be central to our discussion of the conditioning of the stationary distribution vector. Let $B \in \mathbb{R}^{n,n}$ be a matrix with constant row sums, and define the function $\tau(B)$ by

$$\tau(B) = \frac{1}{2} \max_{i,j=1,\ldots,n} \sum_{k=1}^{n} |b_{i,k} - b_{j,k}|. \tag{5.14}$$

Let $T \in \mathcal{AS}_n$, set $Q = I - T$ and define the function $\kappa(T)$ by

$$\kappa(T) = \frac{1}{2} \max_{i,j=1,\ldots,n} (q^{\#}_{j,j} - q^{\#}_{i,j}). \tag{5.15}$$

Lemma 5.3.4 will assist in proving the following result, which shows that both $\tau(Q^{\#})$ and $\kappa(T)$ serve as condition numbers for the stationary distribution. Part a) of the theorem below is due to Seneta (see [108]) while part b) is due to Haviv and Van der Heyden (see [53]).

THEOREM 5.3.5 *Let $T \in \mathcal{AS}_n$ with stationary distribution vector w. Suppose that $\tilde{T} = T + E \in \mathcal{AS}_n$, and denote its stationary distribution vector by $\tilde{w}$. Set $Q = I - T$. Then the following inequalities hold.*
a)

$$||\tilde{w} - w||_1 \leq ||E||_\infty \tau(Q^{\#}). \tag{5.16}$$

b)

$$||\tilde{w} - w||_\infty \leq ||E||_\infty \kappa(T). \tag{5.17}$$

Proof: a) From (5.11) we have $\tilde{w}^t - w^t = \tilde{w}^t E Q^{\#}$, which yields $||\tilde{w} - w||_1 = ||(Q^{\#})^t E^t \tilde{w}||_1$. Observe that $E^t \tilde{w}$ is a vector in $\mathbb{R}^n$ whose entries sum to zero. Applying Lemma 5.3.4 a), we thus have $||\tilde{w} - w||_1 \leq ||E^t \tilde{w}||_1 \tau(Q^{\#})$. Since $||E^t \tilde{w}||_1 \leq ||\tilde{w}||_1 ||E^t||_1 = ||E||_\infty$, the conclusion follows.

b) Again using (5.11) we find that for each $k = 1, \ldots, n$, $|\tilde{w}_k - w_k| = |\tilde{w}^t E Q^{\#} e_k|$. Observing that $\tilde{w}^t E \mathbf{1} = 0$, we find from Lemma 5.3.4 b) that $|\tilde{w}^t E Q^{\#} e_k| \leq \frac{1}{2} ||E^t \tilde{w}||_1 \max_{i,j=1,\ldots,n} (q^{\#}_{i,k} - q^{\#}_{j,k})$. From Proposition 2.5.1, it follows that for each index k, the maximum entry in $Q^{\#} e_k$ is $q^{\#}_{k,k}$; hence we find that $\max_{i,j=1,\ldots,n}(q^{\#}_{i,k} - q^{\#}_{j,k}) = \max_{j=1,\ldots,n}(q^{\#}_{k,k} - q^{\#}_{j,k})$. Consequently, for each index k, $|\tilde{w}_k - w_k| \le \frac{1}{2}||E^t\tilde{w}||_1 \max_{j=1,\ldots,n}(q^{\#}_{k,k} - q^{\#}_{j,k}) \le \frac{1}{2}||E^t\tilde{w}||_1 \max_{i,j=1,\ldots,n}(q^{\#}_{j,j} - q^{\#}_{i,j})$. The conclusion now follows from the fact that $||E^t\tilde{w}||_1 \le ||E||_\infty$. □

REMARK 5.3.6 Observe that the quantities $q^{\#}_{j,j} - q^{\#}_{i,j}, i \ne j$ arise as part of the definition of $\kappa(T)$ in (5.17). These quantities have a natural interpretation, which was observed by Cho and Meyer in [22]. Consider a matrix $T \in \mathcal{IS}_n$, partitioned (as in Proposition 2.5.1) as

$$T = \left[\begin{array}{c|c} T_{1,1} & T_{1,2} \\ \hline T_{2,1} & T_{2,2} \end{array}\right],$$

where $T_{1,1}$ is the leading principal submatrix of T of order $n-1$, and the 'block' $T_{2,2}$ is 1×1. Referring to Proposition 2.5.1, we find that for each $i = 1, \ldots, n-1$,

$$q^{\#}_{n,n} - q^{\#}_{i,n} = \frac{e_i^t(I - T_{1,1})^{-1}\mathbf{1}}{1 + T_{2,1}(I - T_{1,1})^{-1}\mathbf{1}}.$$

In particular, we find that $\kappa(T)$ can be written as

$$\kappa(T) = \frac{1}{2}\max_{j=1,\ldots,n} \frac{||(I - T_{(j)})^{-1}||_\infty}{1 + r_j^t(I - T_{(j)})^{-1}\mathbf{1}}, \tag{5.18}$$

where, for each $j = 1, \ldots, n, T_{(j)}$ is the submatrix of T formed by deleting its j–th row and column, and r_j^t is the row vector in $\mathbb{R}^{n-1}$ formed by deleting the j–th entry of e_j^tT.

Let $M = [m_{i,j}]_{i,j=1,\ldots,n}$ denote the mean first passage matrix for the Markov chain associated with T. From (5.2), we find that $m_{i,n} = e_i^t(I - T_{1,1})^{-1}\mathbf{1}$ for each $i = 1, \ldots, n-1$. Let w denote the stationary distribution vector for T. From the eigen-equation $w^tT = w^t$ and the normalisation constraint $w^t\mathbf{1} = 1$, it is a straightforward exercise to show that

$$w_n = \frac{1}{1 + T_{2,1}(I - T_{1,1})^{-1}\mathbf{1}}.$$

Referring now to (5.3), we see that the mean first return time to vertex n is given by $m_{n,n} = \frac{1}{w_n}$. Consequently, the quantity $q^{\#}_{n,n} - q^{\#}_{i,n}$ can be rewritten as $\frac{m_{i,n}}{m_{n,n}}$. In a similar way we find that for any distinct indices i, j between 1 and n,

$$q^{\#}_{j,j} - q^{\#}_{i,j} = \frac{m_{i,j}}{m_{j,j}} = m_{i,j}w_j. \tag{5.19}$$

In particular, we find that $\kappa(T) = \frac{1}{2}\max_{i,j=1,\ldots,n,i\neq j} \frac{m_{i,j}}{m_{j,j}}$, and so can be interpreted in terms of a maximum ratio of mean first passage times to mean first return times. Alternatively, for distinct indices i and j, (5.19) allows us to express the quantity $q^{\#}_{j,j} - q^{\#}_{i,j}$ in terms of the mean first passage time from i to j and the j-th entry in the stationary distribution vector.

REMARK 5.3.7 Let A be a real matrix of order n with constant row sums. It follows from Lemma 5.3.4 a) that $\tau(A)$ is equal to $\max\{||A^t v||_1 | v \in \mathbb{R}^n, v^t\mathbf{1} = 0, ||v||_1 = 1\}$. Suppose now that B is another $n \times n$ matrix with constant row sums. If $v \in \mathbb{R}^n, v^t\mathbf{1} = 0, ||v||_1 = 1$, then

$$||B^t A^t v||_1 \leq ||A^t v||_1 \tau(B) \leq \tau(A)\tau(B),$$

from which we deduce that $\tau(AB) \leq \tau(A)\tau(B)$.

From Theorem 5.3.5, we find that for a stochastic matrix $T \in \mathcal{AS}_n$, the quantities $\tau((I-T)^{\#})$ and $\kappa(T)$ serve as condition numbers for the stationary distribution vector, providing information on changes in the stationary distribution vector (measured in the 1-norm and ∞-norm, respectively) due to perturbations of T.

EXAMPLE 5.3.8 Let $T = \mathbf{1}w^t$, where $w \in \mathbb{R}^n$ is a positive vector whose entries sum to 1. Evidently w is the stationary distribution vector for T, while setting $Q = I - T$ we have $Q^{\#} = I - \mathbf{1}w^t$. It follows that $\tau(Q^{\#}) = 1$, while $\frac{1}{2}\max_{i,j=1,\ldots,n}(q^{\#}_{j,j} - q^{\#}_{i,j}) = \frac{1}{2}$.

Next, fix a positive number ϵ such that $\epsilon < w_2$, and let $E = \epsilon\mathbf{1}(e_1 - e_2)^t$; observe that $||E||_\infty = 2\epsilon$. The matrix $\tilde{T} = T + E$ is also irreducible and stochastic, and it follows from Lemma 5.3.1 that the stationary distribution vector $\tilde{w}$ for $\tilde{T}$ is given by $w + \epsilon(e_1 - e_2)^t$. Consequently, we have $\tilde{w} - w = \epsilon(e_1 - e_2)$, so that

$$||\tilde{w} - w||_1 = 2\epsilon = ||E||_\infty \tau(Q^{\#}),$$

and

$$||\tilde{w} - w||_\infty = \epsilon = ||E||_\infty \kappa(T).$$

Thus we see that for these choices of T and E, the equality case holds in the inequalities (5.16) and (5.17).

The following result, which appears in [108], relates the conditioning properties of the stationary distribution vector to the eigenvalues of the underlying transition matrix.

COROLLARY 5.3.9 *Suppose that the matrix $T \in \mathcal{IS}_n$ has eigenvalues $1, \lambda_2, \ldots, \lambda_n$, and let $Q = I - T$. Then $\tau(Q^{\#}) \le \sum_{k=2}^{n} \frac{1}{1-\lambda_k}$. In particular, if $\tilde{T} = T + E$ is another irreducible stochastic matrix of order n, with stationary distribution vector $\tilde{w}$, then $||\tilde{w} - w||_1 \le ||E||_\infty \sum_{k=2}^{n} \frac{1}{1-\lambda_k}$.*

Proof: From the definition of $\tau(Q^{\#})$ in (5.16), we need to show that

$$\max_{i,j=1,\ldots,n} \sum_{k=1}^{n} \frac{|q^{\#}_{i,k} - q^{\#}_{j,k}|}{2} \le \sum_{j=2}^{n} \frac{1}{1-\lambda_j}.$$

We begin by observing that for any $a, b \in \mathbb{R}$, $\frac{|a-b|}{2} = \max\{a, b\} - \frac{a+b}{2}$. Hence we find that for any i, j,

$$\sum_{k=1}^{n} \frac{|q^{\#}_{i,k} - q^{\#}_{j,k}|}{2} = \sum_{k=1}^{n} \left(\max\{q^{\#}_{i,k}, q^{\#}_{j,k}\} - \frac{q^{\#}_{i,k} + q^{\#}_{j,k}}{2} \right).$$

Since $\sum_{k=1}^{n} q^{\#}_{i,k} = \sum_{k=1}^{n} q^{\#}_{j,k} = 0$, we thus have that for any i, j, $\frac{1}{2}\sum_{k=1}^{n} |q^{\#}_{i,k} - q^{\#}_{j,k}| = \sum_{k=1}^{n} \max\{q^{\#}_{i,k}, q^{\#}_{j,k}\}$. From Proposition 2.5.1, it follows that for any indices i, j, k, $\max\{q^{\#}_{i,k}, q^{\#}_{j,k}\} \le q^{\#}_{k,k}$, from which we find that for each i, j, $\frac{1}{2}\sum_{k=1}^{n} |q^{\#}_{i,k} - q^{\#}_{j,k}| \le \sum_{k=1}^{n} q^{\#}_{k,k} = \text{trace}(Q^{\#})$. From the discussion of the spectral properties of $Q^{\#}$ in section 2.2, it now follows that $\text{trace}(Q^{\#}) = \sum_{k=2}^{n} \frac{1}{1-\lambda_k}$. □

Given a stochastic matrix $T \in \mathcal{AS}_n$ with eigenvalues $1, \lambda_2, \ldots, \lambda_n$, the quantity $\sum_{k=2}^{n} \frac{1}{1-\lambda_k}$ arising in Corollary 5.3.9 is known as the *Kemeny constant* for the corresponding Markov chain. Observe that if each of $\lambda_2, \ldots, \lambda_n$ is well-separated from 1, then the Kemeny constant is not too large; when this is the case, then from Corollary 5.3.9, we find that the stationary distribution vector for T exhibits good conditioning properties, in the sense that if T is perturbed to yield another stochastic matrix in $\mathcal{AS}_n$, then the difference in the corresponding stationary distribution vectors cannot be too large relative to the size of the perturbing matrix. In section 6.5 we will investigate the Kemeny constant in greater detail.

In our next sequence of results, we will consider a certain structured condition number associated with a particular entry in the stationary distribution. As we shall see, that will eventually lead to further information about the condition number κ. We now define the condition number of interest. Let $T \in \mathcal{AS}_n$, fix an index j between 1 and n, and let

$$C^{(\infty)}_{\mathcal{AS}_n}(T,j) = \limsup_{\epsilon\to 0}\left\{\frac{|\tilde{w}_j - w_j|}{\epsilon}\middle| T+E\in\mathcal{AS}_n,\right.$$
$$\left.\tilde{w}^t(T+E) = \tilde{w}^t, \tilde{w}^t\mathbf{1} = 1, ||E||_\infty \le \epsilon\right\}. \quad (5.20)$$

(Note the similarity between (5.20) and (5.13).) Evidently $C^{(\infty)}_{\mathcal{AS}_n}(T,j)$ serves as a measure of the conditioning of the j–th entry of the stationary distribution vector of T under perturbations of T that yield another matrix in $\mathcal{AS}_n$. As part of our discussion of $C^{(\infty)}_{\mathcal{AS}_n}(T,j)$, we will employ the following useful quantities. Letting $Q = I - T$, we define, for each $1 \le i \le n, \alpha_T(i,j)$ and $\beta_T(i,j)$ by

$$\alpha_T(i,j) = \min\{q^{\#}_{l,j}|t_{i,l} > 0\} \text{ and } \beta_T(i,j) = max\{q^{\#}_{l,j}|t_{i,l} > 0\}. \quad (5.21)$$

The lemma that follows, which appears in a paper of Kirkland [70], will be helpful in the sequel.

LEMMA 5.3.10 *Suppose that* $T \in \mathcal{AS}_n$, *and that* E *is a perturbing matrix such that* $T+E \in \mathcal{AS}_n$. *Set* $Q = I-T$. *Then for each nonnegative vector* p^t *such that* $p^t\mathbf{1} = 1$, *we have*

$$|p^tEQ^{\#}e_j| \le$$
$$\frac{||E||_\infty}{2}\max\left\{q^{\#}_{j,j} - \sum_{i=1}^n p_i\alpha_T(i,j), \sum_{i=1}^n p_i\beta_T(i,j) - \min_{1\le l\le n} q^{\#}_{l,j}\right\}.$$

<u>**Proof**</u>: Since $T+E$ is stochastic, we see that the (l,m) entry of E is negative only if $t_{l,m} > 0$. Fix an i between 1 and n, and note that e_i^tE can be written as $x^t - y^t$ where each of the vectors x and y is nonnegative, where $x^t\mathbf{1} = y^t\mathbf{1} \le ||E||_\infty/2$, and where $y_l > 0$ only if $t_{i,l} > 0$.

Consequently, we have $e_i^tEQ^{\#}e_j = x^tQ^{\#}e_j - y^tQ^{\#}e_j = \sum_{l=1}^n x_lq^{\#}_{l,j} - \sum_{l=1}^n y_lq^{\#}_{l,j}$. From Proposition 2.5.1, $q^{\#}_{l,j}$ is maximised when $l = j$, so we find that $e_i^tEQ^{\#}e_j \le \sum_{l=1}^n x_lq^{\#}_{j,j} - \sum_{l=1}^n y_l\alpha_T(i,j) \le (q^{\#}_{j,j} - \alpha_T(i,j))||E||_\infty/2$. Similarly, we also have $e_i^tEQ^{\#}e_j = \sum_{l=1}^n x_lq^{\#}_{l,j} - \sum_{l=1}^n y_lq^{\#}_{l,j} \ge \sum_{l=1}^n x_l\min_{1\le l\le n} q^{\#}_{l,j} - \sum_{l=1}^n y_l\beta_T(i,j) \ge (\min_{1\le l\le n} q^{\#}_{l,j} - \beta_T(i,j))||E||_\infty/2$.

We now find readily that $p^tEQ^{\#}e_j \le \frac{||E||_\infty}{2}(q^{\#}_{j,j} - \sum_{i=1}^n p_i\alpha_T(i,j))$ and that $p^tEQ^{\#}e_j \ge -\frac{||E||_\infty}{2}(\beta_T(i,j) - \min_{1\le l\le n} Q^{\#}_{l,j})$. The conclusion now follows. □

The following result provides, for a given stochastic matrix T in $\mathcal{AS}_n$, a formula for $C^{(\infty)}_{\mathcal{AS}_n}(T,j)$. The proof adapts techniques developed in [70].

THEOREM 5.3.11 *Suppose that $T \in \mathcal{AS}_n$, and fix an index j between 1 and n. Denote the stationary distribution vector for T by w and let $Q = I - T$. Then*

$$C^{(\infty)}_{\mathcal{AS}_n}(T,j) = \frac{1}{2}\max\left\{q^{\#}_{j,j} - \sum_{i=1}^{n} w_i\alpha_T(i,j), \sum_{i=1}^{n} w_i\beta_T(i,j) - \min_{1\le l\le n} q^{\#}_{l,j}\right\}. \tag{5.22}$$

Proof: Let $\epsilon > 0$ be given, and suppose that E is a perturbing matrix such that $T+E \in \mathcal{AS}_n$ and $||E||_\infty \le \epsilon$. Denote the stationary distribution vector for $T+E$ by $\tilde{w}$. From (5.11), we have $\tilde{w}^t - w^t = \tilde{w}^t E Q^{\#}$, so that $\tilde{w}_j - w_j = \tilde{w}^t E Q^{\#} e_j$. Note that from Lemma 5.3.1 that $\tilde{w} = w + O(\epsilon)$, where, for any $p > 0$, $O(\epsilon^p)$ denotes a vector such that as $\epsilon \to 0^+$, the entries are bounded above in absolute value by a constant times ϵ^p. Applying Lemma 5.3.10, we thus have that

$$|\tilde{w}_j - w_j| \le \frac{||E||_\infty}{2}\max\left\{q^{\#}_{j,j} - \sum_{i=1}^{n} \tilde{w}_i\alpha_T(i,j), \sum_{i=1}^{n} \tilde{w}_i\beta_T(i,j) - \min_{1\le l\le n} q^{\#}_{l,j}\right\}.$$

Since $|E||_\infty \le \epsilon$, we thus have

$$\frac{|\tilde{w}_j - w_j|}{\epsilon} \le \frac{1}{2}\max\left\{q^{\#}_{j,j} - \sum_{i=1}^{n} \tilde{w}_i\alpha_T(i,j), \sum_{i=1}^{n} \tilde{w}_i\beta_T(i,j) - \min_{1\le l\le n} q^{\#}_{l,j}\right\} =$$

$$\frac{1}{2}\max\left\{q^{\#}_{j,j} - \sum_{i=1}^{n} w_i\alpha_T(i,j), \sum_{i=1}^{n} w_i\beta_T(i,j) - \min_{1\le l\le n} q^{\#}_{l,j}\right\} + O(\epsilon). \tag{5.23}$$

Letting $\epsilon \to 0$ in (5.23), it now follows that $C^{(\infty)}_{\mathcal{AS}_n}(T,j) \le \frac{1}{2}\max\left\{q^{\#}_{j,j} - \sum_{i=1}^{n} w_i\alpha_T(i,j), \sum_{i=1}^{n} w_i\beta_T(i,j) - \min_{1\le l\le n} q^{\#}_{l,j}\right\}$.

For each index $i = 1,\dots,n$, select an index $l(i)$ so that $T_{i,l(i)} > 0$ and $q^{\#}_{l(i),j} = \alpha_T(i,j)$. Let $E_1 = \frac{1}{2}\sum_{i=1}^{n} e_i(e_j^t - e_{l(i)}^t)$, and observe that for all sufficiently small $\epsilon > 0, T + \epsilon E_1 \in \mathcal{AS}_n$, while $||\epsilon E_1||_\infty = \epsilon$. Fix a (small) $\epsilon > 0$, and let $\tilde{w}$ denote the stationary distribution vector for $T + \epsilon E_1$. From Lemma 5.3.1 it follows that $\tilde{w}_j - w_j = \epsilon w^t E_1 Q^{\#} e_j + O(\epsilon^2)$. Thus we have $\frac{|\tilde{w}_j - w_j|}{\epsilon} = \frac{1}{2}\left(q^{\#}_{j,j} - \sum_{i=1}^{n} w_i \alpha_T(i,j)\right) + \frac{O(\epsilon^2)}{\epsilon}$. Letting $\epsilon \to 0$, we find that

$$C^{(\infty)}_{\mathcal{AS}_n}(T,j) \geq \frac{1}{2}\left(q^{\#}_{j,j} - \sum_{i=1}^{n} w_i \alpha_T(i,j)\right).$$

Next, for each $i = 1, \ldots, n$,, find an index $p(i)$ such that $T_{i,p(i)} > 0$ and $q^{\#}_{p(i),j} = \beta_T(i,j)$. Let m be an index such that $q^{\#}_{m,j} = \min_{1 \leq l \leq n} q^{\#}_{l,j}$. Now let $E_2 = \frac{1}{2}\sum_{i=1}^{n} e_i(e^t_m - e^t_{l(i)})$. Note that for all sufficiently small $\epsilon > 0, T + \epsilon E_2 \in \mathcal{AS}_n$, and $||\epsilon E_2||_\infty = \epsilon$. Fix a small positive ϵ, and let $\tilde{w}$ denote the stationary distribution vector for $T + E_2$. As above we have $\tilde{w}_j - w_j = \epsilon w^t E_2 Q^{\#} e_j + O(\epsilon^2)$, from which it follows that $\frac{|\tilde{w}_j - w_j|}{\epsilon} = \frac{1}{2}\left(\sum_{i=1}^{n} w_i \beta_T(i,j) - q^{\#}_{m,j}\right) + \frac{O(\epsilon^2)}{\epsilon}$. Now we let $\epsilon \to 0$ to deduce that

$$C^{(\infty)}_{\mathcal{AS}_n}(T,j) \geq \frac{1}{2}\left(\sum_{i=1}^{n} w_i \beta_T(i,j) - q^{\#}_{m,j}\right).$$

Consequently we find that

$$C^{(\infty)}_{\mathcal{AS}_n}(T,j) \geq \frac{1}{2}\max\left\{q^{\#}_{j,j} - \sum_{i=1}^{n} w_i \alpha_T(i,j), \sum_{i=1}^{n} w_i \beta_T(i,j) - \min_{1 \leq l \leq n} q^{\#}_{l,j}\right\},$$

and the conclusion follows. □

REMARK 5.3.12 Suppose that $T \in \mathcal{IS}_n$ with stationary distribution vector w and mean first passage matrix M. Setting $Q = I - T$, and letting $\alpha_T(i,j)$ and $\beta_T(i,j)$ be given by (5.21), we may recast the expressions $q^{\#}_{j,j} - \sum_{i=1}^{n} w_i \alpha_T(i,j)$ and $\sum_{i=1}^{n} w_i \beta_T(i,j) - \min_{1 \leq l \leq n} q^{\#}_{l,j}$ in terms of the mean first passage times as follows.

Observe that

$$\begin{aligned} q^{\#}_{j,j} - \sum_{i=1}^{n} w_i \alpha_T(i,j) &= \sum_{i=1}^{n} w_i (q^{\#}_{j,j} - \min\{q^{\#}_{l,j} | t_{i,l} > 0\}) \\ &= \sum_{i=1}^{n} w_i \max\{q^{\#}_{j,j} - q^{\#}_{l,j} | t_{i,l} > 0\}. \end{aligned}$$

From (5.19), we see that this last expression can be rewritten as

$w_j \sum_{i=1}^n w_i \max\{m_{l,j} | t_{i,l} > 0\}$. Hence

$$q^{\#}_{j,j} - \sum_{i=1}^{n} w_i \alpha_T(i,j) = w_j \sum_{i=1}^{n} w_i \max\{m_{l,j} | t_{i,l} > 0\}.$$

Similarly, we note that

$$\sum_{i=1}^{n} w_i \beta_T(i,j) - \min_{1 \le l \le n} q^{\#}_{l,j} =$$
$$\sum_{i=1}^{n} w_i \left(\max\{q^{\#}_{l,j} | t_{i,l} > 0\} - q^{\#}_{j,j} + q^{\#}_{j,j} - \min_{1 \le l \le n} q^{\#}_{l,j} \right).$$

Again applying (5.19), it now follows that

$$\sum_{i=1}^{n} w_i \beta_T(i,j) - \min_{1 \le l \le n} q^{\#}_{l,j} =$$
$$w_j \sum_{i=1}^{n} w_i \left(\max\{m_{l,j} | l = 1, \ldots, n\} - \min\{m_{l,j} | t_{i,l} > 0\} \right).$$

REMARK 5.3.13 In this remark, we illustrate the fact that in general, the quantities $q^{\#}_{j,j} - \sum_{i=1}^n w_i \alpha_T(i,j)$ and $\sum_{i=1}^n w_i \beta_T(i,j) - \min_{1 \le l \le n} q^{\#}_{l,j}$ of Theorem 5.3.11 are incomparable.

Consider the following stochastic matrices:

$$\hat{T} = \begin{bmatrix} 0 & 1 & 0 & 0 \\ \frac{1}{2} & 0 & \frac{1}{2} & 0 \\ 0 & \frac{1}{2} & 0 & \frac{1}{2} \\ 0 & 0 & 1 & 0 \end{bmatrix}, \tilde{T} = \begin{bmatrix} 0 & 0 & 0 & 1 \\ 0 & 0 & 0 & 1 \\ 0 & 0 & 0 & 1 \\ \frac{1}{3} & \frac{1}{3} & \frac{1}{3} & 0 \end{bmatrix}.$$

We find readily that the stationary distribution vectors for $\hat{T}$ and $\tilde{T}$ are

$$\hat{w} = \begin{bmatrix} \frac{1}{6} \\ \frac{1}{3} \\ \frac{1}{3} \\ \frac{1}{6} \end{bmatrix} \text{ and } \tilde{w} = \begin{bmatrix} \frac{1}{6} \\ \frac{1}{6} \\ \frac{1}{6} \\ \frac{1}{2} \end{bmatrix},$$

respectively. Setting $\hat{Q} = I - \hat{T}$ and $\tilde{Q} = I - \tilde{T}$, we find that

$$\hat{Q}^{\#} = \frac{1}{36} \begin{bmatrix} 35 & 10 & -26 & -19 \\ 5 & 22 & -14 & -13 \\ -13 & -14 & 22 & 5 \\ -19 & -26 & 10 & 35 \end{bmatrix}$$

and

$$\tilde{Q}^{\#} = \frac{1}{12}\begin{bmatrix} 9 & -3 & -3 & -3 \\ -3 & 9 & -3 & -3 \\ -3 & -3 & 9 & -3 \\ -1 & -1 & -1 & 3 \end{bmatrix}.$$

It now follows that

$$\hat{q}^{\#}_{1,1} - \sum_{i=1}^{4} \hat{w}_i \alpha_{\hat{T}}(i,1) = \frac{17}{18} < 1 = \sum_{i=1}^{4} \hat{w}_i \beta_{\hat{T}}(i,1) - \min_{1\le l\le 4} \hat{q}^{\#}_{l,1},$$

while

$$\tilde{q}^{\#}_{2,2} - \sum_{i=1}^{4} \tilde{w}_i \alpha_{\tilde{T}}(i,2) = \frac{11}{12} > \frac{7}{12} = \sum_{i=1}^{4} \tilde{w}_i \beta_{\tilde{T}}(i,2) - \min_{1\le l\le 4} \tilde{q}^{\#}_{l,2}.$$

We thus find that in Theorem 5.3.11, depending on the particular matrix T under consideration, each of the cases $C^{(\infty)}_{\mathcal{AS}_n}(T,j) = \sum_{i=1}^{n} w_i \beta_T(i,j) - \min_{1\le l\le n} q^{\#}_{l,j} > q^{\#}_{j,j} - \sum_{i=1}^{n} w_i \alpha_T(i,j)$ and $C^{(\infty)}_{\mathcal{AS}_n}(T,j) = q^{\#}_{j,j} - \sum_{i=1}^{n} w_i \alpha_T(i,j) > \sum_{i=1}^{n} w_i \beta_T(i,j) - \min_{1\le l\le n} q^{\#}_{l,j}$ may arise.

Theorem 5.3.11 immediately yields the following result, which gives an expression for the structured condition number of (5.13) for the case that $p = q = \infty$.

COROLLARY 5.3.14 *Suppose that $T \in \mathcal{AS}_n$. Then*

$$C^{(\infty,\infty)}_{\mathcal{AS}_n}(T) = \max_{j=1,\dots,n} \frac{1}{2} \max\left\{ q^{\#}_{j,j} - \sum_{i=1}^{n} w_i \alpha_T(i,j), \sum_{i=1}^{n} w_i \beta_T(i,j) - \min_{1\le l\le n} q^{\#}_{l,j} \right\}. \tag{5.24}$$

EXAMPLE 5.3.15 We return once again to the matrix T of Example 1.2.1, and recall that in that example we focused our attention on the behaviour of the twelfth and nineteenth entries in the stationary distribution as T is perturbed. In this example we compute $C^{\infty}_{\mathcal{AS}_{20}}(T,j)$ for the cases $j = 12$ and $j = 19$, thus shedding further light on how those two entries of the stationary distribution behave under perturbation of T.

Set $Q = I - T$. From the combinatorial structure of T it follows that for each $j = 1, \dots, 20$, we have $\alpha_T(1,j) = q^{\#}_{2,j}, \alpha_T(20,j) = q^{\#}_{19,j},$

and $\alpha_T(i,j) = \min\{q^{\#}_{i-1,j}, q^{\#}_{i+1,j}\}, i = 2,\ldots,19$, while $\beta_T(1,j) = q^{\#}_{2,j}, \beta_T(20,j) = q^{\#}_{19,j}$, and $\beta_T(i,j) = \max\{q^{\#}_{i-1,j}, q^{\#}_{i+1,j}\}, i = 2,\ldots,19$. Computing $Q^{\#}$, we find that

$$Q^{\#}e_{12} = \frac{1}{38}\begin{bmatrix} -113 \\ -111 \\ -105 \\ -95 \\ -81 \\ -63 \\ -41 \\ -15 \\ 15 \\ 49 \\ 87 \\ 129 \\ 99 \\ 73 \\ 51 \\ 33 \\ 19 \\ 9 \\ 3 \\ 1 \end{bmatrix} \text{ and } Q^{\#}e_{19} = \frac{1}{38}\begin{bmatrix} -239 \\ -237 \\ -231 \\ -221 \\ -207 \\ -189 \\ -167 \\ -141 \\ -111 \\ -77 \\ -39 \\ 3 \\ 49 \\ 99 \\ 153 \\ 211 \\ 273 \\ 339 \\ 409 \\ 407 \end{bmatrix}.$$

(These expressions for $Q^{\#}e_{12}$ and $Q^{\#}_{19}$ can also be deduced from Example 5.5.1 below.) Applying Theorem 5.3.11, it now follows that $C^{\infty}_{\mathcal{AS}_{20}}(T,12) = \frac{817}{423}$, while $C^{\infty}_{\mathcal{AS}_{20}}(T,19) = \frac{443}{76}$. As we may have expected from Example 1.2.1, $C^{\infty}_{\mathcal{AS}_{20}}(T,19) > C^{\infty}_{\mathcal{AS}_{20}}(T,12)$.

Next we adapt a result from [70] to provide a simple lower bound on $C^{(\infty)}_{\mathcal{AS}_n}(T,j)$

LEMMA 5.3.16 *Suppose that $T \in \mathcal{IS}_n$, with stationary distribution vector w. Fix an index j between 1 and n, and let $Q = I - T$. Then*

$$C^{(\infty)}_{\mathcal{AS}_n}(T,j) \geq \frac{1}{4}\left(q^{\#}_{j,j} - \min_{1\leq i\leq n} q^{\#}_{i,j}\right). \tag{5.25}$$

Further, equality holds in (5.25) if and only if T is a periodic matrix, and one of the diagonal blocks in the periodic normal form for T corresponds to the single index j.

Proof: We have

$$\frac{1}{2}\max\left\{q^{\#}_{j,j}-\sum_{i=1}^{n} w_i\alpha_T(i,j), \sum_{i=1}^{n} w_i\beta_T(i,j)-\min_{1\le l\le n} q^{\#}_{l,j}\right\}\ge$$

$$\frac{1}{4}\left(\left(q^{\#}_{j,j}-\sum_{i=1}^{n} w_i\alpha_T(i,j)\right)+\left(\sum_{i=1}^{n} w_i\beta_T(i,j)-\min_{1\le l\le n} q^{\#}_{l,j}\right)\right)=$$

$$\frac{1}{4}\left(q^{\#}_{j,j}-\min_{1\le l\le n} q^{\#}_{l,j}+\sum_{i=1}^{n} w_i(\beta_T(i,j)-\alpha_T(i,j))\right)\ge$$

$$\frac{1}{4}(q^{\#}_{j,j}-\min_{1\le i\le n} q^{\#}_{i,j}).$$

Next, suppose that equality holds in (5.25). Fix an index i between 1 and n; we necessarily have $\alpha_T(i,j)=\beta_T(i,j)$, so that either there is a single index l such that $t_{i,l}>0$, or for any pair of indices l_1,l_2 such that $t_{i,l_1},t_{i,l_2}>0$, we have $q^{\#}_{l_1,j}=q^{\#}_{l_2,j}$. It now follows that if $t_{i,l}>0$, then $e_i^t(I-\mathbf{1}w^t)e_j=e_i^t(I-T)Q^{\#}e_j=e_i^tQ^{\#}e_j-e_i^tTQ^{\#}e_j=q^{\#}_{i,j}-q^{\#}_{l,j}$. Consequently, if $i\ne j$, then $q^{\#}_{i,j}=q^{\#}_{l,j}-w_j$. Observe also that since $\alpha_T(j,j)=\beta_T(j,j)$, and since $q^{\#}_{l,j}$ is uniquely maximised when $l=j$, it cannot be the case that $t_{j,j}$ is positive. Hence $t_{j,j}=0$.

Consider the directed graph $\mathcal{D}(T)$. We claim that if the distance from vertex i to vertex j in $\mathcal{D}(T)$ is $d\ge 1$, then $q^{\#}_{i,j}=q^{\#}_{j,j}-dw_j$. In order to prove the claim, we proceed by induction on d, and note that the case that $d=1$ has already been established above. Suppose now that the claim holds for vertices whose distance to j is $d\ge 1$, and that the distance from i to j is $d+1$. Necessarily there is a vertex l at distance d from j such that $t_{i,l}>0$. Hence, invoking the induction hypothesis, we have $q^{\#}_{i,j}=q^{\#}_{l,j}-w_j=q^{\#}_{j,j}-dw_j-w_j=q^{\#}_{j,j}-(d+1)w_j$. This completes the induction step, and the proof of the claim.

From the claim, we find that if $i\ne j$ and the distance from vertex i to vertex j is d, then $t_{i,l}>0$ only if the distance from vertex l to vertex j is $d-1$. It follows that we can partition the vertices distinct from j into subsets $S_1,\dots,S_k$ such that
a) $i\in S_l$ if and only if the distance from i to j is l; and
b) if $a\ne j$, then $t_{a,b}>0$ only if for some l between 2 and $k, a\in S_l$ and $b\in S_{l-1}$.

We now find that vertex $a\in S_l$ if and only if $q^{\#}_{a,j}=q^{\#}_{j,j}-lw_j$. Suppose now that there are indices a_1,a_2 such that $t_{j,a_1},t_{j,a_2}>0$ (necessarily, $a_1,a_2\ne j$, since $t_{j,j}=0$.) From the fact that $q^{\#}_{a_1,j}=q^{\#}_{a_2,j}$, we deduce that there is an index l such that $a_1, a_2 \in S_l$. We thus conclude that if i is any index such that $t_{j,i} > 0$, then $i \in S_l$. Since T is irreducible, it must be the case that $l = k$. Consequently, T is a periodic matrix, and one of the diagonal blocks in the periodic normal form for T corresponds to the single index j.

Finally, if T is periodic, with period p say, and has a 1×1 diagonal block in the periodic normal form that corresponds to index j, then by Theorem 5.2.5, we find that $q^{\#}_{j,j} = \frac{p-1}{2p}$ while if i is at distance d from j, then $q^{\#}_{i,j} = \frac{p-1}{2p} - \frac{d}{p}$. It now follows that equality holds in (5.25), with both sides equal to $\frac{p-1}{4p}$. $\square$

Lemma 5.3.16 facilitates the following result from [70], which provides a lower bound on the change in the j-th entry of the stationary distribution vector, under certain perturbations of the underlying stochastic matrix.

THEOREM 5.3.17 *Suppose that $T \in \mathcal{IS}_n$ has stationary distribution vector w. Fix an index $j = 1, \ldots, n$. For each sufficiently small $\epsilon > 0$, there is matrix E such that $||E||_\infty = \epsilon$, $T + E$ is irreducible and stochastic, and further, the stationary distribution vector, $\tilde{w}^t$, of $T + E$ satisfies*

$$|\tilde{w}_j - w_j| > \frac{1}{4}||E||_\infty (q^{\#}_{j,j} - \min_{1 \le i \le n} q^{\#}_{i,j}). \tag{5.26}$$

Proof: Following the proof of Theorem 5.3.11, we find that for all sufficiently small $\epsilon > 0$, there is a perturbing matrix E such that $||E||_\infty = \epsilon$, $T + E$ is irreducible and stochastic, and, denoting the stationary distribution for $T + E$ by $\tilde{w}$, also $|\tilde{w}_j - w_j| = ||E||_\infty C^{(\infty)}_{\mathcal{AS}_n}(T, j) + O(\epsilon^2)$. From Lemma 5.3.16, we have $C^{(\infty)}_{\mathcal{AS}_n}(T, j) \ge \frac{1}{4}(q^{\#}_{j,j} - \min_{1 \le i \le n} q^{\#}_{i,j})$; if that inequality is strict, then the conclusion follows readily.

On the other hand, if equality holds in (5.25), then again appealing to Lemma 5.3.16, we find that T is periodic, and that there is a diagonal block in the periodic normal form for T that corresponds to the index

j. For concreteness, assume that $j = n$. Then T can be written as

$$T = \left[\begin{array}{c|c|c|c|c|c} 0 & T_1 & 0 & 0 & \dots & 0 \\ \hline 0 & 0 & T_2 & 0 & \dots & 0 \\ \hline \vdots & & \ddots & \ddots & & \vdots \\ \hline 0 & 0 & & \dots & T_{p-2} & 0 \\ \hline 0 & 0 & & \dots & 0 & \mathbf{1} \\ \hline t_p^t & 0 & 0 & \dots & 0 & 0 \end{array}\right]. \tag{5.27}$$

We have

$$w^t = \frac{1}{p}\left[\ t_p^t \ \mid \ t_p^t T_1 \ \mid \ t_p^t T_1 T_2 \ \mid \ \dots \ \mid \ t_p^t T_1 \cdots T_{p-2} \ \mid \ 1\ \right].$$

In particular, we have $w_n = \frac{1}{p}$ and from Theorem 5.2.2, we find that $q^{\#}_{n,n} - \min_{1 \le l \le n} q^{\#}_{l,n} = \frac{p-1}{p}$.

Next let $\epsilon > 0$ be given, and consider the following perturbation matrix:

$$E = \epsilon \left[\begin{array}{c|c|c|c|c|c} 0 & -A_1 & 0 & 0 & \dots & \mathbf{1} \\ \hline 0 & 0 & -A_2 & 0 & \dots & \mathbf{1} \\ \hline \vdots & & & \ddots & & \vdots \\ \hline 0 & 0 & & \dots & A_{p-2} & 0 \\ \hline 0 & 0 & & \dots & 0 & 0 \\ \hline -a_p^t & 0 & 0 & \dots & 0 & 1 \end{array}\right], \tag{5.28}$$

where each A_i satisfies $A_i\mathbf{1} = \mathbf{1}$ and has positive entries only in positions for which T_i is positive, and where $a_p^t\mathbf{1} = 1$ with a_p^t positive only where t_p^t is positive. For all sufficiently small ϵ, $T + E \in \mathcal{IS}_n$. Evidently $||E||_\infty = 2\epsilon$, and it is straightforward to show that the stationary distribution vector for $T + E, \tilde{w}$, satisfies

$$\tilde{w}^t = \frac{\epsilon}{1-(1-\epsilon)^p}\left[u^t \ \mid \ u^t S_1 \ \mid \ u^t S_1 S_2 \ \mid \ \dots \ \mid \ u^t S_1 \cdots S_{p-2} \ \mid \ 1\right],$$

where $u^t = t_p^t + \epsilon a_p^t$ and $S_i = T_i + \epsilon A_i$ for each $i = 1, \dots, p-2$.

Consequently, we have

$$\tilde{w}_n - w_n = \frac{p\epsilon + (1-\epsilon)^p - 1}{p(1-(1-\epsilon)^p)}.$$

Since

$$(1-\epsilon)^p = 1 - p\epsilon + \frac{p(p-1)}{2}\epsilon^2 - \frac{p(p-1)(p-2)}{6}\epsilon^3 + O(\epsilon^4),$$

it follows that for all sufficiently small $\epsilon > 0$,

$$\frac{p\epsilon + (1-\epsilon)^p - 1}{p(1-(1-\epsilon)^p)} > \frac{p-1}{2p}\epsilon = \frac{1}{4}||\epsilon E||_\infty(q^{\#}_{n,n} - \min_{1\le l\le n} q^{\#}_{l,n}).$$

□

The following is a direct consequence of Theorem 5.3.17.

COROLLARY 5.3.18 ([70]) *Let T be an irreducible stochastic matrix with stationary distribution vector w. For each sufficiently small positive ϵ, there is matrix E such that $||E||_\infty = \epsilon$, $T+E$ is irreducible and stochastic with stationary distribution vector $\tilde{w}$, say, and $||\tilde{w} - w||_\infty > ||E||_\infty \frac{\kappa(T)}{2}$.*

Suppose that the matrix $T \in \mathcal{IS}_n$ has stationary distribution vector w. For any perturbing matrix E such that $T+E \in \mathcal{AS}_n$, then denoting the stationary distribution vector by $\tilde{w}$, we have from Theorem 5.3.5 b) that $||\tilde{w} - w||_\infty \le ||E||_\infty \kappa(T)$. Moreover, from Corollary 5.3.18, for all sufficiently small $\epsilon > 0$, there is a perturbing matrix $\hat{E}$ such that $T + \hat{E} \in \mathcal{AS}_n, ||\hat{E}|| = \epsilon$, and, denoting the stationary distribution vector for $T + \hat{E}$ by $\hat{w}$, $||\hat{w} - w||_\infty > \frac{1}{2}||E||_\infty \kappa(T)$. Thus, taken together, Theorem 5.3.5 b) and Corollary 5.3.18 show that the function $\kappa(T)$ provides a reasonably accurate measure of the conditioning properties of the stationary distribution, measured in the ∞–norm.

In view of our comments above, we see that lower bounds on κ have implications for the conditioning of the stationary distribution, since such a lower bound then yields, via Corollary 5.3.18, a corresponding lower bound on the size of the change in the stationary distribution vector under certain perturbations of the transition matrix. We now present a couple of lower bounds on κ, the first of which provides a complementary result to Corollary 5.3.9.

THEOREM 5.3.19 *Suppose that $T \in \mathcal{IS}_n$ and has stationary distribution vector w. Denote the eigenvalues of T by $1, \lambda_2, \ldots, \lambda_n$. Then*

$$\kappa(T) \ge \frac{1}{2(n-1)} \sum_{j=1}^{n} \frac{1}{1-\lambda_j}. \tag{5.29}$$

Further, equality holds in (5.29) if and only if there is a scalar $t \in [-\min_{1\le k\le n} w_k, 1)$ such that $T = tI + (1-t)\mathbf{1}w^t$.

Proof: Let $Q = I - T$. Select indices i_0 and j_0 such that $q^{\#}_{j_0,j_0} - q^{\#}_{i_0,j_0} = \max_{i,j=1,\dots,n}\left(q^{\#}_{j,j} - q^{\#}_{i,j}\right)$, so that with this notation, we have $\kappa(T) = \frac{1}{2}\left(q^{\#}_{j_0,j_0} - q^{\#}_{i_0,j_0}\right)$. Next, we note that

$$q^{\#}_{j_0,j_0} - q^{\#}_{i_0,j_0} \geq \frac{1}{n(n-1)}\sum_{i=1}^{n}\sum_{j\neq i}\left(q^{\#}_{j,j} - q^{\#}_{i,j}\right) =$$

$$\frac{1}{n(n-1)}\sum_{i=1}^{n}\sum_{j=1}^{n}\left(q^{\#}_{j,j} - q^{\#}_{i,j}\right) = \frac{1}{n(n-1)}\sum_{i=1}^{n}\sum_{j=1}^{n} q^{\#}_{j,j} = \frac{1}{n-1}\sum_{j=1}^{n}\frac{1}{1-\lambda_j}.$$

The inequality (5.29) now follows.

Suppose now that equality holds in (5.29). Examining the argument above, we see that for each $j = 1, \dots, n$, it must be that case that the off-diagonal entries of $Q^{\#}e_j$ all have the same value. From Proposition 2.5.1 it now follows that for each $j = 1, \dots, n$, there is a scalar a_j such that $(I - T_{(j)})^{-1}\mathbf{1} = a_j\mathbf{1}$, where $T_{(j)}$ denotes the principal submatrix of T obtained by deleting its j-th row and column. Consequently, we find that for each $j = 1, \dots, n, T_{(j)}\mathbf{1} = \left(1 - \frac{1}{a_j}\right)\mathbf{1}$, and since T has row sums all equal to 1, it follows that for each $j = 1, \dots, n$, the off-diagonal entries in the j-th column of T are all equal to $\frac{1}{a_j}$. Thus we find that T can be written as $T = D + \mathbf{1}u^t$, where D is a diagonal matrix and u is a positive vector. Since T is stochastic, we see that $D\mathbf{1} = (1 - u^t\mathbf{1})\mathbf{1}$, so that necessarily D is a multiple of the identity matrix. Hence u must be a scalar multiple of w. Writing $D = tI$ for some scalar t, it now follows from the nonnegativity and irreducibility of T that $-\min_{1\leq k\leq n} w_k \leq t < 1$.

Conversely, if we have $T = tI + (1-t)\mathbf{1}w^t$ for some $-\min_{1\leq k\leq n} w_k \leq t < 1$, then $(I - T)^{\#} = \frac{1}{1-t}(I - \mathbf{1}w^t)$, while $\lambda_j = 1 - t, j = 2, \dots, n$. It is now readily verified that equality holds in (5.29). □

Next, we consider another lower bound (which appears in [69]) on the condition number κ. Recall that a stochastic matrix T is *doubly stochastic* if $\mathbf{1}^tT = \mathbf{1}^t$.

THEOREM 5.3.20 *Suppose that $T \in \mathcal{IS}_n$. Then*

$$\kappa(T) \geq \frac{n-1}{2n}. \tag{5.30}$$

Let M denote the mean first passage matrix for T. Equality holds in (5.30) if and only if

a) T is a doubly stochastic matrix with zero diagonal;
b) for each $j = 1, \ldots, n, \max_{i \neq j} m_{i,j} = n - 1$;
c) for each pair of distinct indices $i, j, t_{j,i} > 0$ only if $m_{i,j} = n - 1$.

Proof: For each $j = 1, \ldots, n$, let $T_{(j)}$ denote the principal submatrix of T formed by deleting its j-th row and column, and let r_j^t denote the row vector formed from $e_j^t T$ by deleting its j-th entry. Letting w denote the stationary distribution vector of T, we find as in Remark 5.3.6 that for each $j = 1, \ldots, n$,

$$w_j = \frac{1}{1 + r_j^t (I - T_{(j)})^{-1} \mathbf{1}}.$$

Since $w^t \mathbf{1} = 1$, we find that

$$1 = \sum_{j=1}^{n} \frac{1}{1 + r_j^t (I - T_{(j)})^{-1} \mathbf{1}} \geq \sum_{j=1}^{n} \frac{1}{1 + ||(I - T_{(j)})^{-1}||_\infty} \geq \frac{n}{1 + \max_{j=1,\ldots,n} ||(I - T_{(j)})^{-1}||_\infty}.$$

Hence

$$\max_{j=1,\ldots,n} ||(I - T_{(j)})^{-1}||_\infty \geq n - 1. \tag{5.31}$$

Next, we recall from (5.18) that

$$2\kappa(T) = \max_{j=1,\ldots,n} \frac{||(I - T_{(j)})^{-1}||_\infty}{1 + r_j^t (I - T_{(j)})^{-1} \mathbf{1}}.$$

Consequently, we find that

$$2\kappa(T) \geq \max_{j=1,\ldots,n} \frac{||(I - T_{(j)})^{-1}||_\infty}{1 + ||(I - T_{(j)})^{-1}||_\infty} \geq \frac{n-1}{n},$$

the last inequality following from (5.31). Consequently, $\kappa(T) \geq \frac{n-1}{2n}$.

Suppose now that equality holds in (5.30). Then from the argument above, it must be the case that for each $j = 1, \ldots, n, ||(I - T_{(j)})^{-1}||_\infty = n - 1$ and $r_j^t (I - T_{(j)})^{-1} \mathbf{1} = n - 1$. Observe that these two conditions together imply that for each $j, w_j = \frac{1}{n}$, so that necessarily T is doubly stochastic. Also, from (5.2) we find that for any index $j, ||(I - T_{(j)})^{-1}||_\infty = \max_{i \neq j} m_{i,j}$, so we see that condition b) also holds. Finally, for each index j, we have $n - 1 = r_j^t (I - T_{(j)})^{-1} \mathbf{1} \leq r_j^t \mathbf{1} ||(I -$

$T_{(j)})^{-1}||_\infty = r_j^t \mathbf{1}(n-1)$. We deduce that $r_j^t\mathbf{1} = 1$ for each $j = 1, \ldots, n$, from which it follows that T has zero diagonal. That fact, together with the fact that T is doubly stochastic, yields condition a). Further, since for each $j = 1, \ldots, n$ we have $n - 1 = r_j^t(I - T_{(j)})^{-1}\mathbf{1} = \sum_{i \neq j} t_{j,i} m_{i,j}$, and since each $m_{i,j} \leq n-1$, whenever $i \neq j$, it must be the case that $t_{j,i} > 0$ only if $m_{i,j} = n-1$. Condition c) now follows.

Finally, if conditions a), b), and c) hold, it is straightforward to show that $\kappa(T) = \frac{n-1}{2n}$. □

REMARK 5.3.21 Suppose that $T \in \mathcal{IS}_n$, and let M denote the corresponding mean first passage matrix. Inspecting the proof of Theorem 5.3.20, we see that in fact we have shown that

$$\max_{i,j=1,\ldots,n, i\neq j} m_{i,j} \geq n-1. \tag{5.32}$$

Further, the proof of Theorem 5.3.20 also reveals that equality holds in (5.32) if and only if conditions a), b) and c) of Theorem 5.3.20 hold.

Notice that the conditions that characterise the case of equality in (5.30) are somewhat opaque, as they involve not only the structure of the stochastic matrix T, but also the corresponding mean first passage matrix M. We now give a few examples of matrices for which equality holds in (5.30).

EXAMPLE 5.3.22 Let T be the $n \times n$ stochastic matrix given by $T = \frac{1}{n-1}(J - I)$, where J is the all ones matrix of order n. Since $Q = I - T$ is equal to $\frac{n}{n-1}\left(I - \frac{1}{n}J\right)$, we find that $Q^\# = \frac{n-1}{n}\left(I - \frac{1}{n}J\right)$; evidently we have $\kappa(T) = \frac{n-1}{2n}$. It is straightforward to show (by computing $(I - T_{(j)})^{-1}\mathbf{1}$ for each j) that the corresponding mean first passage matrix is $M = I + (n-1)J$.

EXAMPLE 5.3.23 Again we consider the stochastic matrix of Example 5.2.3. Specifically, let

$$T = \begin{bmatrix} 0 & 1 & 0 & 0 & \ldots & 0 \\ 0 & 0 & 1 & 0 & \ldots & 0 \\ \vdots & & \ddots & \ddots & & \vdots \\ & & & & & \\ 0 & 0 & \ldots & & 0 & 1 \\ 1 & 0 & 0 & \ldots & & 0 \end{bmatrix}.$$

Referring to Example 5.2.3, we see that

$$(I-T)^{\#} = \begin{bmatrix} \frac{n-1}{2n} & \frac{n-3}{2n} & \frac{n-5}{2n} & \cdots & -\frac{n-3}{2n} & -\frac{n-1}{2n} \\ -\frac{n-1}{2n} & \frac{n-1}{2n} & \frac{n-3}{2n} & \frac{n-5}{2n} & \cdots & -\frac{n-3}{2n} \\ -\frac{n-3}{2n} & -\frac{n-1}{2n} & \frac{n-1}{2n} & \frac{n-3}{2n} & \cdots & -\frac{n-5}{2n} \\ \vdots & & & \ddots & & \vdots \\ \frac{n-3}{2n} & \frac{n-5}{2n} & \cdots & -\frac{n-3}{2n} & -\frac{n-1}{2n} & \frac{n-1}{2n} \end{bmatrix}.$$

In particular, we have $\kappa(T) = \frac{n-1}{2n}$. Computing the mean first passage matrix M corresponding to T, we have

$$M = \begin{bmatrix} n & 1 & 2 & 3 & \cdots & n-1 \\ n-1 & n & 1 & 2 & \cdots & n-2 \\ \vdots & \ddots & \ddots & \ddots & & \vdots \\ 2 & 3 & \cdots & n-1 & n & 1 \\ 1 & 2 & 3 & \cdots & n-1 & n \end{bmatrix}.$$

The following example appears in a paper of Kirkland, Neumann, and Xu [86].

EXAMPLE 5.3.24 Fix a parameter a with $0 < a < 1$, and let $T(a)$ be the 5×5 doubly stochastic matrix given by

$$T(a) = \begin{bmatrix} 0 & 0 & 0 & a & 1-a \\ 1-a & 0 & 0 & 0 & a \\ a & 1-a & 0 & 0 & 0 \\ 0 & a & 1-a & 0 & 0 \\ 0 & 0 & a & 1-a & 0 \end{bmatrix}.$$

Let $T(a)_{(1)}$ denote the trailing 4×4 principal submatrix of $T(a)$. A computation shows that

$$(I - T(a)_{(1)})^{-1}\mathbf{1} = \frac{1}{1-a+a^2-a^3+a^4} \begin{bmatrix} 1+3a-2a^2+a^3 \\ 2+a-4a^2+2a^3 \\ 3-a-a^2+3a^3 \\ 4-3a+2a^2-a^3 \end{bmatrix}.$$

Observing that $T(a)$ is a circulant matrix, it follows that the corresponding mean first passage matrix $M(a)$ is also a circulant matrix. We now deduce that the off-diagonal entries of $M(a)$ are given by

$$\frac{1+3a-2a^2+a^3}{1-a+a^2-a^3+a^4}, \frac{2+a-4a^2+2a^3}{1-a+a^2-a^3+a^4}, \frac{3-a-a^2+3a^3}{1-a+a^2-a^3+a^4},$$

and $\dfrac{4-3a+2a^2-a^3}{1-a+a^2-a^3+a^4}$.

Next, we consider the polynomial $p(x) = 4x^3 - 3x^2 + 2x - 1$. It turns out that $p(x)$ has just one real root, namely

$$a_0 = \frac{1}{4}\left[1 - \frac{5^{\frac{2}{3}}}{(27+12\sqrt{6})^{\frac{1}{3}}} + \frac{(45+20\sqrt{6})^{\frac{1}{3}}}{3^{\frac{2}{3}}}\right].$$

Note that the decimal value of a_0 is about 0.6058.

Using the fact that $p(a_0) = 0$, we find that

$$\frac{4-3a_0+2a_0^2-a_0^3}{1-a_0+a_0^2-a_0^3+a_0^4} = \frac{3-a_0-a_0^2+3a_0^3}{1-a_0+a_0^2-a_0^3+a_0^4} = 4.$$

Further, it is readily shown, again since $p(a_0) = 0$, that $4 - 3a_0 + 2a_0^2 - a_0^3 > 1 + 3a_0 - 2a_0^2 + a_0^3$ and $4 - 3a_0 + 2a_0^2 - a_0^3 > 2 + a_0 - 4a_0^2 + 2a_0^3$. Consequently, we find that the maximum off-diagonal entry in $M(a_0)$ is 4; appealing to Remark 5.3.21, we find that $T(a_0)$ yields equality in (5.30).

In [86], the authors provide several other constructions for stochastic matrices yielding equality in (5.30). However, to date a more transparent characterisation of the equality case in (5.30), (for instance, a characterisation only in terms of the stochastic matrix in question, and not relying on mean first passage times) has yet to be established.

In view of Theorem 5.3.19, it is natural to wonder whether the condition number $\kappa(T)$ can be bounded from above by using the eigenvalues of the stochastic matrix T. In the following sequence of results we will develop an attainable upper bound of this type, and characterise the case that equality holds. Our main result along these lines (Corollary 5.3.28 below) is a sharpening of a bound due to Meyer in [97]. We begin with some preliminary results.

LEMMA 5.3.25 *Let $T \in \mathcal{IS}_n$, written in partitioned form as*

$$T = \left[\begin{array}{c|c} T_{(n)} & \mathbf{1} - T_{(n)}\mathbf{1} \\ \hline r_n^t & 1 - r_n^t\mathbf{1} \end{array}\right],$$

and denote the eigenvalues of T *by* $1, \lambda_2, \ldots, \lambda_n$. *Let* $A = T_{(n)} - \mathbf{1}r_n^t$.
a) We have $(I - A)^{-1}\mathbf{1} = \frac{1}{1+r_n^t(I-T_{(n)})^{-1}\mathbf{1}}(I - T_{(n)})^{-1}\mathbf{1}$.
b) The eigenvalues of A *are* $\lambda_2, \ldots, \lambda_n$.

Proof: a) Using the Sherman–Morrison formula (see [57, section 0.7.4]), we find that

$$(I - A)^{-1} = (I - T_{(n)})^{-1} - \frac{(I - T_{(n)})^{-1}\mathbf{1}r_n^t(I - T_{(n)})^{-1}}{1 + r_n^t(I - T_{(n)})^{-1}\mathbf{1}}.$$

The conclusion now follows.

b) Let $u^t = \left[\begin{array}{c|c} r_n^t & 1 - r_n^t\mathbf{1} \end{array}\right]$, and consider the matrix

$$T - \mathbf{1}u^t = \left[\begin{array}{c|c} A & (r_n^t\mathbf{1})\mathbf{1} - T_{(n)}\mathbf{1} \\ \hline 0^t & 0 \end{array}\right].$$

A result of Brauer ([14]) states that if S is a stochastic matrix of order m with eigenvalues $1, \mu_2, \ldots, \mu_m$, then for any vector $w \in \mathbb{R}^m$, the matrix $S - \mathbf{1}w^t$ has eigenvalues $1 - w^t\mathbf{1}, \mu_2, \ldots, \mu_n$. From this, we find that the eigenvalues of $T - \mathbf{1}u^t$ are given by $0 = 1 - u^t\mathbf{1}, \lambda_2, \ldots, \lambda_n$. We now conclude that the eigenvalues of A are $\lambda_2, \ldots, \lambda_n$. □

We now present a determinantal inequality for the matrix A of Lemma 5.3.25.

LEMMA 5.3.26 *Fix* $k \in \mathbb{N}$, *let* S *be a substochastic matrix of order* k, *and let* $v \in \mathbb{R}^k$ *be a nonnegative vector such that* $0 < v^t\mathbf{1} \equiv \sigma \leq 1$. *Then*

$$\det(I - S + \mathbf{1}v^t) \leq 1 + k\sigma. \tag{5.33}$$

Equality holds in (5.33) if and only if there is a permutation matrix P *such that*

$$PSP^t = \begin{bmatrix} 0 & 0 & 0 & \ldots & 0 \\ 1 & 0 & 0 & \ldots & 0 \\ 0 & 1 & 0 & \ldots & 0 \\ \vdots & \ddots & \ddots & \ddots & \vdots \\ 0 & \ldots & 0 & 1 & 0 \end{bmatrix} \quad \text{and } Pv = \sigma e_k. \tag{5.34}$$

Proof: We begin with the proof of (5.33) and proceed by induction on k. Observe that for the case $k = 1$, we have $\det(I - S + \mathbf{1}v^t) = 1 - S + v_1 = 1 - S + \sigma \leq 1 + \sigma$, as desired. Suppose now that (5.33) holds for order $k - 1 \geq 1$, S is $k \times k$ and substochastic, and that $v \in \mathbb{R}^k$ is nonnegative with entries summing to σ.

Observe that the expression $\det(I - S + \mathbf{1}v^t)$ is linear in each entry of v. It then follows that for some index j between 1 and k, we have $\det(I - S + \mathbf{1}v^t) \le \det(I - S + \sigma\mathbf{1}e_j^t)$. Without loss of generality, we take $j = k$, and note that in order to complete the proof, it suffices to show that $\det(I - S + \sigma\mathbf{1}e_k^t) \le 1 + k\sigma$.

We partition out the last row and column of S, writing S as $S = \left[\begin{array}{c|c} S_{11} & S_{12} \\ \hline S_{21} & S_{22} \end{array}\right]$, where the "block" $S_{2,2}$ is 1×1. Then $I - S + \sigma\mathbf{1}e_k^t = \left[\begin{array}{c|c} I - S_{11} & -S_{12} + \sigma\mathbf{1} \\ \hline -S_{21} & 1 - S_{22} + \sigma \end{array}\right]$. Using the formula for the determinant of a partitioned matrix (see [57, section 0.8.5]), we find that

$$\det(I - S + \sigma\mathbf{1}e_k^t) =$$
$$(1 - S_{22} + \sigma)\det\left(I - S_{11} - \frac{1}{1 - S_{22} + \sigma}(-S_{12} + \sigma\mathbf{1})(-S_{21})\right) =$$
$$(1 - S_{22} + \sigma)\det\left(I - S_{11} - \frac{1}{1 - S_{22} + \sigma}S_{12}S_{21} + \frac{\sigma}{1 - S_{22} + \sigma}\mathbf{1}S_{21}\right).$$

It is straightforward to see that $S_{11} + \frac{1}{1-S_{22}+\sigma}S_{12}S_{21}$ is a substochastic matrix and that $\frac{\sigma}{1-S_{22}+\sigma}S_{21}$ is a nonnegative row vector in $\mathbb{R}^{k-1}$ whose entries sum to $\frac{\sigma S_{21}\mathbf{1}}{1-S_{22}+\sigma} \le 1$. Applying the induction hypothesis, we thus find that

$$\det\left(I - S_{11} - \frac{1}{1 - S_{22} + \sigma}S_{12}S_{21} + \frac{\sigma}{1 - S_{22} + \sigma}\mathbf{1}S_{21}\right) \le 1 + \frac{(k-1)\sigma S_{21}\mathbf{1}}{1 - S_{22} + \sigma}. \quad (5.35)$$

Consequently, we have

$$\det(I - S + \sigma\mathbf{1}v^t) \le \det(I - S + \sigma\mathbf{1}e_k^t) \le$$
$$(1 - S_{22} + \sigma)\left(1 + \frac{(k-1)\sigma S_{21}\mathbf{1}}{1 - S_{22} + \sigma}\right) \le$$
$$1 - S_{22} + \sigma + (k-1)\sigma(1 - S_{22}) = 1 + k\sigma - S_{22}(1 + (k-1)\sigma).$$

The desired inequality now follows, completing the proof of the induction step, and thus establishing (5.33).

Next, we turn to the case of equality in (5.33). We proceed by induction on $k \ge 2$ to show that if equality holds in (5.33), then there is a permutation matrix P such that PSP^t and Pv are as in (5.34). The case $k = 2$ is an uninteresting computation, which we omit. Suppose now that the statement holds for some $k-1 \geq 2$ and that S and v are of orders $k \times k$ and $k \times 1$, respectively, and that equality holds in (5.33). Inspecting the proof above, and adopting its notation, we find that $1 + k\sigma = \det(I - S + \mathbf{1}v^t) \leq (1 - S_{22} + \sigma)(1 + \frac{(k-1)\sigma S_{21}\mathbf{1}}{1-S_{22}+\sigma})$, so that necessarily we must have $S_{22} = 0$ and $S_{21}\mathbf{1} = 1$. Again, inspecting the proof above, we must also have

$$\det\left(I - S_{11} - \frac{1}{1+\sigma}S_{12}S_{21} + \frac{\sigma}{1+\sigma}\mathbf{1}S_{21}\right) = 1 + (k-1)\frac{\sigma}{1+\sigma}.$$

Applying the induction hypothesis to the substochastic matrix $S_{11} + \frac{1}{1+\sigma}S_{12}S_{21}$ and row vector $\frac{1}{1+\sigma}S_{21}$, we find that there is a permutation matrix Q such that $Q(S_{11} + \frac{1}{1+\sigma}S_{12}S_{21})Q^t$ has ones on the first subdiagonal and zeros elsewhere, and $S_{21}Q^t = e^t_{k-1}$. From these two conditions together, it must be the case that S_{12} is the zero vector (otherwise $Q(S_{11} + \frac{1}{1+\sigma}S_{12}S_{21})Q^t$ has a positive entry in the last column, since $S_{21}Q^t = e^t_{k-1}$). Letting P be the permutation matrix whose leading principal submatrix of order $k-1$ is Q, and whose k-th diagonal entry is 1, it follows that PSP^t has the form given in (5.34). The fact that $Pv = \sigma e_k$ now follows easily. This completes the proof of the induction step, as desired.

Finally if there is a permutation matrix P so that (5.34) is satisfied, it is straightforward to determine that equality holds in (5.33). □

Lemma 5.3.25 and 5.3.26 assist in establishing the following result.

THEOREM 5.3.27 *Suppose that $T \in \mathcal{IS}_n$, with eigenvalues $1, \lambda_2, \ldots, \lambda_n$. Write T as*

$$T = \left[\begin{array}{c|c} T_{(n)} & \mathbf{1} - T_{(n)}\mathbf{1} \\ \hline r_n^t & 1 - r_n^t\mathbf{1} \end{array}\right].$$

Then

$$\frac{1}{1 + r_n^t(I - T_{(n)})^{-1}\mathbf{1}}||(I - T_{(n)})^{-1}||_\infty \leq \frac{n-1}{\Pi_{j=2}^n(1-\lambda_j)}. \tag{5.36}$$

Equality holds in (5.36) if and only if the first $n-1$ rows and columns *of T can be simultaneously permuted so that the resulting matrix has the form*

$$\left[\begin{array}{cccccc|c} 0 & 0 & \ldots & 0 & t & & 1-t \\ 1 & 0 & 0 & \ldots & 0 & & 0 \\ 0 & 1 & 0 & \ldots & 0 & & 0 \\ \vdots & & \ddots & \ddots & \vdots & & \vdots \\ 0 & 0 & \ldots & 1 & 0 & & 0 \\ \hline & & u^t & & & & 1-u^t\mathbf{1} \end{array}\right] \tag{5.37}$$

for some scalar $0 \le t < 1$ and some nonnegative, nonzero vector $u \in \mathbb{R}^{n-1}$ such that $u^t\mathbf{1} \le 1$.

Proof: Let $A = T_{(n)} - \mathbf{1}r_n^t$. By Lemma 5.3.25 a),

$$\frac{1}{1 + r_n^t(I - T_{(n)})^{-1}\mathbf{1}}||(I - T_{(n)})^{-1}||_\infty = ||(I - A)^{-1}||_\infty.$$

Thus we need to compute the row sums of $(I - A)^{-1}$. In order to do so, we apply Cramer's rule, noting that for each $i = 1, \ldots, n-1$,

$$e_i^t(I - A)^{-1} = \frac{\det(c_i(I - A))}{\det(I - A)},$$

where $c_i(I - A)$ is the matrix formed from $I - A$ by replacing its i–th column by $\mathbf{1}$. From Lemma 5.3.25 b) it follows that $\det(I - A) = \Pi_{j=2}^n(1 - \lambda_j)$, so in order to establish (5.36), it remains only to prove that $\det(c_i(I - A)) \le n - 1$ for each $i = 1, \ldots, n-1$. For concreteness, we take $i = n-1$; partition out the last row and column of $T_{(n)}$ and the last entry of r_n^t as follows:

$$T_{(n)} = \left[\begin{array}{c|c} S_{11} & S_{12} \\ \hline S_{21} & S_{22} \end{array}\right] \text{ and } r_n^t = \left[\begin{array}{c|c} x_1^t & x_2 \end{array}\right].$$

We then have

$$c_{n-1}(I - A) = \left[\begin{array}{c|c} I - S_{11} + \mathbf{1}x_1^t & \mathbf{1} \\ \hline -S_{21} + x_1^t & 1 \end{array}\right].$$

Applying the formula for the determinant of a partitioned matrix, we thus find that $\det(c_{n-1}(I - A)) = \det(I - S_{11} + \mathbf{1}S_{21})$. By Lemma 5.3.26 we then have

$$\det(c_{n-1}(I - A)) \le 1 + (n-2)S_{21}\mathbf{1} \le n - 1,$$

as desired.

Suppose now that equality holds in (5.36). Then necessarily we find that for some $i = 1, \ldots, n-1$, $\det(c_i(I - A)) = n - 1$. Without loss of generality, we take $i = n - 1$. Referring to the above partitioning for $T_{(n)}$, it must be the case that the entries in S_{21} sum to 1. Further, since $\det(I - S_{11} + \mathbf{1}S_{21}) = n - 1$, from the characterisation of equality in Lemma 5.3.26, there is a permutation matrix Q such that $QS_{11}Q^t$ has ones on the first subdiagonal and zeros elsewhere, and such that $S_{21}Q^t = e_{n-2}^t$. It now follows that the rows and columns of T can be simultaneously permuted so that the resulting matrix has the form in (5.37).

Conversely, it is straightforward to show that if T can be permuted to the form (5.37), then equality holds in (5.36). □

The following is immediate from (5.18) and Theorem 5.3.27.

COROLLARY 5.3.28 *Let $T \in \mathcal{IS}_n$, with eigenvalues $1, \lambda_2, \ldots, \lambda_n$. Then $\kappa(T) \le \frac{n-1}{2\Pi_{j=2}^n(1-\lambda_j)}$, with equality holding if and only if T is permutationally similar to a matrix of the form given in* (5.37).

REMARK 5.3.29 Suppose that T is an irreducible stochastic matrix of the form given in (5.37), for some scalar t and vector $u \in \mathbb{R}^{n-1}$. Denote the eigenvalues of T by $1, \lambda_2, \ldots, \lambda_n$. We can compute $\Pi_{j=2}^n(1-\lambda_j)$ by observing that $(-1)^{n-1}\Pi_{j=2}^n(1-\lambda_j) = (-1)^{n-1}\sum_{j=1}^n \det(I - T_{(j)})$, where as usual, $T_{(j)}$ is formed from T by deleting its j–th row and column. (This follows from the fact that both expressions are the coefficient of the linear term in the characteristic polynomial of $I - T$.) Thus it suffices to find $\sum_{j=1}^n \det(I - T_{(j)})$.

A few straightforward computations reveal that

$$\det(I - T_{(j)}) = \begin{cases} \sum_{k=1}^{n-1} u_k & \text{if } j = 1, \\ \sum_{k=1}^{n-1} u_k - (1-t)\sum_{m=1}^{j-1} u_m & \text{if } j = 2, \ldots, n-1, \\ 1 - t & \text{if } j = n. \end{cases}$$

Consequently we have

$$\Pi_{j=2}^n(1-\lambda_j) = 1 - t + (n-1)\sum_{k=1}^{n-1} u_k - (1-t)\sum_{m=1}^{n-2}(n-m-1)u_m. \quad (5.38)$$

Thus, if T is given by (5.37), we have

$$\kappa(T) = \frac{n-1}{2\left(1-t+(n-1)\sum_{k=1}^{n-1} u_k - (1-t)\sum_{m=1}^{n-2}(n-m-1)u_m\right)}. \tag{5.39}$$

Observe that if $t = 0$ and $u_{n-1} = 1$, then the right side of (5.39) is $\frac{n-1}{2n}$ (the minimum possible value for $\kappa(T)$, by Theorem 5.3.20), while as $t \to 1^-$ and $u \to 0$, the right side of (5.39) diverges to infinity. It now follows that for any $a \geq \frac{n-1}{2n}$, there is a matrix T of the form (5.37) such that $\kappa(T) = a$.

Most of this section has been dedicated to analysing the behaviour of the stationary distribution when the underlying irreducible stochastic matrix is perturbed. We close the section by presenting a couple of results on a related theme, namely analysing the behaviour of the associated group inverse when the underlying irreducible stochastic matrix is perturbed. The following key result is proven in [95].

THEOREM 5.3.30 *Let T and $\tilde{T}$ be in $\mathcal{AS}_n$. Denote the stationary distribution vectors for T by w, set $\tilde{T} = T+E, Q = I-T$ and $\tilde{Q} = I-\tilde{T}$. Then*

$$\tilde{Q}^\# = Q^\#(I-EQ^\#)^{-1} - \mathbf{1}w^t(I-EQ^\#)^{-1}Q^\#(I-EQ^\#)^{-1}.$$

Proof: From Lemma 5.3.1, we find $(I-EQ^\#)^{-1}$ is invertible and that the transpose of the stationary distribution vector for $\tilde{T}$ is $w^t(I-EQ^\#)^{-1}$. Let $G = Q^\#(I-EQ^\#)^{-1} - \mathbf{1}w^t(I-EQ^\#)^{-1}Q^\#(I-EQ^\#)^{-1}$; in order to establish the result, it suffices to show that $G\mathbf{1} = 0, w^t(I-EQ^\#)^{-1}G = 0^t$, and $\tilde{Q}G = G\tilde{Q} = I - \mathbf{1}w^t(I-EQ^\#)^{-1}$.

The first two of these conditions follow readily from the definition of G and the fact that since $(I-EQ^\#)\mathbf{1} = \mathbf{1}$, we have $w^t(I-EQ^\#)^{-1}\mathbf{1} = 1$. Note that $\tilde{Q}G = QQ^\#(I-EQ^\#)^{-1} - EQ^\#(I-EQ^\#)^{-1} = (I-\mathbf{1}w^t)(I-EQ^\#)^{-1} - EQ^\#(I-EQ^\#)^{-1} = I - \mathbf{1}w^t(I-EQ^\#)^{-1}$. A similar calculation shows that $G\tilde{Q} = I - \mathbf{1}w^t(I-EQ^\#)^{-1}$. Consequently, $\tilde{Q} = G$, as desired. □

Next, we consider the special case when the perturbing matrix E in Theorem 5.3.30 has only one nonzero row.

COROLLARY 5.3.31 *Suppose that $T, \tilde{T} \in \mathcal{AS}_n$, and that $\tilde{T}$ differs from T only in the i-th row; that is, $\tilde{T} = T + e_i u^t$ for some vector u*

orthogonal to **1**. *Denote the stationary distribution vector of* T *by* w. *Setting* $\tilde{Q} = I - \tilde{T}$, *we have*

$$\tilde{Q}^{\#} = Q^{\#} + \frac{1}{1 - u^t Q^{\#} e_i} Q^{\#} e_i u^t Q^{\#} - \frac{w_i}{1 - u^t Q^{\#} e_i} \mathbf{1} \left(u^t (Q^{\#})^2 + \frac{u^t (Q^{\#})^2 e_i}{1 - u^t Q^{\#} e_i} u^t Q^{\#} \right).$$

Proof: Set $E = e_i u^t$, and recall from Lemma 5.3.1 that $I - EQ^{\#}$ is invertible. From the Sherman–Morrison formula, we find that $(I - EQ^{\#})^{-1} = I + \frac{1}{1-u^t Q^{\#} e_i} e_i u^t Q^{\#}$. Thus we have $Q^{\#}(I - EQ^{\#})^{-1} = Q^{\#} + \frac{1}{1-u^t Q^{\#} e_i} Q^{\#} e_i u^t Q^{\#}$. It now follows that

$$(I - EQ^{\#})^{-1} Q^{\#} (I - EQ^{\#})^{-1} = Q^{\#} + \frac{1}{1 - u^t Q^{\#} e_i} Q^{\#} e_i u^t Q^{\#} + \frac{1}{1 - u^t Q^{\#} e_i} e_i u^t (Q^{\#})^2 + \frac{u^t (Q^{\#})^2 e_i}{(1 - u^t Q^{\#} e_i)^2} e_i u^t Q^{\#}.$$

Thus we see that

$$w^t (I - EQ^{\#})^{-1} Q^{\#} (I - EQ^{\#})^{-1} = \frac{w_i}{1 - u^t Q^{\#} e_i} \left(u^t (Q^{\#})^2 + \frac{u^t (Q^{\#})^2 e_i}{1 - u^t Q^{\#} e_i} u^t Q^{\#} \right),$$

and (5.40) now follows by applying Theorem 5.3.30. □

REMARK 5.3.32 Here we maintain the notation of Corollary 5.3.31, and consider the special case that our perturbing matrix has the form $ae_i(e_i^t - e_i^t T)$ for some $0 < a < 1$. Observe that such a perturbation has the effect of multiplying the off-diagonal entries in the i-th row of T by $1 - a$, and adding $a(1 - t_{i,i})$ to the diagonal entry in that row. In the notation of Corollary 5.3.31 we have $u^t = ae_i^t Q$, so that $u^t Q^{\#} = a(e_i - w)^t$. Setting $\tilde{T} = T + ae_i e_i^t Q$, $Q = I - T$ and $\tilde{Q} = I - \tilde{T}$, it now follows from (5.40) that

$$\tilde{Q}^{\#} = Q^{\#} + \frac{a}{1 - a(1 - w_i)} Q^{\#} e_i (e_i - w)^t - \frac{aw_i}{1 - a(1 - w_i)} \mathbf{1} \left(e_i^t Q^{\#} + \frac{a q^{\#}_{i,i}}{1 - a(1 - w_i)} (e_i^t - w^t) \right).$$

5.4 Bounds on the subdominant eigenvalue

Suppose that we have a primitive $n \times n$ stochastic matrix T. As mentioned in section 5.1, the sequence $T^m, m \in \mathbb{N}$ converges as $m \to \infty$, and its limit is $\mathbf{1}w^t$, where w is the stationary distribution for T. This in turn implies that for any initial vector $v_1 \geq 0$ with $v_1^t\mathbf{1} = 1$, the sequence of iterates of the corresponding Markov chain $v_1^tT^m$, necessarily converges to w^t. Evidently the convergence of both the sequence of powers of T, and of the sequence of iterates of the associated Markov chain, is governed by the non–Perron eigenvalues of T. In particular, we find that if the eigenvalues of T are denoted $1, \lambda_2, \ldots, \lambda_n$, then asymptotically, the convergence of the sequence T^m to its limit $\mathbf{1}w^t$ is geometric with ratio $\gamma(T) \equiv \max_{j=2,\ldots,n} |\lambda_j|$.

This last observation has lead to a great deal of interest in localising the non–Perron eigenvalues for a stochastic matrix T in general, and in estimating $\gamma(T)$ in particular. We begin this section with a brief discussion of results in that area. Our intention here is not to give a comprehensive survey of the area; rather we aim to give the reader the flavour of some of the different types of approaches that have been taken to address the issue of localising the spectrum of a stochastic matrix. We extend the definition of γ slightly as follows: given a matrix $T \in \mathcal{IS}_n$ with eigenvalues $1, \lambda_2, \ldots, \lambda_n$, we set $\gamma(T) = \max_{j=2,\ldots,n} |\lambda_j|$. An eigenvalue λ_j of T is said to be *subdominant* if $|\lambda_j| = \gamma(T)$. We begin with a result, due to Hopf, in [56].

PROPOSITION 5.4.1 *Let T be a positive stochastic matrix, and denote its minimum and maximum entries by m and M, respectively. Then $\gamma(T) \leq \frac{M-m}{M+m}$.*

Observe that the conclusion of Proposition 5.4.1 still holds if we relax the hypothesis to allow T to be nonnegative (and stochastic); however, in that case the conclusion is that $\gamma(T) \leq 1$, which yields no new information, since the spectral radius of T is already known to be 1.

The next result is found in Pokarowski's paper [104]. It is motivated in part by considering so-called nearly decoupled Markov chains, i.e., those whose transition matrix is close to being block diagonal.

PROPOSITION 5.4.2 *Let T be an $n \times n$ primitive stochastic matrix.*

Let $\lambda_2, \ldots, \lambda_n$ be the non–Perron eigenvalues of T. Let λ be a non–Perron eigenvalue of T that is closest to 1. i.e., such that $|1-\lambda| = \min_{j=2,\ldots,n} |1-\lambda_j|$. Set

$$\sigma = \min_{C_1 \cap C_2 = \emptyset} \left(\sum_{i \in C_1, j \notin C_1} t_{i,j} + \sum_{i \in C_2, j \notin C_2} t_{i,j} \right).$$

Then

$$\frac{|1-\lambda|}{n-1} \le \sigma \le n^2 \left(\frac{|1-\lambda|}{2} \right)^{\frac{1}{n-1}}.$$

For a directed graph D containing at least one cycle, the *girth* of D is defined as being the length of the shortest cycle in D. Given a stochastic matrix T, the following result in Kirkland's paper [71] provides a lower bound on $\gamma(T)$ in terms of the girth of $\mathcal{D}(T)$.

PROPOSITION 5.4.3 *Fix $g, n \in \mathbb{N}$ with $1 \le g \le n$ and $n \ge 2$. Set*

$$\underline{\gamma}(g,n) = \inf\{\gamma(T) | T \in \mathcal{IS}_n, \text{ and } \mathcal{D}(T) \text{ has girth } g\}.$$

For any $g \ge 2$, we have $\underline{\gamma}(g,n) \ge \left(\frac{1}{n-1}\right)^{\frac{1}{g-1}}$. Further we have the following values for small g:

$$\underline{\gamma}(1,n) = 0, \underline{\gamma}(2,n) = \frac{1}{n-1}, \underline{\gamma}(3,n) = \begin{cases} \frac{\sqrt{n+1}}{n-1} & \text{if } n \text{ is odd,} \\ \sqrt{\frac{n+2}{n^2-2n}} & \text{if } n \text{ is even.} \end{cases}$$

The following result of Bauer, Deutsch, and Stoer in [9] will inform some of our work in this section.

PROPOSITION 5.4.4 *Let A be a real matrix of order n such that for some $a \in \mathbb{R}, A\mathbf{1} = a\mathbf{1}$. Let $\lambda \ne a$ be an eigenvalue of A. Then*

$$|\lambda| \le \tau(A) = \frac{1}{2} \max_{i,j=1,\ldots,n} \sum_{k=1}^{n} |a_{i,k} - a_{k,j}|. \tag{5.40}$$

We note that in the context of Proposition 5.4.4, the expression $\tau(A)$ can be rewritten either as $a - \min_{i,j} \sum_{k=1}^{n} \min\{a_{i,k}, a_{j,k}\}$ or as $\max_{i,j} \sum_{k=1}^{n} \max\{a_{i,k}, a_{j,k}\} - a$. These expressions are established by noting that for any $p, q \in \mathbb{R}, \min\{p,q\} = \frac{1}{2}(p+q-|p-q|)$, while $\max\{p,q\} = \frac{1}{2}(p+q+|p-q|)$. In the context of a stochastic matrix T, the function $\tau(T)$ is known as a coefficient of ergodicity, and [107, section 4.3] discusses the application of such functions to nonhomogeneous Markov chains.

In Kirkland and Neumann's paper [84], it is shown that if equality holds in (5.40) for an $n \times n$ stochastic matrix T with eigenvalue $\lambda \neq 1$, then either:
i) there is a $k \in \mathbb{N}$ with $k \leq n$, and a k–th root of 1, ω, say, such that $\lambda = \tau(T)\omega$, or
ii) there is an odd $k \in \mathbb{N}$ with $k \leq n$, and a k–th root of -1, $\tilde{\omega}$, say, such that $\lambda = \tau(T)\tilde{\omega}$.

Observe that for any stochastic matrix T, we have $\tau(T) \leq 1$, with equality holding if and only if there is a pair of rows of T, say rows i_0 and j_0, such that for each index k, one (or both) of $a_{i_0,k}$ and $a_{j_0,k}$ is zero. A stochastic matrix T is called *scrambling* if $\tau(T) < 1$, or equivalently, if $\mathcal{D}(T)$ has the property that for any pair of distinct vertices i, j of $\mathcal{D}(T)$, there is a vertex k such that $i \to k$ and $j \to k$ in $\mathcal{D}(T)$. The *scrambling index* for a primitive stochastic matrix T is defined as

$$\mathsf{si}(T) = \min\{m \in \mathbb{N} | T^m \text{ is a scrambling matrix}\}.$$

The following result, in Akelbek's thesis [2], uses the combinatorial information contained in the scrambling index $\mathsf{si}(T)$ to bound $\gamma(T)$.

PROPOSITION 5.4.5 *Let T be a primitive stochastic matrix, and let μ denote the smallest positive entry in T. Then*

$$\gamma(T) \leq \left(1 - \mu^{\mathsf{si}(T)}\right)^{\frac{1}{\mathsf{si}(T)}}.$$

One of the main results of Akelbek and Kirkland's paper [3] is that for a primitive stochastic matrix T of order n, the corresponding scrambling index is at most $\left\lceil \frac{(n-1)^2+1}{2} \right\rceil$. From this it now follows from Proposition 5.4.5 that

$$\gamma(T) \leq \left(1 - \mu^{\left\lceil \frac{(n-1)^2+1}{2} \right\rceil}\right)^{\frac{1}{\left\lceil \frac{(n-1)^2+1}{2} \right\rceil}},$$

where again μ is the smallest positive entry in T.

The results in Propositions 5.4.1–5.4.5 all proceed by working directly with the transition matrix T. Our objective in this section is to work with $(I - T)^{\#}$ in order to localise the non–Perron eigenvalues of T. We begin with a straightforward consequence of Proposition 5.4.4.

THEOREM 5.4.6 *Suppose that $T \in \mathcal{IS}_n$, and set $Q = I - T$. For each non–Perron eigenvalue λ of T, we have*

$$\frac{1}{|1-\lambda|} \leq \tau(Q^{\#}). \tag{5.41}$$

Proof: From our remarks in section 2.2, we find that $Q^{\#}\mathbf{1} = 0$ (so that $Q^{\#}$ has constant row sums) and that each eigenvalue $\lambda \neq 1$ of T, yields a nonzero eigenvalue $\frac{1}{1-\lambda}$ of $Q^{\#}$. The conclusion now follows from Proposition 5.4.4. □

Next we show that, under certain circumstances, if equality holds in (5.40) for some eigenvalue, then it also holds in (5.41) for some eigenvalue. The following result appears in [84].

THEOREM 5.4.7 *Suppose that T is primitive stochastic matrix, and that $\gamma(T)$ is an eigenvalue of T. If equality holds in (5.40) for the eigenvalue $\gamma(T)$, then equality must also hold in (5.41) for the eigenvalue $\gamma(T)$.*

Proof: Since $\gamma(T)$ is an eigenvalue of T by hypothesis, we have $\frac{1}{1-\gamma(T)} \leq \tau(Q^{\#})$ by (5.41). Thus it suffices to show that $\tau(Q^{\#}) \leq \frac{1}{1-\gamma(T)}$. Let w^t denote the stationary distribution for T. Consider the matrix $T - \mathbf{1}w^t$, and note that its eigenvalues consist of 0, along with the non–Perron eigenvalues of T. Since T is primitive, we thus find that the spectral radius of $T - \mathbf{1}w^t$ is less than 1.

We claim that $\tau(Q^{\#}) \leq \frac{1}{1-\tau(T)}$. To see the claim, first we note that $Q^{\#}$ can be written as $Q^{\#} = I - \mathbf{1}w^t + \sum_{k=1}^{\infty}(T - \mathbf{1}w^t)^k$ (the series converges because the spectral radius of $T - \mathbf{1}w^t$ is less than 1). Thus

$$\tau(Q^{\#}) \leq \tau(I - \mathbf{1}w^t) + \sum_{k=1}^{\infty} \tau((T - \mathbf{1}w^t)^k).$$

From Remark 5.3.7, it follows that $\tau((T-\mathbf{1}w^t)^k) \leq (\tau(T-\mathbf{1}w^t))^k$. Thus we find that

$$\tau(Q^{\#}) \leq 1 + \sum_{k=1}^{\infty}(\tau(T - \mathbf{1}w^t))^k = 1 + \sum_{k=1}^{\infty}(\tau(T))^k = \frac{1}{1-\tau(T)},$$

as claimed.

By hypothesis, $\gamma(T) = \tau(T)$. Hence we have $\tau(Q^{\#}) \leq \frac{1}{1-\gamma(T)}$, so that equality holds in (5.40) for the eigenvalue $\gamma(T)$ of T. □

The following illustrative example appears in [84].

EXAMPLE 5.4.8 Consider the matrix

$$T = \begin{bmatrix} \frac{7}{8} & \frac{1}{16} & 0 & \frac{1}{16} & 0 \\ \frac{1}{2} & 0 & \frac{1}{2} & 0 & 0 \\ \frac{1}{2} & \frac{1}{2} & 0 & 0 & 0 \\ \frac{1}{2} & 0 & 0 & 0 & \frac{1}{2} \\ \frac{1}{2} & 0 & 0 & \frac{1}{2} & 0 \end{bmatrix}.$$

Evidently $\tau(T) = \frac{1}{2}$, while the eigenvalues of T are $1, \frac{3}{8}, \frac{1}{2}$, and $-\frac{1}{2}$, the last with multiplicity two. Thus we have $\gamma(T) = \frac{1}{2}$, which is an eigenvalue of T. From Theorem 5.4.7, it must be the case that $\tau((I-T)^{\#}) = \frac{1}{1-\gamma(T)} = 2$. Indeed, a computation yields

$$(I-T)^{\#} = \begin{bmatrix} \frac{8}{25} & -\frac{19}{225} & -\frac{17}{225} & -\frac{19}{225} & -\frac{17}{225} \\ -\frac{32}{25} & \frac{251}{225} & \frac{118}{225} & -\frac{49}{225} & -\frac{32}{225} \\ -\frac{32}{25} & \frac{101}{225} & \frac{268}{225} & -\frac{49}{225} & -\frac{32}{225} \\ -\frac{32}{25} & -\frac{49}{225} & -\frac{32}{225} & \frac{251}{225} & \frac{118}{225} \\ -\frac{32}{25} & -\frac{49}{225} & -\frac{32}{225} & \frac{101}{225} & \frac{268}{225} \end{bmatrix},$$

from which it can be verified directly that $\tau((I-T)^{\#}) = 2$.

Our next result is a consequence of Corollary 5.3.28.

THEOREM 5.4.9 *Suppose that $T \in \mathcal{IS}_n$. Then*

$$\gamma(T) \geq 1 - \left(\frac{n-1}{2\kappa(T)}\right)^{\frac{1}{n-1}}. \tag{5.42}$$

Proof: Denote the non–Perron eigenvalues of T by $\lambda_2, \ldots, \lambda_n$. From Corollary 5.3.28 we have $\kappa(T) \leq \frac{n-1}{2\Pi_{j=2}^{n}(1-\lambda_j)}$, and since $\frac{1}{\Pi_{j=2}^{n}(1-\lambda_j)} \leq \left(\frac{1}{1-\gamma(T)}\right)^{n-1}$, (5.42) follows readily. □

Observe that the lower bound on $\gamma(T)$ in (5.42) is nonnegative only in the case that $\kappa(T) \geq \frac{n-1}{2}$.

Next, we apply Theorem 5.3.19 to generate an upper bound on positive non–Perron eigenvalues for a stochastic matrix.

THEOREM 5.4.10 *Suppose that $T \in \mathcal{IS}_n$, and that T has a positive non–Perron eigenvalue λ. Then*

$$\lambda \le 1 - \frac{1}{2(n-1)\kappa(T)}. \tag{5.43}$$

Proof: Denote the non–Perron eigenvalues of T by $\lambda_j, j = 2, \ldots, n$. From Theorem 5.3.19, we have

$$\kappa(T) \ge \frac{1}{2(n-1)} \sum_{j=2}^{n} \frac{1}{1-\lambda_j}.$$

Evidently $\frac{1}{1-\lambda_j} > 0$ for any real eigenvalue λ_j, while for a complex conjugate pair of eigenvalues $\lambda_j, \overline{\lambda_j}$, we have

$$\frac{1}{1-\lambda_j} + \frac{1}{1-\overline{\lambda_j}} = \frac{2(1 - Re(\lambda_j))}{|1-\lambda_j|^2} > 0.$$

It now follows that $\sum_{j=2}^{n} \frac{1}{1-\lambda_j} \ge \frac{1}{1-\lambda}$, so that $\kappa(T) \ge \frac{1}{2(n-1)(1-\lambda)}$. This last inequality can be rearranged to yield (5.43). □

Our next sequence of results focuses on bounding the non–Perron eigenvalues of symmetric nonnegative matrices. Here we follow the development in Kirkland, Neumann and Shader's paper [79].

THEOREM 5.4.11 *Suppose that $A \in \Phi^{n,n}$ is symmetric with Perron value 1. Denote the remaining eigenvalues of A by $\lambda_2 \ge \lambda_3 \ge \ldots \ge \lambda_n$. Let x denote the Perron vector for A, normalised so that $x^t x = 1$, and let $Q = I - A$. Then*

$$\lambda_2 \ge 1 - \frac{1 - \min_{i=1,\ldots,n} x_i^2}{\max_{i=1,\ldots,n} q^{\#}_{i,i}} \tag{5.44}$$

and

$$\lambda_n \le 1 - \frac{1 - \max_{i=1,\ldots,n} x_i^2}{\min_{i=1,\ldots,n} q^{\#}_{i,i}} \tag{5.45}$$

In particular, if A is stochastic, we have

$$\lambda_2 \ge 1 - \frac{n-1}{n \max_{i=1,\ldots,n} q^{\#}_{i,i}}$$

and

$$\lambda_n \le 1 - \frac{n-1}{n \min_{i=1,\ldots,n} q^{\#}_{i,i}}.$$

Proof: For each $j = 2, \ldots, n$, let $v^{(j)}$ denote an eigenvector of A corresponding to the eigenvalue λ_j. Without loss of generality we will assume that the vectors $x, v^{(2)}, \ldots, v^{(n)}$ form an orthonormal basis of $\mathbb{R}^n$. We have $Q = \sum_{j=2}^n (1-\lambda_j) v^{(j)} (v^{(j)})^t$, from which we find that $Q^\# = \sum_{j=2}^n \frac{1}{1-\lambda_j} v^{(j)} (v^{(j)})^t$.

Fix an index i between 1 and n. From our expression for $Q^\#$, we have

$$q_{i,i}^\# = \sum_{j=2}^n \frac{(v_i^{(j)})^2}{1-\lambda_j} \le \sum_{j=2}^n \frac{(v_i^{(j)})^2}{1-\lambda_2} = \frac{1-x_i^2}{1-\lambda_2}, \tag{5.46}$$

the last relation following from the fact that $xx^t + \sum_{j=2}^n v^{(j)}(v^{(j)})^t = I$. Rearranging (5.46), we find that for each $i = 1, \ldots, n$,

$$\lambda_2 \ge 1 - \frac{1-x_i^2}{q_{i,i}^\#}.$$

Select an index j so that $q_{j,j}^\# = \max_{i=1,\ldots,n} q_{i,i}^\#$; noting that $1 - \frac{1-x_j^2}{q_{j,j}^\#} \ge 1 - \frac{1-\min_{i=1,\ldots,n} x_i^2}{q_{j,j}^\#}$, we find that $\lambda_2 \ge 1 - \frac{1-\min_{i=1,\ldots,n} x_i^2}{\max_{i=1,\ldots,n} q_{i,i}^\#}$.

The proof of (5.45) proceeds along analogous lines, and is omitted. □

REMARK 5.4.12 Consider a symmetric matrix $A \in \Phi^{n,n}$. Adopting the notation of Theorem 5.4.11, we can use some of the proof technique of that result in order to establish some further inequalities on λ_2 and λ_n. Recall that within the proof of Theorem 5.4.11, it is established that for each $i = 1, \ldots, n$,

$$\lambda_2 \ge 1 - \frac{1-x_i^2}{q_{i,i}^\#}.$$

Select an index k so that $x_k = \max_{i=1,\ldots,n} x_i$. We find that

$$\frac{1-x_k^2}{q_{k,k}^\#} \le \frac{1-x_k^2}{\min_{i=1,\ldots,n} q_{i,i}^\#},$$

and so we deduce that

$$\lambda_2 \ge 1 - \frac{1-x_k^2}{q_{k,k}^\#} \ge 1 - \frac{1-x_k^2}{\min_{i=1,\ldots,n} q_{i,i}^\#} = 1 - \frac{1-\max_{i=1,\ldots,n} x_i^2}{\min_{i=1,\ldots,n} q_{i,i}^\#}.$$

Consequently,

$$\lambda_2 \geq 1 - \frac{1 - \max_{i=1,\ldots,n} x_i^2}{\min_{i=1,\ldots,n} q_{i,i}^{\#}}.$$

An analogous line of argumentation shows that

$$\lambda_n \leq 1 - \frac{1 - \min_{i=1,\ldots,n} x_i^2}{\max_{i=1,\ldots,n} q_{i,i}^{\#}}.$$

Our next result provides a characterisation of the case of equality in (5.44).

THEOREM 5.4.13 ([79]) *Suppose that $A \in \Phi^{n,n}$ is symmetric. Suppose further that the Perron value for A is 1, that the other eigenvalues of A are $\lambda_2 \geq \ldots \geq \lambda_n$. Let x denote the Perron vector for A, normalised so that $x^t x = 1$, and let $Q = I - A$. We have*

$$\lambda_2 = 1 - \frac{1 - \min_{i=1,\ldots,n} x_i^2}{\max_{i=1,\ldots,n} q_{i,i}^{\#}}$$

if and only if there is a permutation matrix P such that PAP^t can be written as

$$PAP^t = \left[\begin{array}{c|c} 1 - \frac{u^t u}{\alpha} & u^t \\ \hline u & (1-\alpha)Y \end{array}\right], \tag{5.47}$$

where $1 \geq \alpha > 0$,

$$Yu = u, \tag{5.48}$$

$$u \geq \alpha \mathbf{1}, \tag{5.49}$$

and, where, denoting the eigenvalues of Y by $1 \equiv \mu_1 \geq \mu_2 \geq \ldots \geq \mu_{n-1}$, we also have

$$1 - \alpha - \frac{u^t u}{\alpha} \geq (1-\alpha)\mu_2. \tag{5.50}$$

<u>**Proof**</u>: For each $j = 2, \ldots, n$, let $v^{(j)}$ denote an eigenvector of A corresponding to the eigenvalue λ_j. Suppose first that equality holds in (5.44), and denote the multiplicity of λ_2 as an eigenvalue of A by m. Without loss of generality we assume that $q_{1,1}^{\#} = \max_{i=1,\ldots,n} q_{i,i}^{\#}$. Inspecting the proof of Theorem 5.4.11, we see that since equality holds in (5.44), it must be the case that $e_1^t v^{(j)} = 0$ for each $j = m+2, \ldots, n$. Further, by taking linear combinations if necessary, we may also assume that for each $j = 3, \ldots, m+1, e_1^t v^{(j)} = 0$. Finally, we may also take it to be the case that the vectors $x, v^{(2)}, \ldots, v^{(n)}$ are orthonormal.

Write A in partitioned form as

$$\left[\begin{array}{c|c} a & u^t \\ \hline u & M \end{array}\right],$$

and for each $j = 3, \ldots, n$, partition $v^{(j)}$ conformally as

$$v^{(j)} = \left[\begin{array}{c} 0 \\ \hline \widehat{v}^{(j)} \end{array}\right].$$

From the fact $Av^{(j)} = \lambda_j v^{(j)}, j = 3, \ldots, n$, we find that $\widehat{v}^{(j)}$ is an eigenvector of M for each such j. As $u^t v^{(j)} = 0, j = 3, \ldots, n$, we find that necessarily u must also be an eigenvector of M, say $Mu = (1-\alpha)u$ for some $\alpha > 0$. We further find that since x and $v^{(2)}$ are both orthogonal to each of $v^{(3)}, \ldots, v^{(n)}$, it must be the case that for some scalars s, t, the vectors x and $v^{(2)}$ are scalar multiples of the vectors

$$\left[\begin{array}{c} s \\ \hline u \end{array}\right] \text{ and } \left[\begin{array}{c} t \\ \hline u \end{array}\right],$$

respectively. In particular, we note that necessarily $u > 0$. Applying the eigen–equation for the Perron vector, we find that $s = \alpha$, which in turn yields the fact that $a = 1 - \frac{u^t u}{\alpha}$. It now follows that $t = -\frac{u^t u}{\alpha}$. A computation now shows that $\lambda_2 = 1 - \alpha - \frac{u^t u}{\alpha}$.

Write $M = (1-\alpha)Y$; we find that $Yu = u$, (yielding (5.48)) and since $u > 0$, it follows that the spectral radius of Y is 1. Denote the eigenvalues of Y by $1 \equiv \mu_1 \geq \mu_2 \geq \ldots \geq \mu_{n-1}$. Since $\lambda_2 = 1 - \alpha - \frac{u^t u}{\alpha}$ and $(1-\alpha)\mu_2$ is an eigenvalue of A, (5.50) now follows.

Using Proposition 2.5.1, we find that $Q^\#$ can be written as

$$Q^\# = \left[\begin{array}{c|c} \frac{\alpha u^t u}{(\alpha^2 + u^t u)^2} & -\frac{\alpha^2}{(\alpha^2+u^t u)^2} u^t \\ \hline -\frac{\alpha^2}{(\alpha^2+u^t u)^2} u & [I - (1-\alpha)Y]^{-1} - \frac{(2\alpha^2 + u^t u)}{\alpha(\alpha^2+u^t u)^2} uu^t \end{array}\right]. \tag{5.51}$$

By hypothesis we have $\max_{i=1,\ldots,n} q^\#_{i,i} = q^\#_{1,1} = \frac{\alpha u^t u}{(\alpha + u^t u)^2}$. Observe that the Perron vector for A, normalised to have 2-norm 1 is

$$x = \frac{1}{\alpha^2 + u^t u} \left[\begin{array}{c} \alpha \\ \hline u \end{array}\right].$$

Since

$$\lambda_2 = 1 - \alpha - \frac{u^t u}{\alpha} = 1 - \frac{1 - \min_{i=1,\ldots,n} x_i^2}{q^\#_{1,1}} = 1 - \frac{1 - \min_{i=1,\ldots,n} x_i^2}{\frac{\alpha u^t u}{(\alpha+u^t u)^2}},$$

we deduce that $\min_{i=1,\ldots,n} x_i^2 = \alpha$. Hence (5.49) holds.

Next, we consider the converse. So, suppose that there is a permutation matrix P so that (5.47), (5.48), (5.49), and (5.50) hold. As above we find that necessarily $\min_{i=1,\ldots,n} x_i^2 = \alpha$ and $\lambda_2 = 1 - \alpha - \frac{u^t u}{\alpha}$. Further, (5.51) also holds, and so it suffices to show that $\max_{i=1,\ldots,n} q^{\#}_{i,i} = \frac{\alpha u^t u}{(\alpha^2+u^t u)^2}$.

Let $z^{(j)}, j = 2, \ldots, n-1$ denote an orthonormal collection of eigenvectors of Y corresponding, respectively, to $\mu_j, j = 2, \ldots, n-1$. For each $i = 1, \ldots, n-1$ we have

$$\begin{aligned}([I-(1-\alpha)Y]^{-1})_{i,i} &= \frac{u_i^2}{\alpha u^t u} + \sum_{j=2}^{n-1} \frac{1}{1-(1-\alpha)\mu_j}(z_i^{(j)})^2 \\ &\leq \frac{u_i^2}{\alpha u^t u} + \frac{1}{1-(1-\alpha)\mu_2}\left(1 - \frac{u_i^2}{u^t u}\right).\end{aligned}$$

Consequently we find that for each $i = 1, \ldots, n-1$,

$$\begin{aligned}q^{\#}_{i+1,i+1} &= ([I-(1-\alpha)Y]^{-1})_{i,i} - \frac{(2\alpha^2+u^t u)u_i^2}{\alpha(\alpha^2+u^t u)^2} \\ &\leq \frac{u_i^2}{\alpha u^t u} + \frac{1}{1-(1-\alpha)\mu_2}\left(1-\frac{u_i^2}{u^t u}\right) - \frac{(2\alpha^2+u^t u)u_i^2}{\alpha(\alpha^2+u^t u)^2} \\ &= \frac{\alpha}{\alpha^2+u^t u} - \frac{\alpha u_i^2}{(\alpha^2+u^t u)^2} \leq \frac{\alpha u^t u}{(\alpha^2+u^t u)^2},\end{aligned}$$

the last inequality following from (5.49). Hence $\max_{i=1,\ldots,n} q^{\#}_{i,i} = \frac{\alpha u^t u}{(\alpha^2+u^t u)^2}$, and so we conclude that

$$\lambda_2 = 1 - \frac{1 - \min_{i=1,\ldots,n} x_i^2}{\max_{i=1,\ldots,n} q^{\#}_{i,i}}.$$

□

The following result from [79] provides a characterisation of the equality case in (5.45). The proof, which is analogous to that of Theorem 5.4.13, is omitted.

THEOREM 5.4.14 *Let $A \in \Phi^{n,n}$ be symmetric. Suppose that the Perron value for A is 1, that the other eigenvalues of A are $\lambda_2 \geq \ldots \geq \lambda_n$.*

Let x denote the Perron vector for A, normalised so that $x^t x = 1$, and let $Q = I - A$. We have

$$\lambda_n = 1 - \frac{1 - \max_{i=1,\ldots,n} x_i^2}{\min_{i=1,\ldots,n} q_{i,i}^{\#}}$$

if and only if there is a permutation matrix P such that PAP^t can be written as

$$PAP^t = \left[\begin{array}{c|c} 1 - \frac{u^t u}{\alpha} & u^t \\ \hline u & (1-\alpha)Y \end{array}\right],$$

where $\alpha > 0$,

$$Yu = u,$$

$$u \leq \alpha \mathbf{1},$$

and, where, denoting the eigenvalues of Y by $1 \equiv \mu_1 \geq \mu_2 \geq \ldots \geq \mu_{n-1}$, we also have

$$1 - \alpha - \frac{u^t u}{\alpha} \leq (1-\alpha)\mu_{n-1}.$$

REMARK 5.4.15 In this remark, we maintain the notation of Theorems 5.4.13 and 5.4.14. Consider the special case that A is an $n \times n$ symmetric stochastic matrix. In that case note we have $x = \frac{1}{\sqrt{n}}\mathbf{1}$, so that $\min_{i=1,\ldots,n} x_i^2 = \max_{i=1,\ldots,n} x_i^2 = \frac{1}{n}$. Consequently, the quantities $1 - \frac{1-\min_{i=1,\ldots,n} x_i^2}{\max_{i=1,\ldots,n} q_{i,i}^{\#}}$ and $1 - \frac{1-\max_{i=1,\ldots,n} x_i^2}{\min_{i=1,\ldots,n} q_{i,i}^{\#}}$ of Theorems 5.4.13 and 5.4.14 and given by $1 - \frac{n-1}{n \max_{i=1,\ldots,n} q_{i,i}^{\#}}$ and $1 - \frac{n-1}{n \min_{i=1,\ldots,n} q_{i,i}^{\#}}$, respectively.

It follows from the proof of Theorem 5.4.13 that we then have $\lambda_2 = 1 - \frac{n-1}{n \max_{i=1,\ldots,n} q_{i,i}^{\#}}$ if and only if there is a permutation matrix P and a scalar $\alpha > 0$ such that

$$PAP^t = \left[\begin{array}{c|c} 1-(n-1)\alpha & \alpha\mathbf{1}^t \\ \hline \alpha\mathbf{1} & (1-\alpha)Y \end{array}\right],$$

where Y is stochastic, and in addition, denoting the eigenvalues of Y by $1 \geq \mu_2 \geq \ldots \geq \mu_{n-1}$, we also have $1 - n\alpha \geq (1-\alpha)\mu_2$.

Similarly, we have $\lambda_n = 1 - \frac{n-1}{n \min_{i=1,\ldots,n} q_{i,i}^{\#}}$ if and only if there is a permutation matrix P and a scalar $\alpha > 0$ such that

$$PAP^t = \left[\begin{array}{c|c} 1-(n-1)\alpha & \alpha\mathbf{1}^t \\ \hline \alpha\mathbf{1} & (1-\alpha)Y \end{array}\right],$$

where Y is stochastic, and in addition, denoting the eigenvalues of Y by $1 \geq \mu_2 \geq \ldots \geq \mu_{n-1}$, we also have $1 - n\alpha \leq (1-\alpha)\mu_{n-1}$.

Let T be an irreducible stochastic matrix of order n, with stationary distribution vector w. The Markov chain with transition matrix T is said to be *reversible* if the matrix $W^{\frac{1}{2}}TW^{-\frac{1}{2}}$ is symmetric, where $W = \text{diag}(w)$. Evidently the transition matrix for a reversible Markov chain has real eigenvalues. We now use Remark 5.4.12 to localise the eigenvalues of transition matrices for reversible Markov chains.

THEOREM 5.4.16 *Suppose that $T \in \mathcal{IS}_n$, and that the corresponding Markov chain is reversible. Denote the non–Perron eigenvalues of T by $\lambda_2 \geq \lambda_3 \geq \ldots \geq \lambda_n$. Let $Q = I - T$. Then*

$$\lambda_2 \geq 1 - \frac{n-1}{n \min_{i=1,\ldots,n} q^{\#}_{i,i}} \text{ and } \lambda_n \leq 1 - \frac{n-1}{n \max_{i=1,\ldots,n} q^{\#}_{i,i}}.$$

Proof: Let w be the stationary distribution for T, and let $W = \text{diag}(w)$. Letting $A = W^{\frac{1}{2}}TW^{-\frac{1}{2}}$, we see that A is symmetric. Is straightforward to see that the Perron vector for A, normalised to have 2-norm equal to 1, is $W^{\frac{1}{2}}\mathbf{1}$. Note also that $Q^{\#} = W^{-\frac{1}{2}}(I-A)^{\#}W^{\frac{1}{2}}$. In particular we find that $q^{\#}_{i,i} = (I-A)^{\#}_{i,i}, i = 1, \ldots, n$.

Applying Remark 5.4.12 to A, it follows that

$$\lambda_2 \geq 1 - \frac{1 - \max_{i=1,\ldots,n} w_i}{\min_{i=1,\ldots,n}(I-A)^{\#}_{i,i}}.$$

Since the entries of w sum to 1, we see that $\max_{i=1,\ldots,n} w_i \geq \frac{1}{n}$. Hence we find that

$$\lambda_2 \geq 1 - \frac{n-1}{n \min_{i=1,\ldots,n}(I-A)^{\#}_{i,i}} = 1 - \frac{n-1}{n \min_{i=1,\ldots,n} q^{\#}_{i,i}}.$$

Similarly, we find from Remark 5.4.12 that

$$\lambda_n \leq 1 - \frac{1 - \min_{i=1,\ldots,n} w_i}{\max_{i=1,\ldots,n}(I-A)^{\#}_{i,i}}.$$

Since $\min_{i=1,\ldots,n} w_i \leq \frac{1}{n}$, we find as above that

$$\lambda_n \leq 1 - \frac{n-1}{n \max_{i=1,\ldots,n} q^{\#}_{i,i}}.$$

□

EXAMPLE 5.4.17 Let $w \in \mathbb{R}^n$ with $w > 0$ and $w^t\mathbf{1} = 1$, and fix a scalar $a \in [0, 1)$. Let $T = aI + (1-a)\mathbf{1}w^t$. It is readily seen that the stochastic matrix T has stationary distribution vector w, and that the Markov chain corresponding to T is reversible. The eigenvalues of T are a (with multiplicity $n-1$) and 1, so that in the notation of Theorem 5.4.16, we have $\lambda_2 = a = \lambda_n$.

Since $I - T = (1-a)(I - \mathbf{1}w^t)$, we find that $(I-T)^\# = \frac{1}{1-a}(I - \mathbf{1}w^t)$. So for this example, our bounds of Theorem 5.4.16 are $\lambda_2 \geq 1 - \frac{(n-1)(1-a)}{n(1-\max_{i=1,\ldots,n} w_i)}$ and $\lambda_n \leq 1 - \frac{(n-1)(1-a)}{n(1-\min_{i=1,\ldots,n} w_i)}$. Two short computations reveal that the gap between λ_2 and the lower bound above is equal to $\frac{(1-a)(n\max_{i=1,\ldots,n} w_i - 1)}{n(1-\max_{i=1,\ldots,n} w_i)}$, while the gap between λ_n and the upper bound above is $\frac{(1-a)(1-n\min_{i=1,\ldots,n} w_i)}{n(1-\min_{i=1,\ldots,n} w_i)}$. Evidently the estimates of Theorem 5.4.16 for λ_2 and λ_n perform well if a is close to 1 or if w is close to $\frac{1}{n}\mathbf{1}$.

5.5 Examples

In this section we present a few examples of specific classes of stochastic matrices, and use them to illustrate some of the ideas and techniques developed in this chapter.

EXAMPLE 5.5.1 In this example, we investigate the group inverse associated with the transition matrix for the simple random walk on the undirected path on n vertices. Specifically, we consider the $n \times n$ stochastic matrix T given by

$$T = \begin{bmatrix} 0 & 1 & 0 & 0 & \ldots & 0 \\ \frac{1}{2} & 0 & \frac{1}{2} & 0 & \ldots & 0 \\ 0 & \frac{1}{2} & 0 & \frac{1}{2} & \ldots & 0 \\ \vdots & & \ddots & \ddots & \ddots & \vdots \\ & & & & & \\ 0 & 0 & \ldots & \frac{1}{2} & 0 & \frac{1}{2} \\ 0 & \ldots & 0 & 0 & 1 & 0 \end{bmatrix}. \tag{5.52}$$

Partition out the last row and column of T as

$$T = \left[\begin{array}{c|c} T_{1,1} & T_{1,2} \\ \hline T_{2,1} & T_{2,2} \end{array}\right].$$

Setting $Q = I - T$, we use the approach of Proposition 2.5.1 in order to find $Q^{\#}$.

It is straightforward to show that

$$(I - T_{1,1})^{-1} =$$

$$\begin{bmatrix} n-1 & n-2 & n-3 & \dots & 2 & 1 \\ n-2 & n-2 & n-3 & \dots & 2 & 1 \\ n-3 & n-3 & n-3 & n-4 & \dots & 1 \\ \vdots & & & & & \vdots \\ 1 & 1 & \dots & & 1 & 1 \end{bmatrix} \begin{bmatrix} 1 & & & & \\ & 2 & & & \\ & & 2 & & \\ & & & \ddots & \\ & & & & 2 \end{bmatrix}.$$

Consequently we find that

$$T_{2,1}(I - T_{1,1})^{-1} = \begin{bmatrix} 1 & 2 & 2 & \dots & 2 \end{bmatrix},$$

which in turn yields $1 + T_{2,1}(I - T_{1,1})^{-1}\mathbf{1} = 2(n-1)$. We also find that for each $k = 1, \dots, n-1$, $e_k^t(I - T_{1,1})^{-1}\mathbf{1} = (n-k)(n+k-2)$.

Note further that

$$\begin{aligned} T_{2,1}(I - T_{1,1})^{-2}e_k &= \begin{bmatrix} 1 & 2 & 2 & \dots & 2 \end{bmatrix}(I - T_{1,1})^{-1}e_k \\ &= \begin{cases} (n-1)^2 & \text{if } k = 1, \\ 2(n-k)(n+k-2) & \text{if } 2 \le k \le n-1 \end{cases}. \end{aligned}$$

From this last it follows that $T_{2,1}(I - T_{1,1})^{-2}\mathbf{1} = \frac{(n-1)(4n^2-8n+3)}{3}$.

Having computed all of the quantities needed to implement the formula for $Q^{\#}$ in Proposition 2.5.1, it is now simply a matter of substituting (and simplifying) in order to find $Q^{\#}$. Applying Proposition 2.5.1, it follows that

$$\begin{aligned} q^{\#}_{i,1} &= \frac{4n^2-8n+3}{12(n-1)} - \frac{(i-1)(2n-i-1)}{2(n-1)} \text{ for } i = 1, \dots, n, \\ q^{\#}_{i,n} &= \frac{4n^2-8n+3}{12(n-1)} - \frac{(n-i)(n+i-2)}{2(n-1)} \text{ for } j = n, i = 1, \dots, n, \text{ and} \\ q^{\#}_{i,j} &= \frac{4n^2-8n+3}{6(n-1)} + 2(n - \max\{i,j\}) - \frac{(n-i)(n+i-2)}{(n-1)} \\ &\quad - \frac{(n-j)(n+j-2)}{(n-1)} \text{ for } 2 \le j \le n-1, i = 1, \dots, n. \end{aligned} \tag{5.53}$$

In particular, we find from (5.53) that

$$q^{\#}_{1,1} = \frac{4n^2 - 8n + 3}{12(n-1)} = q^{\#}_{n,n},$$

while for $k = 2, \ldots, n-1$, we have

$$\begin{aligned} q^{\#}_{k,k} &= \frac{4n^2 - 8n + 3}{6(n-1)} + 2(n-k) - \frac{2(n-k)(n+k-2)}{n-1} \\ &= \frac{4n^2 - 8n + 3}{6(n-1)} + \frac{2(k-1)(k-n)}{n-1}. \end{aligned}$$

It follows that $q^{\#}_{k,k}$ is maximised for $k = 2$ and $k = n-1$ and minimised when $k = \lfloor \frac{n+1}{2} \rfloor$ (and, if n is even, $k = \lceil \frac{n+1}{2} \rceil$). Consequently, we have $\max_{k=1,\ldots,n} q^{\#}_{k,k} = \frac{4n^2-20n+27}{6(n-1)}$ and $\min_{k=1,\ldots,n} q^{\#}_{k,k} = \frac{4n^2-8n+3}{6(n-1)} - 2\frac{\lfloor \frac{n+1}{2} \rfloor \lceil \frac{n+1}{2} \rceil}{n-1}$.

It is not difficult to see that T is the transition matrix for a reversible Markov chain. In view of Theorem 5.4.16, we find that the largest non–Perron eigenvalue λ_2 of T satisfies

$$\lambda_2 \geq 1 - \frac{n-1}{n \min_{k=1,\ldots,n} q^{\#}_{k,k}} = 1 - \frac{n-1}{n\left(\frac{4n^2-8n+3}{6(n-1)} - 2\frac{\lfloor \frac{n+1}{2} \rfloor \lceil \frac{n+1}{2} \rceil}{n-1}\right)}.$$

Observe that for large values of n, this lower bound is roughly $1 - \frac{6}{n}$. We note in passing that the eigenvalues of T are known to be given by $\cos\left(\frac{(k-1)\pi}{n-1}\right), k = 1, \ldots, n$ (with the $n \times n$ matrix $U = \left[\cos\left(\frac{(i-1)(j-1)\pi}{n-1}\right)\right]_{i,j=1,\ldots,n}$ serving as a diagonalising matrix for T). Hence $\lambda_2 = \cos\left(\frac{\pi}{n-1}\right)$; for large values of n, we see that λ_2 is approximately $1 - \frac{\pi^2}{2n^2}$.

Note that as we descend each column of $Q^{\#}$, the entries increase up to the diagonal element, then decrease afterwards. We thus find that $\kappa(T) = \frac{1}{2}\max_{j=1,\ldots,n}(q^{\#}_{j,j} - \min\{q^{\#}_{1,j}, q^{\#}_{n,j}\})$. From (5.53), it follows that $q^{\#}_{1,1} - q^{\#}_{n,1} = \frac{n-1}{2} = q^{\#}_{n,n} - q^{\#}_{1,n}$, while for each $j = 2, \ldots, n-1$, $q^{\#}_{j,j} - q^{\#}_{1,j} = \frac{(j-1)^2}{n-1}$ and $q^{\#}_{j,j} - q^{\#}_{n,j} = \frac{(n-j)^2}{n-1}$. We now readily deduce that $\kappa(T) = \frac{(n-2)^2}{2(n-1)}$. From our computation of the diagonal entries of $Q^{\#}$ above, we find that $\mathrm{trace}(Q^{\#}) = \frac{2n^2-4n+3}{6}$. Observe that $\frac{(n-2)^2}{2(n-1)} \geq \frac{1}{2(n-1)}\frac{2n^2-4n+3}{6}$, which illustrates the conclusion of Theorem 5.3.19, namely that $\kappa(T) \geq \frac{1}{2(n-1)}\mathrm{trace}(Q^{\#})$.

Example 5.5.1 allows us to (at last) address Example 1.2.1 in some detail.

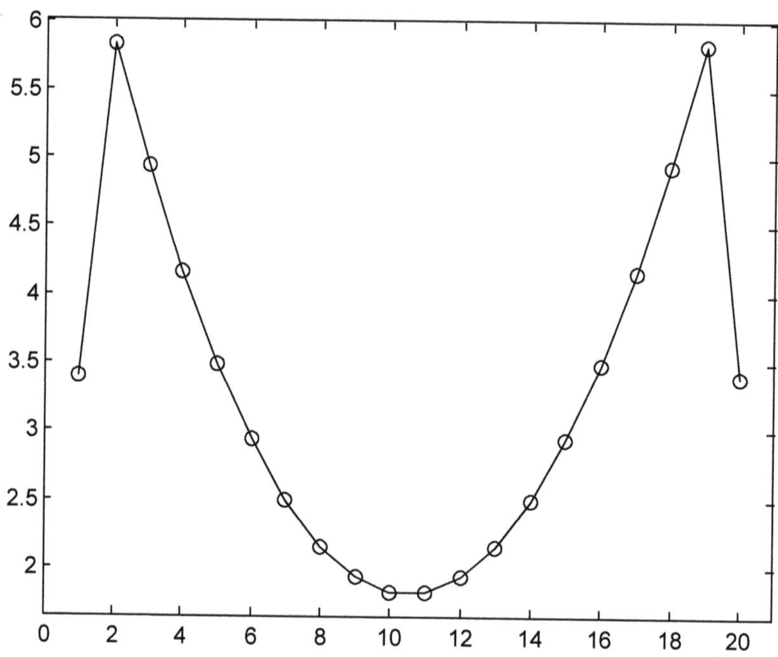

FIGURE 5.2
The structured condition numbers $C^{(\infty)}_{\mathcal{AS}_{20}}(T, j), j = 1, \ldots, 20$ for Example 5.5.2

EXAMPLE 5.5.2 Let T and B be as in Example 1.2.1, and for each $s \in [0, \frac{1}{2})$, denote the stationary distribution vector of $T + sB$ by $w(s)$. Recall that our interest in Example 1.2.1 centred on the difference in behaviours between $w_{12}(s)$ and $w_{19}(s)$ for small values of s. It follows from Corollary 5.3.2 that

$$\left.\frac{dw_{12}(s)}{ds}\right|_{s=0} = w(0)^t B(I - T)^{\#} e_{12}$$

and

$$\left.\frac{dw_{19}(s)}{ds}\right|_{s=0} = w(0)^t B(I - T)^{\#} e_{19}.$$

Observe that we may use Example 5.5.1 (with the parameter n set to

20) in order to compute $(I-T)^{\#}$.

The vector $w(0)$ is given by

$$w(0) = \frac{1}{38}\begin{bmatrix} 1 \\ 2 \\ 2 \\ \vdots \\ 2 \\ 1 \end{bmatrix},$$

from which we find that

$$w(0)^t B = \frac{1}{38} = \begin{bmatrix} 0 & -1 & -2 & -2 & \ldots & -2 & 35 & -2 \end{bmatrix}.$$

Using (5.53) to compute $(I-T)^{\#}e_{12}$ and $(I-T)^{\#}e_{19}$, it now follows that

$$\left.\frac{dw_{12}(s)}{ds}\right|_{s=0} = -\frac{3}{38}, \text{ while } \left.\frac{dw_{19}(s)}{ds}\right|_{s=0} = \frac{375}{38}.$$

These values serve to explain, for values of s near 0, the qualitative difference between the two curves shown in Figure 1.2.

In a related direction, we computed, for the matrix T of Example 1.2.1, the quantities $C^{(\infty)}_{\mathcal{AS}_{20}}(T,j), j = 1,\ldots,20$; recall that each $C^{(\infty)}_{\mathcal{AS}_{20}}(T,j)$ is a structured condition number for the j–th entry in the stationary distribution. Figure 5.2 shows a plot of $C^{(\infty)}_{\mathcal{AS}_{20}}(T,1),\ldots,C^{(\infty)}_{\mathcal{AS}_{20}}(T,20)$, and we see that for this example, $C^{(\infty)}_{\mathcal{AS}_{20}}(T,j)$ is maximised for $j = 2, 19$ (with a maximum value of $\frac{443}{76}$) while $C^{(\infty)}_{\mathcal{AS}_{20}}(T,j)$ is minimised for $j = 10, 11$ (with a minimum value of $\frac{2635}{1444}$). Again this reinforces the notion that for this example, the entries of the stationary distribution vector in positions 2 and 19 are the most sensitive to small perturbations of T.

In our next example, we return to the class of stochastic Leslie matrices, and consider a highly structured instance.

EXAMPLE 5.5.3 Fix $t \in (0,1)$, and consider the following simple stochastic Leslie matrix of order n:

$$T = \begin{bmatrix} 0 & 0 & 0 & \dots & 0 & t & 0 & \dots & 0 & 1-t \\ 1 & 0 & 0 & & & \dots & & & & 0 \\ 0 & 1 & 0 & & & \dots & & & & 0 \\ \vdots & & \ddots & \ddots & & & & & & \vdots \\ 0 & 0 & & & \dots & & & 0 & 1 & 0 \end{bmatrix},$$

where the nonzero entry t in the first row is in the k-th position, say. In this example we compute a few of the quantities arising in Corollary 5.3.9 and Theorem 5.3.19.

Note that the stationary distribution for T is given by

$$w = \frac{1}{n - t(n-k)} \begin{bmatrix} \frac{\mathbf{1}_k}{(1-t)\mathbf{1}_{n-k}} \end{bmatrix},$$

where the subscripts on the all ones vectors indicate the orders of the corresponding vectors. Set $Q = I - T$, and note that we can use (4.13) and (4.14) to compute $Q^{\#}$. Observe that in our setting, the matrix R of (4.13) is the lower triangular matrix of order $n-1$ with ones on and below the diagonal, and zeros elsewhere. It now follows that for any indices i, j with $1 \le i < j \le n$, we have

$$e_i^t Q^{\#} - e_j^t Q^{\#} = (j-i)w^t - \sum_{l=i+1}^{j} e_l^t. \tag{5.54}$$

From (5.54) we deduce readily that $\max_{i,j}\left(q^{\#}_{j,j} - q^{\#}_{i,j}\right) = \frac{n-1}{n-t(n-k)}$; in particular, we have $\kappa(T) = \frac{n-1}{2n-2t(n-k)}$.

Suppose that the eigenvalues of T are $1, \lambda_2, \dots, \lambda_n$. Then $\sum_{j=2}^{n} \frac{1}{1-\lambda_j} = \text{trace}(Q^{\#})$, and the latter can be found from (4.14). Letting $\hat{w}$ denote the trailing subvector of w of order $n-1$, we find from (4.14) that $\text{trace}(Q^{\#}) = \text{trace}(R) - \hat{w}^t R\mathbf{1}$. A computation now yields

$$\sum_{j=2}^{n} \frac{1}{1-\lambda_j} = \text{trace}(Q^{\#}) = n - 1 - \frac{(1-t)n(n-1) + tk(k-1)}{2n - 2t(n-k)}.$$

We now turn our attention to $\tau(Q^{\#})$, and for ease of exposition we introduce the simplifying assumption that $k \geq \lceil \frac{n+1}{2} \rceil$.Note that for any pair of indices i, j, the 1-norm of $\frac{1}{2}(e_i^t Q^\# - e_j^t Q^\#)^t$ is the same as the sum of the positive entries in $(e_i^t Q^\# - e_j^t Q^\#)$ (this is because $(e_i^t Q^\# - e_j^t Q^\#)\mathbf{1} = 0$). Fix a pair of indices i, j with $1 \leq i < j \leq n$, and let $j - i = p$. From (5.54) we have $e_i^t Q^\# - e_j^t Q^\# = pw^t - \sum_{l=i+1}^{j} e_l^t$; now observe that the sum of the positive entries in $e_i^t Q^\# - e_j^t Q^\#$ is no greater than the sum of the positive entries in the vector $pw^t - \sum_{l=n-p+1}^{n} e_l^t$. In other words, for $j - i = p$, we have that

$$|| \left(Q^\#\right)^t e_i - \left(Q^\#\right)^t e_j||_1 \leq || \left(Q^\#\right)^t e_{n-p} - \left(Q^\#\right)^t e_n||_1.$$

It is straightforward to determine that

$$\frac{1}{2}|| \left(Q^\#\right)^t e_{n-p} - \left(Q^\#\right)^t e_n||_1 =$$

$$\begin{cases} \frac{p(n-p)-tp(n-k-p)}{n-t(n-k)} & \text{if} 1 \leq p \leq n-k, \\ \frac{p(n-p)}{n-t(n-k)} & \text{if} n-k+1 \leq p \leq n-1. \end{cases}$$

It now follows that for each $p = 1, \ldots, n-1$,

$$\frac{1}{2}|| \left(Q^\#\right)^t e_{n-p} - \left(Q^\#\right)^t e_n||_1 \leq \frac{\lfloor \frac{n}{2} \rfloor \lceil \frac{n}{2} \rceil}{n - t(n-k)}.$$

Invoking our simplifying assumption that $k \geq \lceil \frac{n+1}{2} \rceil$, we see that $n - k + 1 \leq \lfloor \frac{n+1}{2} \rfloor$. We then find that

$$\frac{1}{2}|| \left(Q^\#\right)^t e_{n-\lfloor \frac{n+1}{2} \rfloor} - \left(Q^\#\right)^t e_n||_1 = \frac{\lfloor \frac{n}{2} \rfloor \lceil \frac{n}{2} \rceil}{n - t(n-k)}.$$

Consequently, we have

$$\tau(Q^\#) = \frac{\lfloor \frac{n}{2} \rfloor \lceil \frac{n}{2} \rceil}{n - t(n-k)}.$$

The conclusions of Corollary 5.3.9 and Theorem 5.3.19 thus correspond to the inequalities

$$\frac{\lfloor \frac{n}{2} \rfloor \lceil \frac{n}{2} \rceil}{n - t(n-k)} \leq n - 1 - \frac{(1-t)n(n-1) + tk(k-1)}{2n - 2t(n-k)} \tag{5.55}$$

(for $k \geq \lceil \frac{n+1}{2} \rceil$), and

$$\frac{1}{2(n-1)} \left(n - 1 - \frac{(1-t)n(n-1) + tk(k-1)}{2n - 2t(n-k)} \right) \leq \frac{n-1}{2n - 2t(n-k)}, \tag{5.56}$$

respectively. Observe that when n is large and t is close to zero, the right sides of (5.55) and (5.56) are roughly twice the corresponding left sides.

We continue with stochastic Leslie matrices in the next example, this time focusing on the irreducible and periodic case.

EXAMPLE 5.5.4 Suppose that we have an irreducible stochastic Leslie matrix of order n, say

$$L = \begin{bmatrix} x_1 & x_2 & \dots & x_{n-1} & x_n \\ 1 & 0 & 0 & \dots & 0 \\ 0 & 1 & 0 & \dots & 0 \\ \vdots & & \ddots & \ddots & \vdots \\ 0 & 0 & \dots & 1 & 0 \end{bmatrix},$$

where $x_j \geq 0, j = 1, \dots, n-1, x_n > 0$, and $\sum_{j=1}^{n} x_j = 1$. Considering the directed graph $\mathcal{D}(L)$, we find that it contains a cycle of length k if and only if $x_k > 0$. Consequently, L is primitive if and only if $\gcd\{j | x_j > 0\} = 1$, and periodic with period $d = \gcd\{j | x_j > 0\}$ otherwise.

In this example, we suppose that L is periodic with period $d \geq 2$; observe that necessarily $n = kd$ for some $k \in \mathbb{N}$. Our aim is to provide an expression for $(I - L)^{\#}$ by using the machinery of Theorem 5.2.2. We begin by permuting L to periodic normal form. Let P be the permutation matrix such that

$$p_{ak+b, bd-a} = 1, \text{ for } a = 0, \dots, d-1, b = 1, \dots, k.$$

We then find that

$$PLP^t = \begin{bmatrix} 0 & I & 0 & 0 & \dots & 0 \\ 0 & 0 & I & 0 & \dots & 0 \\ \vdots & & \ddots & \ddots & \vdots & \\ 0 & 0 & \dots & 0 & 0 & I \\ \hat{L} & 0 & 0 & \dots & 0 & 0 \end{bmatrix}, \tag{5.57}$$

where each of the identity matrix blocks is $k \times k$, and where $\hat{L}$ is the $k \times k$ stochastic Leslie matrix

$$\hat{L} = \begin{bmatrix} x_d & x_{2d} & \dots & x_{(k-1)d} & x_{kd} \\ 1 & 0 & 0 & \dots & 0 \\ 0 & 1 & 0 & \dots & 0 \\ \vdots & & \ddots & \ddots & \vdots \\ 0 & 0 & \dots & 1 & 0 \end{bmatrix}.$$

Note that since d is the greatest common divisor of the set of ks so that x_k is positive, it must be the case that $\gcd\{j|x_{jd} > 0\} = 1$. We conclude that necessarily $\hat{L}$ is primitive. Let $\hat{w}$ denote the stationary distribution vector for $\hat{L}$.

Adopting the language and notation of Theorem 5.2.2, we find that $A_j = \hat{L}$ and $w_j = \hat{w}$ for each $j = 1, \ldots, d$. Setting $G = (I - PLP^t)^{\#}$, we find from Theorem 5.2.2 that G can be written as a $d \times d$ block matrix

$$\left[\begin{array}{c|c|c} G_{1,1} & \cdots & G_{1,d} \\ \hline \vdots & & \vdots \\ \hline G_{d,1} & \cdots & G_{d,d} \end{array}\right],$$

where the $k \times k$ blocks $G_{i,j}$ are given by

$$G_{i,j} = \begin{cases} (I - \hat{L})^{\#} + \left(\frac{d-1}{2d} - \frac{j-i}{d}\right) \mathbf{1}\hat{w}^t & \text{if } i \le j, \\ (I - \hat{L})^{\#}\hat{L} + \left(\frac{d-1}{2d} - \frac{d+j-i}{d}\right) \mathbf{1}\hat{w}^t & \text{if } i > j. \end{cases} \tag{5.58}$$

Since $(I - \hat{L})^{\#}(I - \hat{L}) = I - \mathbf{1}\hat{w}^t$, it follows that $(I - \hat{L})^{\#}\hat{L} = (I - \hat{L})^{\#} - I + \mathbf{1}\hat{w}^t$. Thus we may rewrite (5.58) as

$$G_{i,j} = \begin{cases} (I - \hat{L})^{\#} + \left(\frac{d-1}{2d} - \frac{j-i}{d}\right) \mathbf{1}\hat{w}^t & \text{if } i \le j, \\ (I - \hat{L})^{\#} - I + \left(\frac{d-1}{2d} - \frac{j-i}{d}\right) \mathbf{1}\hat{w}^t & \text{if } i > j. \end{cases} \tag{5.59}$$

Observe that the blocks of G have the appealing property that they can be expressed in terms of the stochastic Leslie matrix of lower order, namely $\hat{L}$.

Finally, we may recover $(I - L)^{\#}$ from the block matrix G by observing that $(I - L)^{\#} = P^tGP$. Given indices i, j between 1 and n, write $i = a_ik + b_i, j = a_jk + b_k$, where $a_i, a_j \in \{0, 1, \ldots, d-1\}$ and $b_i, b_j \in \{1, \ldots, k\}$. It is straightforward to see that $a_i = \lceil \frac{i}{k} \rceil - 1$ and $a_j = \lceil \frac{j}{k} \rceil - 1$, so that $b_i = i + k - k\lceil \frac{i}{k} \rceil$ and $b_j = j + k - k\lceil \frac{j}{k} \rceil$. We then find that $e_i^tP^t = e^t_{db_i - a_i}$ and $Pe_j = e_{db_j - a_j}$. Observe here that since $b_i = i - a_ik$, we have $db_i - a_i = di - a_i(n+1)$, and similarly $db_j - a_j = dj - a_j(n+1)$. Next we write $db_i - a_i = p_ik + q_i$ and $db_j - a_j = p_jk + q_j$, where $p_i, p_j \in \{0, 1, \ldots, d-1\}, q_i, q_j \in \{1, \ldots, k\}$. (As above, $p_i = \lceil \frac{db_i - a_i}{k} \rceil - 1, p_j = \lceil \frac{db_j - a_j}{k} \rceil - 1, q_i = db_i - a_i + k - k\lceil \frac{db_i - a_i}{k} \rceil$ and $q_j = db_j - a_j + k - k\lceil \frac{db_j - a_j}{k} \rceil$.)

With these preliminary computations, we then have:

$$e_i^t(I - L)^{\#}e_j = e_i^tP^tGPe_j = e^t_{db_i - a_i}Ge_{db_j - a_j} = e^t_{p_ik + q_i}Ge_{p_jk + q_j}. \tag{5.60}$$

From the partitioning of G as a $d \times d$ block matrix (where all of the blocks are $k \times k$), it follows that

$$e^t_{p_ik+q_i} G e_{p_jk+q_j} = e^t_{q_i}(G_{p_i+1,p_j+1})e_{q_j},$$

where the block G_{p_i+1,p_j+1} is given by (5.59). It now follows from (5.60) that the (i,j) entry of $(I-L)^{\#}$ coincides with the (q_i,q_j) entry of the block G_{p_i+1,p_j+1}.

We close the chapter by examining a highly structured class of stochastic circulant matrices.

EXAMPLE 5.5.5 Suppose that $n \in \mathbb{N}$ is even, and write $n = 2k$. We begin with the following motivating example appearing in [95], which we will subsequently generalise.

Let M_k be the $n \times n$ matrix given by

$$M_k = \frac{1}{4k}\begin{bmatrix} 3 & 1 & 3 & 1 & \dots & 1 & 3 & 1 \\ 1 & 3 & 1 & 3 & \dots & 3 & 1 & 3 \\ 3 & 1 & 3 & 1 & \dots & 1 & 3 & 1 \\ 1 & 3 & 1 & 3 & \dots & 3 & 1 & 3 \\ & \ddots & \ddots & \ddots & & \ddots & \ddots & \\ 3 & 1 & 3 & 1 & \dots & 1 & 3 & 1 \\ 1 & 3 & 1 & 3 & \dots & 3 & 1 & 3 \end{bmatrix}.$$

It is straightforward to check then that

$$(I-M_k)^{\#} = I - \frac{1}{k}\begin{bmatrix} 0 & 1 & 0 & 1 & \dots & 1 & 0 & 1 \\ 1 & 0 & 1 & 0 & \dots & 0 & 1 & 0 \\ 0 & 1 & 0 & 1 & \dots & 1 & 0 & 1 \\ 1 & 0 & 1 & 0 & \dots & 0 & 1 & 0 \\ & \ddots & \ddots & \ddots & & \ddots & \ddots & \\ 0 & 1 & 0 & 1 & \dots & 1 & 0 & 1 \\ 1 & 0 & 1 & 0 & \dots & 0 & 1 & 0 \end{bmatrix}.$$

We note with interest that in particular, $(I-M_k)^{\#}$ is an M-matrix.

Following the approach of Chen, Kirkland, and Neumann in [21], we now generalise Meyer's example. Fix a parameter $\alpha \in [0,1)$, and consider the $n \times n$ stochastic circulant matrix $T_k(\alpha)$ given by

$$T_k(\alpha) = \frac{1}{k}\begin{bmatrix} \alpha & 1-\alpha & \alpha & 1-\alpha & \dots & 1-\alpha & \alpha & 1-\alpha \\ 1-\alpha & \alpha & 1-\alpha & \alpha & \dots & \alpha & 1-\alpha & \alpha \\ \alpha & 1-\alpha & \alpha & 1-\alpha & \dots & 1-\alpha & \alpha & 1-\alpha \\ 1-\alpha & \alpha & 1-\alpha & \alpha & \dots & \alpha & 1-\alpha & \alpha \\ & \ddots & \ddots & \ddots & & \ddots & \ddots & \\ \alpha & 1-\alpha & \alpha & 1-\alpha & \dots & 1-\alpha & \alpha & 1-\alpha \\ 1-\alpha & \alpha & 1-\alpha & \alpha & \dots & \alpha & 1-\alpha & \alpha \end{bmatrix}. \quad (5.61)$$

Evidently Meyer's example above corresponds to the case $\alpha = \frac{3}{4}$.

Observe that we can write $T_k(\alpha)$ as $T_k(\alpha) = AB$, where

$$A = \frac{1}{k}\begin{bmatrix} \alpha & 1-\alpha \\ 1-\alpha & \alpha \\ \alpha & 1-\alpha \\ 1-\alpha & \alpha \\ \vdots & \vdots \\ \alpha & 1-\alpha \\ 1-\alpha & \alpha \end{bmatrix} \text{ and } B = \begin{bmatrix} 1 & 0 & 1 & 0 & \dots & 1 & 0 \\ 0 & 1 & 0 & 1 & \dots & 0 & 1 \end{bmatrix}.$$

Observe that

$$BA = \begin{bmatrix} \alpha & 1-\alpha \\ 1-\alpha & \alpha \end{bmatrix},$$

so that

$$(I - BA)^{\#} = \begin{bmatrix} \frac{1}{4(1-\alpha)} & -\frac{1}{4(1-\alpha)} \\ -\frac{1}{4(1-\alpha)} & \frac{1}{4(1-\alpha)} \end{bmatrix}.$$

Evidently the stationary distribution vector for $T_k(\alpha)$ is $\frac{1}{2k}\mathbf{1}$, and so applying Lemma 5.2.1, it follows that

$$(I - T_k(\alpha))^{\#} = I - \frac{1}{2k}J + A(I - BA)^{\#}B =$$

$$= I + \frac{1}{4k(1-\alpha)}\begin{bmatrix} 4\alpha-3 & -1 & 4\alpha-3 & -1 & \dots & -1 \\ -1 & 4\alpha-3 & -1 & 4\alpha-3 & \dots & 4\alpha-3 \\ 4\alpha-3 & -1 & 4\alpha-3 & -1 & \dots & -1 \\ -1 & 4\alpha-3 & -1 & 4\alpha-3 & \dots & 4\alpha-3 \\ \vdots & \ddots & \ddots & \ddots & & \vdots \\ 4\alpha-3 & -1 & 4\alpha-3 & -1 & \dots & -1 \\ -1 & 4\alpha-3 & -1 & 4\alpha-3 & \dots & 4\alpha-3 \end{bmatrix}.$$

We see in particular that $(I - T_k(\alpha))^{\#}$ is an M-matrix if and only if $0 \le \alpha \le \frac{3}{4}$.

Next we turn our attention to computing $\tau((I - T_k(\alpha))^{\#})$. Note that if indices i and j have the same parity, then $(e_i - e_j)^t(I - T_k(\alpha))^{\#} = (e_i - e_j)^t$, while if i is odd and j is even,

$$(e_i - e_j)^t(I - T_k(\alpha))^{\#} =$$
$$(e_i - e_j)^t + \frac{4\alpha - 2}{4k(1-\alpha)}\left[\,[1 \quad -1 \quad 1 \quad -1 \quad \dots \quad 1 \quad -1\,\right].$$

It now follows that if indices i and j have opposite parities, then

$$\frac{1}{2}\|\left((I - T_k(\alpha))^{\#}\right)^t(e_i - e_j)\|_1 = 1 + \frac{4\alpha-2}{4k(1-\alpha)} + (k-1)\frac{|4\alpha-2|}{4k(1-\alpha)}.$$

We thus find that

$$\tau((I - T_k(\alpha))^{\#}) = \begin{cases} \frac{1}{2(1-\alpha)} & \text{if } \frac{1}{2} \le \alpha < 1, \\ 1 + (k-2)\frac{1-2\alpha}{2k(1-\alpha)} & \text{if } 0 \le \alpha < \frac{1}{2}. \end{cases}$$

It is readily determined that the eigenvalues of $T_k(\alpha)$ are $1, 2\alpha - 1$, and 0, the last with multiplicity $n - 2$. In particular, note that for the eigenvalue $\lambda \equiv 2\alpha - 1$, we have $\frac{1}{|1-\lambda|} = \frac{1}{2(1-\alpha)}$. Thus we see that if $\alpha \ge \frac{1}{2}$, then $\frac{1}{|1-\lambda|} = \tau((I - T_k(\alpha))^{\#})$, so that equality holds in (5.41). Observe also that $T_k(\alpha)$ satisfies the hypotheses of Theorem 5.4.7 when $\alpha \ge \frac{1}{2}$, which provides an alternate way of concluding that equality holds in (5.41). When $\alpha < \frac{1}{2}$, we have $\tau((I - T_k(\alpha))^{\#}) = 1 + (k-2)\frac{1-2\alpha}{2k(1-\alpha)} = \frac{1}{2(1-\alpha)} + \frac{(k-1)(1-2\alpha)}{k(1-\alpha)}$, so that the bound on $\frac{1}{|1-\lambda|}$ in (5.41) is strict, though still reasonably accurate for values of α close to (but less than) $\frac{1}{2}$.

6

Mean First Passage Times for Markov Chains

In this chapter, we consider the mean first passage matrix associated with a Markov chain on n states having an irreducible transition matrix T. In section 6.1, we give a basic result showing how the mean first passage matrix can be computed from the group inverse of $I-T$, and discuss some of its consequences. A triangle inequality for the entries in the mean first passage matrix is derived in section 6.2; we then go on to extend that inequality to the entries in certain symmetric inverse M-matrices. Section 6.3 addresses the problem of determining which positive matrices are realisable as mean first passage matrices, and links that problem to a special case of the inverse M-matrix problem. In section 6.4, we consider a mean first passage matrix in partitioned form as a 2×2 block matrix, and show how those blocks can be computed from the mean first passage matrices corresponding to certain stochastic complements of the original transition matrix. Finally, in section 6.5, we discuss some of the properties of the Kemeny constant associated with an irreducible stochastic matrix; in particular, we find the minimum value of the Kemeny constant over all transition matrices whose directed graphs are subordinate to a given directed graph.

6.1 Mean first passage matrix via the group inverse

In section 5.1, we briefly introduced the notion of the mean first passage time from one state to another in a Markov chain whose transition matrix is irreducible. Here we develop the connection between mean first passage times and the group inverse associated with the transition matrix. To fix the notation, suppose that we have a matrix $T \in \mathcal{IS}_n$. Recall that for the Markov chain associated with T, the first passage time from state i to state j, $f_{i,j}$, say, is the random variable taking on the value given by the smallest $k \geq 1$ such that the chain is in state j after k steps, given that the chain started in state i. Let $m_{i,j}$ denote the mean first passage time from state i to state j — i.e., the expected value of $f_{i,j}$. The following conditional expectation argument establishes a connection between the $m_{i,j}$s and the entries in T.

Fix indices i and j between 1 and n, and suppose that the Markov chain is initially in state i. We may compute $m_{i,j}$ by conditioning on the state of the Markov chain after one step has been taken. After one step, the chain is in state j (with probability $t_{i,j}$) or it is in some state $k \neq j$ (with probability $t_{i,k}$). We thus find that

$$m_{i,j} = t_{i,j} + \sum_{k=1,\ldots,n,k\neq j} t_{i,k}(m_{k,j}+1) = 1 + \sum_{k=1,\ldots,n,k\neq j} t_{i,k}m_{k,j}. \quad (6.1)$$

Let $M = [m_{i,j}]_{i,j=1,\ldots,n}$ denote the mean first passage matrix for the Markov chain. Then (6.1) can be rewritten as

$$M = T(M - M_{dg}) + J, \quad (6.2)$$

where for any $n \times n$ matrix A, we define A_{dg} as

$$A_{dg} = \operatorname{diag}\left(\begin{bmatrix} a_{1,1} & \ldots & a_{n,n} \end{bmatrix}\right).$$

The following result, which appears in [94], shows how we can write M in terms of the group inverse of $I - T$.

THEOREM 6.1.1 *Suppose that $T \in \mathcal{IS}_n$, let w denote its stationary vector, let $Q = I-T$, and denote the mean first passage matrix associated with T by M. Then*

$$M = (I - Q^{\#} + JQ^{\#}_{dg})W^{-1}, \quad (6.3)$$

where $W = \operatorname{diag}(w)$ and where $Q^{\#}_{dg}$ is the diagonal matrix whose i-th diagonal entry is $q^{\#}_{i,i}, i = 1,\ldots,n$.

Proof: We begin by noting that from (6.2), we find that $w^tM = w^tT(M - M_{dg}) + \mathbf{1}^t$, so that $w^tM_{dg} = \mathbf{1}^t$. Hence we have $M_{dg} = W^{-1}$. Thus we can rewrite (6.2) as $QM = -TW^{-1} + J$. Multiplying on the left by $Q^{\#}$ now yields $(I - \mathbf{1}w^t)M = -Q^{\#}TW^{-1}$. Since $-Q^{\#}T = I - \mathbf{1}w^t - Q^{\#}$, we find that for some suitable vector v, we have $M = (I - Q^{\#} + \mathbf{1}v^t)W^{-1}$. Since $m_{i,i} = \frac{1}{w_i}$ for each $i = 1,\ldots,n$, it must be the case that $v_i = q^{\#}_{i,i}, i = 1,\ldots,n$, i.e., $v = Q^{\#}_{dg}\mathbf{1}$. The expression (6.3) now follows. □

We note that in [66, sections 4.3, 4.4] the mean first passage matrix is approached in a somewhat different manner, by use of the so–called *fundamental matrix* for the Markov chain. To be specific, if T is an irreducible stochastic matrix with stationary vector w, the associated fundamental matrix is given by $Z = (I - T + \mathbf{1}w^t)^{-1}$ (it is readily seen that $I - T + \mathbf{1}w^t$ is invertible). The mean first passage matrix is then computed in [66] via the formula

$$M = (I - Z + JZ_{dg})W^{-1}, \tag{6.4}$$

where as in Theorem 6.1.1, $W = \text{diag}(w)$. The relationship between (6.4) and (6.3) becomes apparent upon observing that $(I - T)^{\#} = Z - \mathbf{1}w^t$, a fact readily verified by considering the eigendecomposition for both $(I - T)^{\#}$ and $Z - \mathbf{1}w^t$.

The following observation is made in the paper of Neumann and Sze [102].

REMARK 6.1.2 Suppose that we have an irreducible stochastic matrix with stationary vector w, and let $Q = I - T$. We see from Theorem 6.1.1 that if we happen to know $Q^{\#}$ then may compute the mean first passage matrix M from (6.3). It turns out that the converse holds – i.e., if we happen to know M, then we can recover $Q^{\#}$ from (6.3). The following argument shows how this can be accomplished.

As in Theorem 6.1.1, we let $W = \text{diag}(w)$. We find readily from (6.3) that

$$MW - I = -Q^{\#} + J(Q^{\#})_{dg}.$$

Multiplying both sides on the left by w^t and using the facts that $w^tQ^{\#} = 0^t$ and $w^t\mathbf{1} = 1$, we find that $\mathbf{1}^t(Q^{\#})_{dg} = w^t(MW - I)$. Consequently, we find from (6.3) that $Q^{\#} = I + J(Q^{\#})_{dg} - MW = I + \mathbf{1}w^t(MW - I) - MW$ and simplifying this last yields

$$Q^{\#} = (I - \mathbf{1}w^t)(I - MW).$$

Since the stationary distribution vector w can be recovered from the diagonal entries of M, we thus find that we can reconstruct $Q^{\#}$ from the mean first passage matrix M.

The following is immediate from Theorem 6.1.1.

COROLLARY 6.1.3 *Suppose that $T \in \mathcal{IS}_n$, with stationary vector w*

and mean first passage matrix M. Let $Q = I - T$. Then for each pair of distinct indices i, j between 1 and n, we have

$$m_{i,j} = \frac{q^{\#}_{j,j} - q^{\#}_{i,j}}{w_j}. \tag{6.5}$$

EXAMPLE 6.1.4 Here we revisit the simple random walk on the path on n vertices, discussed in Example 5.5.1. Let T be the stochastic matrix given in (5.52); it is straightforward to determine that the corresponding stationary vector w is given by

$$w_j = \begin{cases} \frac{1}{2(n-1)} & \text{if } j = 1, n, \\ \frac{1}{n-1} & \text{if } 2 \le j \le n-1. \end{cases}$$

Our goal is to use (6.5) in order to find the mean first passage time $m_{i,j}$ between distinct states i and j.

Fix an index j between 2 and n, and an index $i \neq j$. From (5.53) it follows that

$$q^{\#}_{j,j} - q^{\#}_{i,j} = 2(\max\{i,j\} - j) + \frac{(j-i)(j+i-2)}{n-1},$$

which it turn yields

$$m_{i,j} = 2(n-1)(\max\{i,j\} - j) + (j-i)(j+i-2).$$

Similarly, for an index $i \neq 1$, it follows from (5.53) that $m_{i,1} = (i-1)(2n-i-1)$.

It is straightforward to determine from the formulas above that the maximum entries in the mean first passage matrix M are in the $(1,n)$ and $(n,1)$ positions, with $\max_{i,j=1,\ldots,n} m_{i,j} = (n-1)^2$. Evidently the minimum entry in M is 1, attained at positions $(1,2)$ and $(n, n-1)$.

Recall that in Chapter 2, we took an interest in the signs of the off-diagonal entries of $(I - T)^{\#}$ for an irreducible stochastic matrix T. In particular, Question 3.3.8 asks for a characterisation of the case that all off-diagonal entries of $(I - T)^{\#}$ are nonpositive. The following result, which appears in a paper of Catral, Neumann and Xu [18], provides an interpretation of the situation that $(I - T)^{\#}_{i,j} \le 0$ in terms of mean first passage times.

COROLLARY 6.1.5 *Suppose that the matrix $T \in \mathcal{IS}_n$ has stationary vector w and mean first passage matrix M. Let $Q = I - T$. Fix a pair of distinct indices i and j between 1 and n. Then $q^{\#}_{i,j} \le 0$ if and only if $m_{i,j} \ge \sum_{k=1,\ldots,n, k\neq j} w_k m_{k,j}$.*

Proof: From Corollary 6.1.3 we have $m_{i,j} = \frac{q^{\#}_{j,j} - q^{\#}_{i,j}}{w_j}$, so $q^{\#}_{i,j} \le 0$ if and only if $m_{i,j} \ge \frac{q^{\#}_{j,j}}{w_j}$. Next, we find from Theorem 6.1.1 that $w^t M e_j = w^t(I - Q^{\#} + JQ^{\#}_{dg})W^{-1}e_j$, where $W = \text{diag}(w)$. Simplifying this last, we see that $\sum_{k=1}^{n} w_k m_{k,j} = w^t M e_j = \frac{q^{\#}_{j,j}}{w_j} + 1$. Recalling that $m_{j,j} = \frac{1}{w_j}$, we thus find that $\frac{q^{\#}_{j,j}}{w_j} = \sum_{k=1,\ldots,n,k\ne j} w_k m_{k,j}$.

Assembling these observations, we thus find that $q^{\#}_{i,j} \le 0$ if and only if $m_{i,j} \ge \sum_{k=1,\ldots,n,k\ne j} w_k m_{k,j}$, as desired. □

Our next result describes the behaviour of the mean first passage matrix when the underlying stochastic matrix is perturbed. The same problem is addressed from an algorithmic perspective by Hunter in [59].

THEOREM 6.1.6 *Suppose that T is an irreducible stochastic matrix with stationary vector w, and set $Q = I - T$. Suppose that $\tilde{T}$ is another irreducible stochastic matrix, and write $\tilde{T} = T + E$. Let $\tilde{M}$ denote the mean first passage matrix corresponding to $\tilde{T}$. Then*

$$\tilde{M} = (I - Q^{\#}(I - EQ^{\#})^{-1} + J(Q^{\#}(I - EQ^{\#})^{-1})_{dg})\tilde{W}^{-1},$$

where $\tilde{W}$ is the diagonal matrix whose diagonal entries are the corresponding entries of $w^t(I - EQ^{\#})^{-1}$.

Proof: From Theorem 5.3.1 we find that the stationary vector for $\tilde{T}$ is given by $w^t(I - EQ^{\#})^{-1}$. Letting $\tilde{Q} = I - \tilde{T}$, it follows from Theorem 5.3.30 that

$$-\tilde{Q}^{\#} + J(\tilde{Q}^{\#})_{dg} = -Q^{\#}(I - EQ^{\#})^{-1} + J(Q^{\#}(I - EQ^{\#})^{-1})_{dg}.$$

The conclusion now follows readily from Theorem 6.1.1. □

In the special case that only one row of T is perturbed, we have the following result.

PROPOSITION 6.1.7 *Let T be a matrix in $\mathcal{IS}_n$ with stationary vector w and mean first passage matrix M. Suppose that $\tilde{T} = T + e_i u^t$ is another irreducible stochastic matrix, and denote its mean first passage matrix by $\tilde{M}$. Let $Q = I - T$. Fix a pair of distinct indices k, j between 1 and n, with $k, j \ne i$, and let*

$$\theta = \frac{u^t Q^{\#} e_j}{1 - u^t Q^{\#} e_i}.$$

Then

$$\tilde{m}_{k,j} = \frac{w_j}{w_j + \theta w_i} m_{k,j} + \frac{\theta w_i}{w_j + \theta w_i}(m_{k,i} - m_{k,j}).$$

Also, for any index $k \neq i$, $\tilde{m}_{k,i} = m_{k,i}$.

Proof: Letting $\tilde{Q} = I - \tilde{T}$, we find from Corollary 6.1.3 that

$$\tilde{m}_{k,j} = \frac{\tilde{q}^{\#}_{j,j} - \tilde{q}^{\#}_{k,j}}{\tilde{w}_j}, \tag{6.6}$$

where $\tilde{w}$ denotes the stationary vector for $\tilde{T}$. Observing that

$$(I - e_i u^t Q^{\#})^{-1} = I + \frac{1}{1 - u^t Q^{\#} e_i} e_i u^t Q^{\#},$$

we find from Theorem 5.3.1 that $\tilde{w}_j = w_j + \theta w_i$. Similarly from Theorem 5.3.30 it follows that $\tilde{q}^{\#}_{j,j} - \tilde{q}^{\#}_{k,j} = q^{\#}_{j,j} - q^{\#}_{k,j} + \frac{u^t Q^{\#} e_j}{1 - u^t Q^{\#} e_i}(q^{\#}_{j,i} - q^{\#}_{k,i})$. Observe that $q^{\#}_{j,j} - q^{\#}_{k,j} = w_j m_{k,j}$, while

$$q^{\#}_{j,i} - q^{\#}_{k,i} = w_i(q^{\#}_{i,i} - q^{\#}_{k,i} - (q^{\#}_{i,i} - q^{\#}_{j,i})) = w_i(m_{k,i} - m_{j,i}).$$

Substituting the various expressions into (6.6) now yields the desired expression.

Finally suppose that $k \neq i$. Since the submatrix formed from $\tilde{T}$ by deleting row and column i coincides with the corresponding submatrix of T, it follows from (5.2) that $\tilde{m}_{k,i} = m_{k,i}$. □

EXAMPLE 6.1.8 Let $T = \frac{1}{n-1}(J - I)$, where J is the $n \times n$ all ones matrix. We find readily that the stationary distribution vector for T is $\frac{1}{n}\mathbf{1}$, and that $(I - T)^{\#} = \frac{n}{n-1}(I - \frac{1}{n}J)$; applying Theorem 6.1.1 we find that the mean first passage matrix for T is $M = I + (n-1)J$. Now let $\tilde{T} = T + \frac{1}{n-1} e_1(e_1 - e_n)^t$, and let $\tilde{M}$ denote the corresponding mean first passage matrix. Observe that $\tilde{T}$ is formed from T by taking the mass on the $(1, n)$ entry and shifting it to the $(1, 1)$ entry.

Fix a pair of distinct indices k, j with $k, j \neq 1$, and note that $\frac{1}{n-1}(e_1 - e_n)^t(I - T)^{\#} e_j = \frac{1}{n}(e_1 - e_n)^t e_j$. In particular, for $j \neq n$, we have $\frac{1}{n-1}(e_1 - e_n)^t(I - T)^{\#} e_j = 0$, and it now follows from Proposition 6.1.7 and $\tilde{m}_{k,j} = m_{k,j} = n - 1$. Also, we find from Proposition 6.1.7 that for each $k = 2, \ldots, n$, $\tilde{m}_{k,1} = m_{k,1}$. Consequently we see that for any index $j \in \{1, \ldots, n-1\}$ and any index $k \neq j$, the perturbation taking

T to $\tilde{T}$ does not affect the mean first passage time from state k to state j.

In the case that $j = n$ and $k \neq 1, n$, we find that $\frac{1}{n-1}(e_1 - e_n)^t(I - T)^\# e_j = -\frac{1}{n}$. Again appealing to Proposition 6.1.7, it follows that $\tilde{m}_{k,n} = \frac{n-1}{n-2} m_{k,n} = \frac{(n-1)^2}{n-2}$. Further, using (6.1) (with $i = 1, j = n$) in conjunction with the values $\tilde{m}_{k,n}, k = 2 \ldots, n-1$ determined above, we find readily that $\tilde{m}_{1,n} = \frac{n(n-1)}{n-2} > m_{1,n}$. Thus, as might be expected, the perturbation taking T to $\tilde{T}$ has the effect of increasing the mean first passage time from any state $k \in \{1, \ldots, n-1\}$ to state n.

Our next result involves the Kemeny constant, which we saw in section 5.3.

THEOREM 6.1.9 ([66, section 4.4]) *Suppose that $T \in \mathcal{IS}_n$, with stationary vector w and mean first passage matrix M. Set $Q = I - T$. Then*

$$Mw = (\text{trace}(Q^\#) + 1)\mathbf{1}. \tag{6.7}$$

In particular, denoting the eigenvalues of T by $1, \lambda_2, \ldots, \lambda_n$, we have

$$Mw = \left(1 + \sum_{k=2}^{n} \frac{1}{1-\lambda_j}\right)\mathbf{1}.$$

Proof: From Theorem 6.1.1, we have $M = (I - Q^\# + J(Q^\#)_{dg})W^{-1}$, where $W = \text{diag}(w)$. Hence, $Mw = (I - Q^\# + J(Q^\#)_{dg})W^{-1}w = (I - Q^\# + J(Q^\#)_{dg})\mathbf{1} = (1 + \mathbf{1}^t(Q^\#)_{dg}\mathbf{1})\mathbf{1}$. Since $\mathbf{1}^t(Q^\#)_{dg}\mathbf{1} = \text{trace}(Q^\#)$, the conclusions follow. □

In view of Theorem 5.2.2, one might suspect that for an irreducible and periodic stochastic matrix, the corresponding mean first passage matrix will have a tractable structure. Our next result, which appears in [68], confirms that suspicion.

THEOREM 6.1.10 *Let $T \in \mathcal{IS}_n$ be periodic with period $d \geq 2$. Suppose that T is in periodic normal form, with*

$$T = \left[\begin{array}{c|c|c|c|c|c} 0 & T_1 & 0 & 0 & \cdots & 0 \\ \hline 0 & 0 & T_2 & 0 & \cdots & 0 \\ \hline \vdots & & \ddots & \ddots & & \vdots \\ \hline 0 & \cdots & & 0 & T_{d-2} & 0 \\ \hline 0 & 0 & \cdots & & 0 & T_{d-1} \\ \hline T_d & 0 & 0 & \cdots & 0 & 0 \end{array}\right]. \tag{6.8}$$

Write $S_i = T_iT_{i+1}\dots T_dT_1\dots T_{i-1}, i = 1,\dots,d,$ *let* w_1 *denote the stationary distribution vector for* S_1*, and for each* $j = 2,\dots,d,$ *let* w_j *be given by* $w_j^t = w_1^tT_1T_2\dots T_{j-1}$. *For each* $j = 1,\dots,d,$ *let* $W_j = \mathrm{diag}(w_j)$.

Consider the mean first passage matrix M *associated with* T*, and partition* M *conformally with* T *as a* $d \times d$ *block matrix, with blocks* $M_{i,j}, i,j = 1,\dots,d$. *Then we have the following:*

$$\begin{aligned} M_{i,j} &= d(-T_i\dots T_{j-1}(I-S_j)^{\#} + J((I-S_j)^{\#})_{dg})W_j^{-1} \\ &\qquad +(j-i)J \text{ if } i<j, \\ M_{i,j} &= d(I-(I-S_i)^{\#} + J((I-S_i)^{\#})_{dg})W_i^{-1} \text{ if } i=j, \\ M_{i,j} &= d(-T_i\dots T_dT_1\dots T_{j-1}(I-S_j)^{\#} + J((I-S_j)^{\#})_{dg})W_j^{-1} \\ &\qquad +(d+j-i)J \text{ if } i>j. \end{aligned} \quad (6.9)$$

Alternatively, denoting the mean first passage matrix for S_i *by* $M_{S_i}, i = 1,\dots,d,$ *we have*

$$M_{i,j} = \begin{cases} dT_i\dots T_{j-1}(M_{S_j} - W_j^{-1}) + (j-i)J & \text{if } i<j, \\ dM(S_i) & \text{if } i=j, \\ dT_i\dots T_dT_1\dots T_{j-1}(M_{S_j} - W_j^{-1}) + (d+j-i)J & \text{if } i>j. \end{cases} \quad (6.10)$$

Proof: First, observe that the stationary distribution vector for T is given by

$$w = \frac{1}{d}\begin{bmatrix} w_1 \\ \hline w_2 \\ \hline \vdots \\ \hline w_d \end{bmatrix}.$$

Partition $(I-T)^{\#}$ conformally with T, denoting the blocks by $(I-T)^{\#}_{i,j}, i,j = 1,\dots,d$. Suppose for concreteness that $i<j$. Applying Theorem 6.1.1, it follows that

$$M_{i,j} = (-(I-T)^{\#}_{i,j} + J((I-T)^{\#}_{j,j})_{dg})\left(\frac{1}{d}W_j\right)^{-1}. \quad (6.11)$$

From the formula (5.7) for the blocks of $(I-T)^{\#}$ in Theorem 5.2.2, we find that $(I-T)^{\#}_{i,j} = (I-S_i)^{\#}T_i\dots T_{j-1} + (\frac{d-1}{2d} - \frac{j-i}{d})\mathbf{1}w_j^t$, and applying Lemma 5.2.1 yields $(I-S_i)^{\#}T_i\dots T_{j-1} = T_i\dots T_{j-1}(I-S_j)^{\#}$. Similarly, $(I-T)^{\#}_{j,j} = (I-S_j)^{\#} + \frac{d-1}{2d}\mathbf{1}w_j^t$. Substituting in these expressions into (6.11) and simplifying now yields the fact that

$$M_{i,j} = d(-T_i\dots T_{j-1}(I-S_j)^{\#} + J((I-S_j)^{\#})_{dg})W_j^{-1} + (j-i)J.$$

The cases that $i = j$ and $i > j$ are handled similarly, thus establishing (6.9).

Applying Theorem 6.1.1 to the matrix S_j, we find that $M_{S_j} = (I - (I - S_j)^{\#} + J((I - S_j)^{\#})_{dg})W_j^{-1}$; substituting that expression into (6.9) now yields (6.10). □

REMARK 6.1.11 As noted in [68], (6.10) admits a natural probabilistic interpretation in terms of the underlying periodic Markov chain. For concreteness, suppose that the partitioning of T as in (6.8) corresponds to a partitioning of the set $\{1, \ldots, n\}$ into subsets $U_1, \ldots, U_d$, where $t_{i,j} > 0$ only if for some l between 1 and d we have $i \in U_l, j \in U_{(l+1) \bmod d}$.

If we have indices $p, q \in U_i$, observe that the Markov chain can make a k-step transition from state p to state q only in the case that k is a multiple of d. Note also that the d-step transition matrix for states in U_i is S_i. From these two observations, it follows that the mean first passage time from p to q is given by the corresponding entry in dM_{S_i}.

Next suppose that we have indices $p \in U_i$ and $q \in U_j$ with $i < j$. The Markov chain can make a k-step transition from p to q only if k is congruent to $(j - i) \mod d$, and as above, the d-step transition matrix for states in U_j is S_j. It follows that the mean first passage time from state p to state q can be written as

$$\begin{aligned}
&(j-i)(\text{probability of a transition from } p \text{ to } q \text{ in } j-i \text{ steps}) + \\
&\sum_{k \in U_j \setminus \{q\}} [(\text{probability of a transition from } p \text{ to } k \text{ in } j-i \text{ steps}) \\
&\qquad \times (j - i + \text{ mean first passage time from } k \text{ to } q)]. \qquad (6.12)
\end{aligned}$$

It is now straightforward to verify that (6.12) coincides with the entry of $dT_i \ldots T_{j-1}(M(S_j) - W_j^{-1}) + (j - i)J$ in the position corresponding to (p, q). A similar interpretation of (6.10) goes through if $p \in U_i$ and $q \in U_j$ with $i > j$.

Our next example revisits the simple random walk on the graph depicted in Figure 5.1.

EXAMPLE 6.1.12 Let T denote the stochastic matrix of Example

5.2.4, and recall that we have

$$(I-T)^{\#} = \left[\begin{array}{c|c|c} 2I - \frac{7}{4k}J & I - \frac{9}{8k}J & -\frac{1}{8}\mathbf{1} \\ \hline 2I - \frac{9}{4k}J & 2I - \frac{11}{8k}J & -\frac{3}{8}\mathbf{1} \\ \hline -\frac{1}{4k}\mathbf{1}^t & -\frac{3}{8k}\mathbf{1}^t & \frac{5}{8} \end{array}\right],$$

where the diagonal blocks are of orders k, k, and 1, respectively. It is readily determined that the stationary vector w is given by

$$w = \frac{1}{4k}\left[\begin{array}{c} 21 \\ \hline 1 \\ \hline k \end{array}\right],$$

where here our partitioning is conformal with that of $(I-T)^{\#}$.

Next, we apply Theorem 6.1.1 to find that the mean first passage matrix M is as follows:

$$M = \left[\begin{array}{c|c|c} 4kJ - 2kI & (8k-1)J - 4kI & 3\mathbf{1} \\ \hline (4k+1)J - 4kI & 8kJ - 4kI & 4\mathbf{1} \\ \hline (4k-3)\mathbf{1}^t & (8k-4)\mathbf{1}^t & 4 \end{array}\right]. \tag{6.13}$$

Observe that the smallest entry in M is 1, occurring in positions $(k+i, i), i = 1, \dots, k$; this comes as no surprise in view of the fact that $t_{k+i,i} = 1$ for each $i = 1, \dots, k$. We also note that the maximum entry in M is $8k$, occurring in positions $(k+i, k+j)$, where $i, j = 1, \dots, k$ and $i \neq j$. Again, in view of the fact that such pairs of indices correspond to distinct pendent vertices of the graph shown in Figure 5.1, this is not so surprising.

EXAMPLE 6.1.13 In this example we consider the Ehrenfest model for diffusion. Suppose that $n \in \mathbb{N}$ is given, and that we have a collection of n balls, which have been distributed between two urns. We consider the following Markov chain on $n+1$ states, where the state of the chain is the number of balls in the left–hand urn: at each time step, a ball is chosen at random from the collection, and moved from one urn to the other. We index the states of the Markov chain with the numbers from 1 to $n+1$, with index j corresponding to the case that there are $j-1$ balls in the left-hand urn. See the paper of Krafft and Schaefer [88] for further details.

The transition matrix T for our Markov chain is $(n+1) \times (n+1)$, and can be written as

$$T = \begin{bmatrix} 0 & 1 & 0 & 0 & \dots & 0 & 0 & 0 \\ \frac{1}{n} & 0 & \frac{n-1}{n} & 0 & 0 & \dots & 0 & 0 \\ 0 & \frac{2}{n} & 0 & \frac{n-2}{n} & 0 & \dots & 0 & 0 \\ 0 & 0 & \frac{3}{n} & 0 & \frac{n-3}{n} & 0 & \dots & 0 \\ \vdots & & & \ddots & \ddots & \ddots & & \vdots \\ 0 & 0 & \dots & 0 & \frac{n-2}{n} & 0 & \frac{2}{n} & 0 \\ 0 & 0 & \dots & 0 & 0 & \frac{n-1}{n} & 0 & \frac{1}{n} \\ 0 & 0 & 0 & \dots & 0 & 0 & 1 & 0 \end{bmatrix}.$$

It is readily verified that the stationary vector for T is given by

$$w = \frac{1}{2^n} \begin{bmatrix} \binom{n}{0} \\ \binom{n}{1} \\ \binom{n}{2} \\ \vdots \\ \binom{n}{n-1} \\ \binom{n}{n} \end{bmatrix}.$$

From the expression for w, we find that for each $j = 1, \dots, n+1$, the mean first return time to state $j, m_{j,j}$, is given by

$$m_{j,j} = \frac{2^n}{\binom{n}{j-1}}.$$

Our goal in this example is to use Corollary 6.1.3 in order to determine the mean first passage times $m_{i,j}, i, j = 1, \dots, n+1$, when $i \neq j$.

Partition off the last row and column of T, so that T is written as

$$T = \left[\begin{array}{c|c} T_{1,1} & T_{1,2} \\ \hline T_{2,1} & T_{2,2} \end{array} \right].$$

For each $j = 1, \dots, n$, set $\delta_j = \sum_{l=j}^{n} \prod_{k=1}^{l-1} \left(\frac{k}{n-k} \right)$, where we define the empty product (corresponding to an upper index of 0) to be 1. We note in passing that the δ_js are decreasing with j. Next, we construct the $n \times n$ matrix U such that for each pair of indices i, j, between 1 and n, $u_{i,j} = \binom{n}{j-1} \delta_{\max\{i,j\}}$. It is readily verified that $(I - T_{1,1})U = I$, so that $(I - T_{1,1})^{-1} = U$. From Proposition 2.5.1, it follows that for any $i, j = 1, \dots, n$,

$$q^{\#}_{j,j} - q^{\#}_{i,j} = u_{j,j} - u_{i,j} - \frac{1}{1 + T_{2,1} U \mathbf{1}} (e_j - e_i)^t U \mathbf{1} T_{2,1} U e_j.$$

Since $T_{2,1} = e_n^t$ and $u_{n,k} = \binom{n}{k-1}, k = 1, \ldots, n,$ we find that $1 + T_{2,1}U\mathbf{1} = 2^n$. We also have $T_{2,1}Ue_j = u_{n,j} = \binom{n}{j-1}, j = 1, \ldots, n$. Observe also that for each $k = 1, \ldots, n, (e_j - e_i)^t Ue_k = \binom{n}{k-1}\left(\delta_{\max\{j,k\}} - \delta_{\max\{i,k\}}\right)$. Consequently, we find that if $i < j$,

$$(e_j - e_i)^t Ue_k = \binom{n}{k-1}\begin{cases} 0 & \text{if } i < j \le k, \\ \delta_j - \delta_k & \text{if } i \le k \le j, \\ \delta_j - \delta_i & \text{if } k \le i < j, \end{cases}$$

while if $j < i$,

$$(e_j - e_i)^t Ue_k = \binom{n}{k-1}\begin{cases} 0 & \text{if } j < i \le k, \\ \delta_k - \delta_i & \text{if } j \le k \le i, \\ \delta_j - \delta_i & \text{if } k \le j < i. \end{cases}$$

It now follows that

$$(e_j - e_i)^t U\mathbf{1} = \begin{cases} \sum_{k=1}^{i-1}\binom{n}{k-1}(\delta_j - \delta_i) + \sum_{k=i}^{j}\binom{n}{k-1}(\delta_j - \delta_k) & \text{if } i < j, \\ \sum_{k=1}^{j-1}\binom{n}{k-1}(\delta_j - \delta_i) + \sum_{k=j}^{i}\binom{n}{k-1}(\delta_k - \delta_i) & \text{if } j < i. \end{cases}$$

Assembling the above quantities, we now find that for $i, j = 1, \ldots, n$ and $i \ne j$,

$$\begin{aligned} m_{i,j} &= \frac{q^{\#}_{j,j} - q^{\#}_{i,j}}{w_j} \\ &= \begin{cases} \sum_{k=1}^{i-1}\binom{n}{k-1}(\delta_i - \delta_j) + \sum_{k=i}^{j}\binom{n}{k-1}(\delta_k - \delta_j) & \text{if } i < j, \\ 2^n(\delta_j - \delta_i) - \sum_{k=1}^{j-1}\binom{n}{k-1}(\delta_j - \delta_i) - \sum_{k=j}^{i}\binom{n}{k-1}(\delta_k - \delta_i) & \text{if } j < i. \end{cases} \end{aligned} \tag{6.14}$$

In a similar way, we find that for each $i = 1, \ldots, n$,

$$m_{i,n+1} = \sum_{k=1}^{i}\binom{n}{k-1}\delta_i + \sum_{k=i+1}^{n}\binom{n}{k-1}\delta_k \tag{6.15}$$

and for each $j = 1, \ldots, n$,

$$m_{n+1,j} = 2^n\delta_j - \sum_{k=1}^{j}\binom{n}{k-1}\delta_j - \sum_{k=i+1}^{n}\binom{n}{k-1}\delta_k. \tag{6.16}$$

We note that expressions for these mean first passage times have been developed using other techniques in [88], for instance.

As a small concrete example, we find that for $n = 6$, the mean first passage matrix for the Ehrenfest model is given by

$$M = \begin{bmatrix} 64 & 1 & 2.4 & 4.6 & 8.8 & 20.2 & 83.2 \\ 63 & 10.\overline{6} & 1.4 & 3.6 & 7.8 & 19.2 & 82.2 \\ 74.4 & 11.4 & 4.2\overline{6} & 2.2 & 6.4 & 17.8 & 80.8 \\ 78.6 & 15.6 & 4.2 & 3.2 & 4.2 & 15.6 & 78.6 \\ 80.8 & 17.8 & 6.4 & 2.2 & 4.2\overline{6} & 11.4 & 74.4 \\ 82.2 & 19.2 & 7.8 & 3.6 & 1.4 & 10.\overline{6} & 63 \\ 83.2 & 20.2 & 8.8 & 4.6 & 2.4 & 1 & 64 \end{bmatrix}.$$

Observe that as we descend each column of M, the entries are decreasing above the main diagonal, and increasing below the main diagonal. Indeed it is straightforward to show from (6.14), (6.15) and (6.16), that this property for the mean first passage matrix holds for the Ehrenfest model on $n+1$ states for any $n \in \mathbb{N}$. In view of the structure of the corresponding transition matrix T, this comes as no surprise.

Note that in Examples 6.1.4 and 6.1.13, the transition matrices in question are tridiagonal. In [88], the authors provide an approach for computing the mean first passage matrix associated with any irreducible stochastic tridiagonal matrix. Thus the techniques in that paper provide an alternate approach to Examples 6.1.4 and 6.1.13.

6.2 A proximity inequality for the group inverse

In this section we investigate a natural relationship between first passage times, and its consequences for entries in the group inverse associated with an irreducible singular M-matrix. We begin with a triangle inequality for mean first passage times found in [59], where the inequality is derived by probabilistic reasoning. Here we provide a matrix-theoretic proof, which also facilitates a characterisation of the equality case. Our approach borrows some of the ideas used in [69].

THEOREM 6.2.1 *Let $T \in \mathcal{IS}_n$, and denote the corresponding mean first passage matrix by M. Then for any triple of indices i, j, k, we have*

$$m_{i,j} + m_{j,k} \geq m_{i,k}. \tag{6.17}$$

Equality holds in (6.17) if and only if j is distinct from both i and k, and in addition, every path in $\mathcal{D}(T)$ from vertex i to vertex k passes through vertex j.

Proof: It is straightforward to see that (6.17) holds with strict inequality if either $j = i$ or $j = k$, so henceforth we suppose that j is distinct from both i and k.

We first consider the case that $i \neq k$, and for concreteness we take $j = n-1, k = n$ and $1 \leq i \leq n-2$. Partition out the last two rows and columns of T, and write T as

$$T = \left[\begin{array}{c|cc} S & y_1 & y_2 \\ \hline x_1^t & a_1 & b_1 \\ x_2^t & b_2 & a_2 \end{array}\right].$$

Referring (5.2), we see that we need expressions for $(I - T_{(l)})^{-1}\mathbf{1}$ for $l = n-1, n$, where we recall that $T_{(l)}$ denotes the principal submatrix of T formed by deleting the l-th row and column. For $p = 1, 2$, set

$$z_p = \frac{1}{1 - a_p - x_p^t(I-S)^{-1}y_p}.$$

From the formula for the inverse of a partitioned matrix (see [57, section 0.7.3]), we find that

$$(I - T_{(n)})^{-1} = \left[\begin{array}{c|c} (I-S)^{-1} + z_1(I-S)^{-1}y_1x_1^t(I-S)^{-1} & z_1(I-S)^{-1}y_1 \\ \hline z_1x_1^t(I-S)^{-1} & z_1 \end{array}\right]$$

and

$$(I - T_{(n-1)})^{-1} = \left[\begin{array}{c|c} (I-S)^{-1} + z_2(I-S)^{-1}y_2x_2^t(I-S)^{-1} & z_2(I-S)^{-1}y_2 \\ \hline z_2x_2^t(I-S)^{-1} & z_2 \end{array}\right].$$

Setting $c_p = x_p^t(I-S)^{-1}\mathbf{1} + 1, p = 1, 2$, it now follows that

$$(I - T_{(n)})^{-1}\mathbf{1} = \left[\begin{array}{c} (I-S)^{-1}\mathbf{1} + z_1c_1(I-S)^{-1}y_1 \\ \hline z_1c_1 \end{array}\right] \text{ and}$$

$$(I - T_{(n-1)})^{-1}\mathbf{1} = \left[\begin{array}{c} (I-S)^{-1}\mathbf{1} + z_2c_2(I-S)^{-1}y_2 \\ \hline z_2c_2 \end{array}\right].$$

With this notation in hand, we see that $m_{i,n-1} + m_{n-1,n} \geq m_{i,n}$ if and only if

$$e_i^t(I-S)^{-1}\mathbf{1}+z_2c_2e_i^t(I-S)^{-1}y_2+z_1c_1 \geq e_i^t(I-S)^{-1}\mathbf{1}+z_1c_1e_i^t(I-S)^{-1}y_1,$$

i.e., if and only if

$$z_2c_2e_i^t(I-S)^{-1}y_2 + z_1c_1 \geq z_1c_1e_i^t(I-S)^{-1}y_1.$$

Since $S\mathbf{1} + y_1 + y_2 = \mathbf{1}$, we find readily that

$$(I-S)^{-1}y_1 = \mathbf{1} - (I-S)^{-1}y_2,$$

so that the inequality $m_{i,n-1} + m_{n-1,n} \geq m_{i,n}$ is equivalent to the inequality

$$z_2c_2e_i^t(I-S)^{-1}y_2 + z_1c_1 \geq z_1c_1(1 - e_i^t(I-S)^{-1}y_2).$$

As this last inequality certainly holds, we conclude that $m_{i,n-1} + m_{n-1,n} \geq m_{i,n}$.

From our analysis above, we find that $m_{i,n-1}+m_{n-1,n} = m_{i,n}$ if and only if $e_i^t(I-S)^{-1}y_2 = 0$. From the fact that $(I-S)^{-1} = \sum_{p=0}^{\infty} S^p$, it follows that $(I-S)^{-1}$ has a positive entry in the position (i,l) if and only if there is a path from vertex i to vertex l in $\mathcal{D}(T)$ that does not pass through vertices $n-1$ or n. Since the l–th entry of y_2 is positive if and only if $\mathcal{D}(T)$ contains the arc $l \to n$, we deduce that $e_i^t(I-S)^{-1}y_2 > 0$ if and only if there is a path in $\mathcal{D}(T)$ from vertex i to vertex n that does not pass through vertex $n-1$. Hence, $e_i^t(I-S)^{-1}y_2 = 0$ if and only if every path in $\mathcal{D}(T)$ from i to n passes through $n-1$.

Finally we consider the case that $i = k \neq j$, and without loss of generality we take $i = k = n$ and $j = n-1$. Recall that $m_{n,n}$ is the reciprocal of the n–th entry in the stationary distribution for T. It now follows that, with the notation above, $m_{n,n} = \frac{z_1c_1+z_2c_2}{z_2}$. The inequality $m_{n,n-1} + m_{n-1,n} \geq m_{n,n}$ is then equivalent to

$$z_2c_2 + z_1c_1 \geq \frac{z_1c_1 + z_2c_2}{z_2};$$

as $z_2 = \frac{1}{1-a_2-x_2^t(I-S)^{-1}y_2} \geq 1$, it now follows that $m_{n,n-1} + m_{n-1,n} \geq m_{n,n}$. Observe that from this analysis, we see that $m_{n,n-1} + m_{n-1,n} = m_{n,n}$ if and only if $a_2 = 0$ and $x_2^t(I-S)^{-1}y_2 = 0$. Note that $a_2 = 0$ if and only if $\mathcal{D}(T)$ does not contain a loop at vertex n, while $x_2^t(I-S)^{-1}y_2 = 0$ if and only if there is no path in $\mathcal{D}(T)$ from vertex n back to itself that does not pass through vertex $n-1$. Hence we find that

$m_{n,n-1} + m_{n-1,n} = m_{n,n}$ if and only if every path $\mathcal{D}(T)$ from vertex n back to vertex n must pass through vertex $n-1$. □

Theorem 6.2.1 has the following immediate consequence, which appears in [18].

COROLLARY 6.2.2 *Let $T \in \mathcal{IS}_n$, and denote its stationary vector by w. Setting $Q = I - T$, we have, for any triple of indices $i, j, k = 1, \ldots, n$, that*

$$\frac{q^{\#}_{j,j}}{w_j} - \frac{q^{\#}_{i,j}}{w_j} + \frac{q^{\#}_{i,k}}{w_k} - \frac{q^{\#}_{j,k}}{w_k} \geq 0. \tag{6.18}$$

Equality holds in (6.18) if and only if either $i = j$, or $j = k$, or j is distinct from both i and k and in addition, every path in $\mathcal{D}(T)$ from vertex i to vertex k passes through vertex j.

Proof: Observe that both sides of (6.18) are zero if $i = j$ or $j = k$. We suppose henceforth that $j \neq i, k$, and denote the mean first passage matrix for T by M.

If $i = k$, then from Corollary 6.1.3, we have

$$\frac{q^{\#}_{j,j}}{w_j} - \frac{q^{\#}_{i,j}}{w_j} + \frac{q^{\#}_{i,k}}{w_k} - \frac{q^{\#}_{j,k}}{w_k} = \frac{q^{\#}_{j,j} - q^{\#}_{k,j}}{w_j} + \frac{q^{\#}_{k,k} - q^{\#}_{j,k}}{w_k} = m_{j,k} + m_{k,j} \geq 2.$$

If the indices i, j, k are all distinct, then again from Corollary 6.1.3, we have

$$\frac{q^{\#}_{j,j}}{w_j} - \frac{q^{\#}_{i,j}}{w_j} + \frac{q^{\#}_{i,k}}{w_k} - \frac{q^{\#}_{j,k}}{w_k} = \frac{q^{\#}_{j,j} - q^{\#}_{i,j}}{w_j} + \frac{q^{\#}_{j,j} - q^{\#}_{j,k}}{w_k} - \frac{q^{\#}_{j,j} - q^{\#}_{i,k}}{w_k} = m_{i,j} + m_{j,k} - m_{i,k}.$$

The inequality (6.18), as well as the characterisation of the equality case, now follow from Theorem 6.2.1. □

The inequality (6.18) may be recast as a statement for irreducible nonnegative matrices in two different ways, as the following result (from [18]) shows.

PROPOSITION 6.2.3 *Suppose that $A \in \Phi^{n,n}$, with right and left Perron vectors x and y, respectively, normalised so that $y^t x = 1$. Denote* *the Perron value of A by $r(A)$, and let $Q = r(A)I - A$. Then for indices $i, j, k = 1, \dots, n$, we have*

$$\frac{q^{\#}_{j,j}}{x_j y_j} - \frac{q^{\#}_{i,j}}{x_i y_j} + \frac{q^{\#}_{i,k}}{x_i y_k} - \frac{q^{\#}_{j,k}}{x_j y_k} \geq 0, \tag{6.19}$$

and

$$\frac{1}{x_j y_j}\frac{\partial^2 r(A)}{\partial^2_{j,j}} - \frac{1}{x_i y_j}\frac{\partial^2 r(A)}{\partial^2_{j,i}} + \frac{1}{x_i y_k}\frac{\partial^2 r(A)}{\partial^2_{k,i}} - \frac{1}{x_j y_k}\frac{\partial^2 r(A)}{\partial^2_{k,j}} \geq 0. \tag{6.20}$$

Equality holds in either (6.19) or (6.20) if and only if either $i = j$, or $j = k$, or j is distinct from both i and k and in addition, every path in $\mathcal{D}(T)$ from vertex i to vertex k passes through vertex j.

Proof: Let $X = \text{diag}(x)$. Then the matrix $\tilde{A} = \frac{1}{r(A)} X^{-1}AX$ is irreducible and stochastic, with stationary vector Xy. From Remark 2.5.3 we find that $Q^{\#} = \frac{1}{r(A)} X(I - \tilde{A})^{\#} X^{-1}$. Hence for any pair of indices p, q we have

$$\frac{q^{\#}_{p,q}}{x_p y_q} = \frac{1}{r(A) x_q y_q}\left((I - \tilde{A})^{\#}\right)_{p,q}.$$

The inequality (6.19), along with the characterisation of the equality case, now follows from Corollary 6.2.2.

Similarly, for any pair of indices p, q, we find from Corollary 3.3.1 that

$$q^{\#}_{p,q} = \frac{2}{x_p y_q}\frac{\partial^2 r(A)}{\partial^2_{q,p}}.$$

Inequality (6.20) and the equality characterisation now follow from (6.19). □

Next we show how the ideas above can be used to generalise a result in a paper of Chebotarev and Shamis [20] concerning the so–called proximity between two vertices in a connected undirected graph. In order to frame that result, we need to introduce a little notation. Suppose that we have a connected undirected graph $\mathcal{G}$ on vertices labelled $1, \dots, n$. Suppose further that we have a function w that assigns each edge of $\mathcal{G}$ a positive number known as the *weight* of the edge. For each pair of adjacent vertices i, j in $\mathcal{G}$ we denote the weight of the corresponding edge by $w_{i,j}$, and for each $i = 1, \dots, n$, we let d_i denote the sum of the weights of the edges incident with vertex i. The pair $(\mathcal{G}, w)$ is called a *weighted graph.* The *Laplacian matrix for* the weighted graph $(\mathcal{G}, w)$ is the matrix L given by

$$L = \begin{cases} -w_{i,j} & \text{if } i \text{ is adjacent to } j \text{ in } \mathcal{G}, \\ 0 & \text{if } i \neq j \text{ and } i \text{ is not adjacent to } j \text{ in } \mathcal{G}, \\ d_i & \text{if } i = j. \end{cases}$$

We refer the reader to the book of Molitierno [101] for a comprehensive introduction to Laplacian matrices; our own investigation of Laplacian matrices for weighted graphs will continue in Chapter 7.

Since $\mathcal{G}$ is connected, it follows readily that its Laplacian matrix is a singular irreducible M–matrix, and hence the matrix $I + L$ is a nonsingular M–matrix. In [20], the authors consider the matrix $H = (I+L)^{-1}$, and interpret the entries of H as proximities between corresponding vertices. Specifically, if $h_{i,j}$ is small, then this is given the interpretation that vertex i is far from vertex j. In [20] it is asserted that for any triple of indices i, j, k,

$$h_{j,j} - h_{i,j} + h_{i,k} - h_{j,k} \geq 0, \tag{6.21}$$

with strict inequality holding if $j \neq i, k$. There is an obvious resemblance between (6.21) and the inequalities in Corollary 6.2.2 and Proposition 6.2.3; our goal now is to relate (6.21) to those results. Our first result in that direction concerns the group inverse of the Laplacian matrix for a connected weighted graph.

PROPOSITION 6.2.4 ([18]) *Let $(\mathcal{G}, w)$ be a connected weighted graph on n vertices, and denote its Laplacian matrix by L. Then for any indices $i, j, k = 1, \ldots, n$, we have*

$$l^{\#}_{j,j} - l^{\#}_{i,j} + l^{\#}_{i,k} - l^{\#}_{j,k} \geq 0. \tag{6.22}$$

Equality holds in (6.22) *if and only if either $j = i$ or $j = k$, or every path from vertex i to vertex k in $\mathcal{G}$ passes through vertex j.*

<u>**Proof**</u>: Let s denote the maximum diagonal entry in L, and note that L can be written as $L = s(I - T)$, where T is an irreducible symmetric and stochastic matrix whose stationary vector is $\frac{1}{n}\mathbf{1}$. Further, $L^{\#} = \frac{1}{s}(I - T)^{\#}$. Applying Corollary 6.2.2 to T, the inequality (6.22), as well as the characterisation of the equality case, now follow immediately. □

We now apply Proposition 6.2.4 in order to generalise (6.21).

COROLLARY 6.2.5 ([18]) *Let P be an $n \times n$ irreducible symmetric nonsingular M–matrix that is row diagonally dominant, and let $B = P^{-1}$. Then for each $i, j, k = 1, \ldots, n$, we have*

$$b_{j,j} - b_{i,j} + b_{i,k} - b_{j,k} \geq 0. \tag{6.23}$$

Set $S = \{l | e_l^t P\mathbf{1} > 0\}$, and let $\mathcal{G}$ denote the graph on vertices $1, \ldots, n+1$ constructed as follows: for each pair i, j between 1 and n, vertices i and j are adjacent in $\mathcal{G}$ if and only if $p_{i,j} < 0$; for each l between 1 and n, vertex l is adjacent to vertex $n+1$ if and only if $l \in S$. Then equality holds in (6.23) if and only if either $j = i$, or $j = k$, or if $j \neq i, k$ and every path from vertex i to vertex k in $\mathcal{G}$ passes through vertex j.

Proof: Let $u = P\mathbf{1}$, and note that $u_l > 0$ if and only if $l \in S$. Observe also that since P is invertible by hypothesis, it must be the case that $S \neq \emptyset$. Next we construct the matrix L of order $n+1$ via

$$L = \left[\begin{array}{c|c} P & -u \\ \hline -u^t & u^t\mathbf{1} \end{array}\right].$$

Observe that L is an irreducible symmetric singular M–matrix whose row sums are all zero. Consequently, L can be thought of as the Laplacian matrix of a weighted graph. Indeed, the corresponding graph is $\mathcal{G}$, where the weights of edges between vertices in $\{1, \ldots, n\}$ are the corresponding entries of P, and where, for each $l \in S$, the weight of the edge between vertex l and vertex $n+1$ is u_l.

From Observation 2.3.4, the leading principal submatrix of order n of $L^{\#}$ is given by

$$\frac{\mathbf{1}^t B\mathbf{1}}{n+1}J + B - \frac{1}{n+1}BJ - \frac{1}{n+1}JB.$$

Consider a triple of indices i, j, k between 1 and n. Then

$$\begin{aligned}
&l^{\#}_{j,j} - l^{\#}_{i,j} + l^{\#}_{i,k} - l^{\#}_{j,k} = \\
&(e_j - e_i)^t \left(\frac{\mathbf{1}^t B\mathbf{1}}{n+1}J + B - \frac{1}{n+1}BJ - \frac{1}{n+1}JB\right) e_j + \\
&(e_i - e_j)^t \left(\frac{\mathbf{1}^t B\mathbf{1}}{n+1}J + B - \frac{1}{n+1}BJ - \frac{1}{n+1}JB\right) e_k = \\
&\left(b_{j,j} - b_{i,j} - \frac{1}{n+1}(e_j - e_i)^t B\mathbf{1}\right) + \\
&\left(b_{i,k} - b_{j,k} - \frac{1}{n+1}(e_i - e_j)^t B\mathbf{1}\right) = \\
&b_{j,j} - b_{i,j} + b_{i,k} - b_{j,k}.
\end{aligned}$$

The inequality (6.23), as well as the characterisation of equality, now follows from Proposition 6.2.4. □

The following example illustrates the conclusion of Corollary 6.2.5.

EXAMPLE 6.2.6 Consider the matrix P given by

$$P = \begin{bmatrix} 1 & -1 & 0 & 0 & 0 \\ -1 & 2 & -1 & 0 & 0 \\ 0 & -1 & 3 & -1 & 0 \\ 0 & 0 & -1 & 2 & -1 \\ 0 & 0 & 0 & -1 & 2 \end{bmatrix}.$$

It is readily determined that P is an irreducible symmetric M–matrix that is diagonally dominant. Further, just two rows of P have positive row sums, namely rows 3 and 5. It follows that the graph $\mathcal{G}$ of Corollary 6.2.5 is that depicted in Figure 6.1.

Setting $B = P^{-1}$, we find that

$$B = \frac{1}{4}\begin{bmatrix} 11 & 7 & 3 & 2 & 1 \\ 7 & 7 & 3 & 2 & 1 \\ 3 & 3 & 3 & 2 & 1 \\ 2 & 2 & 2 & 4 & 2 \\ 1 & 1 & 1 & 2 & 3 \end{bmatrix}.$$

Selecting indices i, j, k as $2, 3$ and 5, respectively, we have

$$b_{j,j} - b_{i,j} + b_{i,k} - b_{j,k} = \frac{3}{4} - \frac{3}{4} + \frac{1}{4} - \frac{1}{4} = 0;$$

observe from Figure 6.1 that there are just two paths in $\mathcal{G}$ from vertex 2 to vertex 5, and that both paths pass through vertex 3. On the other hand, if we select indices i, j, k as $2, 4$ and 5, respectively, we have

$$b_{j,j} - b_{i,j} + b_{i,k} - b_{j,k} = 1 - \frac{1}{2} + \frac{1}{4} - \frac{1}{2} = \frac{1}{4}.$$

Referring to Figure 6.1, we see that only one of the paths in $\mathcal{G}$ from vertex 2 to vertex 5 passes through vertex 4.

6.3 The inverse mean first passage matrix problem

Suppose that we are given a square matrix A with positive entries. Under what circumstances can A be realised as the mean first passage matrix

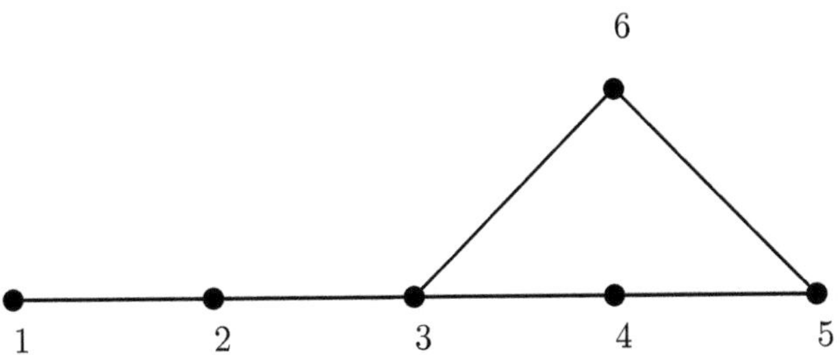

FIGURE 6.1
The graph $\mathcal{G}$ for Example 6.2.6

for some Markov chain? This is the so-called *inverse mean first passage matrix problem*, which is considered (and solved) in [66]. It will turn out that if A happens to be the mean first passage matrix for some Markov chain, then we may work with (6.2) in order to recover the associated transition matrix.

The statement of the inverse mean first passage matrix problem bears a resemblance to that of the *inverse M-matrix problem*, which can be phrased as follows: given a square matrix A with positive entries, under what circumstances can A be realised as the inverse of an irreducible nonsingular M-matrix? There is a wealth of literature on that problem, see for example the survey papers of Johnson [63], and Johnson and Smith [64]. In this section we discuss the inverse mean first passage matrix problem and establish a connection with the inverse M-matrix problem. Our approach follows that of [102]. We begin with a solution to the inverse mean first passage matrix problem which is essentially found in [66, section 4.4].

THEOREM 6.3.1 *Let M be a square matrix with positive entries. Then M is the mean first passage matrix for a Markov chain with an irreducible transition matrix if and only if all of the following conditions hold:*
i) the matrix $N = M - M_{dg}$ is invertible;
ii) $\mathbf{1}^t(M_{dg})^{-1}\mathbf{1} = 1$;
iii) the vector $M(M_{dg})^{-1}\mathbf{1}$ is a scalar multiple of $\mathbf{1}$;
iv) the matrix $I + (M_{dg} - J)N^{-1}$ is entrywise nonnegative.

In the event that conditions i)–iv) hold, the transition matrix associated with M is equal to $I + (M_{dg} - J)N^{-1}$.

Proof: Suppose first that M is the mean first passage matrix for a Markov chain with an irreducible transition matrix T, of order n, say. Set $Q = I - T$, denote the stationary vector for T by w, and let $W = \text{diag}(w)$. From (6.3), we find that $M = (I - Q^\# + JQ^\#_{dg})W^{-1}$. Note that $M_{dg} = W^{-1}$, so that $N = M - M_{dg} = M - W^{-1} = (-Q^\# + JQ^\#_{dg})W^{-1}$. Evidently N is invertible if and only if $-Q^\# + JQ^\#_{dg}$ is invertible. Denoting the eigenvalues of T by $1, \lambda_2, \ldots, \lambda_n$, we find from a result in [14] that for each $j = 2, \ldots, n$, $\frac{-1}{1-\lambda_j}$ is an eigenvalue of $-Q^\# + JQ^\#_{dg}$. Further, since $\text{trace}(-Q^\# + JQ^\#_{dg}) = 0$, it follows that the remaining eigenvalue of $-Q^\# + JQ^\#_{dg}$ is given by $\sum_{j=2}^n \frac{1}{1-\lambda_j}$. As none of the eigenvalues of $-Q^\# + JQ^\#_{dg}$ is zero, we deduce that N is invertible, establishing i).

To see that ii) holds, we simply observe that $\mathbf{1}^t(M_{dg})^{-1}\mathbf{1} = \mathbf{1}^t W \mathbf{1} = \mathbf{1}^t w = 1$. To see that iii) holds, note that $M(M_{dg})^{-1}\mathbf{1} = Mw$; from Theorem 6.1.9, we find that $Mw = \left(1 + \sum_{j=2}^n \frac{1}{1-\lambda_j}\right)\mathbf{1}$.

In order to establish iv), we first recall (6.2), namely

$$M = T(M - M_{dg}) + J.$$

Since N is invertible, we find that $T = (M - J)N^{-1}$. Writing M as $N + M_{dg}$, we have $T = (N + M_{dg} - J)N^{-1} = I + (M_{dg} - J)N^{-1}$. Condition iv) now follows from the fact that T is a nonnegative matrix.

Next, we consider the converse, so suppose that conditions i)–iv) hold, and let $T = I + (M_{dg} - J)N^{-1}$. Observe that $T(M - M_{dg}) = M - J$, so that $M = T(M - M_{dg}) + J$. If we can show that T is an irreducible and stochastic matrix, then from the fact that there is a unique solution to (6.2) (which is necessarily given by (6.3)) it will follow that M is the mean first passage matrix for the Markov chain with transition matrix T. Thus it suffices to show that T is irreducible and stochastic.

Set $w = (M_{dg})^{-1}\mathbf{1}$. From iii) we have that Mw is a scalar multiple of $\mathbf{1}$, from which we find that $Nw = t\mathbf{1}$ for some scalar t; from i) we find that necessarily $t \neq 0$. As a result, $(M_{dg} - J)N^{-1}\mathbf{1} = \frac{1}{t}(M_{dg} - J)w = \frac{1}{t}(1 - \mathbf{1}^t w)\mathbf{1}$. From ii) we have $\mathbf{1}^t w = \mathbf{1}^t(M_{dg})^{-1}\mathbf{1} = 1$, so that $(M_{dg} - J)N^{-1}\mathbf{1} = 0$. It now follows that $T\mathbf{1} = \mathbf{1}$, and since T is nonnegative by iv), we see that T is stochastic.

It remains only to show that T is irreducible. Suppose to the contrary that T is reducible; by simultaneously permuting its rows and columns, we may assume that T has the form

$$T = \left[\begin{array}{c|c} T_{1,1} & 0 \\ \hline T_{2,1} & T_{2,2} \end{array}\right],$$

where $T_{1,1}$ is both irreducible and stochastic. Partition M conformally with T as

$$M = \left[\begin{array}{c|c} M_{1,1} & M_{1,2} \\ \hline M_{2,1} & M_{2,2} \end{array}\right].$$

By considering the $(1,2)$ block of the equation $M = T(M - M_{dg}) + J$, we find that $M_{1,2} = T_{1,1}M_{1,2} + J$. Letting $\hat{w}$ denote the stationary vector for $T_{1,1}$, we then have $\hat{w}^t M_{1,2} = \hat{w}^t(TM_{1,2} + J) = \hat{w}^t M_{1,2} + \mathbf{1}^t$, a contradiction. We conclude that necessarily T is irreducible, as desired.

Finally, we observe that in the event that M satisfies conditions i)–iv), the corresponding transition matrix is given by $T = I + (M_{dg} - J)N^{-1}$. □

We have the following consequence of Theorem 6.3.1.

COROLLARY 6.3.2 *Suppose that M is a positive matrix that is the mean first passage for some Markov chain. Then for any $t \geq 1$, the matrix $tM + (1-t)M_{dg}$ is also the mean first passage matrix for some Markov chain.*

Fix a $t \geq 1$ and let $\tilde{M} = tM + (1-t)M_{dg}$. It suffices to verify that $\tilde{M}$ satisfies conditions i)–iv) of Theorem 6.3.1. Observing that $\tilde{M}_{dg} = M_{dg}$, the fact that $\tilde{M}$ satisfies ii) follows from the fact that M satisfies ii). Also, since $\tilde{M} - \tilde{M}_{dg} = t(M - M_{dg})$, we see that $\tilde{M} - \tilde{M}_{dg}$ is invertible, since $M - M_{dg}$ is. Further, $\tilde{M}(\tilde{M}_{dg})^{-1}\mathbf{1} = \tilde{M}(M_{dg})^{-1}\mathbf{1} = tM(M_{dg})^{-1}\mathbf{1} - (1-t)\mathbf{1}$. Since iii) holds for M, it must be the case that $M(M_{dg})^{-1}\mathbf{1}$ is a multiple of $\mathbf{1}$, and it now follows that $\tilde{M}$ satisfies iii). Finally, we observe that $I + (\tilde{M}_{dg} - J)(\tilde{M} - \tilde{M}_{dg})^{-1} = \frac{1}{t}\left(tI + (M_{dg} - J)(M - M_{dg})^{-1}\right)$. As $I + (M_{dg} - J)(M - M_{dg})^{-1}$ is nonnegative, so too is $I + (\tilde{M}_{dg} - J)(\tilde{M} - \tilde{M}_{dg})^{-1}$. Thus $\tilde{M}$ satisfies iv), and so by Theorem 6.3.1, $\tilde{M}$ is the mean first passage matrix for a Markov chain. □

REMARK 6.3.3 Suppose that M is the mean first passage matrix for a Markov chain with transition matrix T, and that $t \geq 1$. As in Corollary 6.3.2, we form $\tilde{M}$ as $\tilde{M} = tM + (1-t)M_{dg}$. Then $\tilde{M}$ is the mean first passage matrix for a Markov chain. Denoting the corresponding transition matrix by $\tilde{T}$, it is readily verified that $\tilde{T} = \frac{1}{t}T + \left(1 - \frac{1}{t}\right)I$.

EXAMPLE 6.3.4 Suppose that $n \in \mathbb{N}$ with $n \geq 4$, and let $t \in \mathbb{R}$. Consider the matrix $M(t) = nJ + t(e_1 - e_2)(e_{n-1} - e_n)^t$. Evidently $M(t)$ is a positive matrix if and only if $t \in (-n, n)$. For which such values of t is $M(t)$ the mean first passage matrix for some Markov chain?

Referring to Theorem 6.3.1, we see that since $(M(t))_{dg} = nI$, conditions ii) and iii) of Theorem 6.3.1 hold for any $t \in (-n, n)$. Note also that $N(t) \equiv M(t) - (M(t))_{dg} = n(J - I) + t(e_1 - e_2)(e_{n-1} - e_n)^t$; it is readily verified that $N(t)$ is invertible for any $t \in (-n, n)$, and that

$$(N(t))^{-1} = \frac{1}{n}\left(-I + \frac{1}{n-1}J - \frac{t}{n}(e_1 - e_2)(e_{n-1} - e_n)^t\right).$$

Consequently, condition i) of Theorem 6.3.1 holds for each $t \in (-n, n)$. Finally, a straightforward computation shows that

$$I + ((M(t))_{dg} - J)(N(t))^{-1} = \frac{1}{n}\left(J - t(e_1 - e_2)(e_{n-1} - e_n)^t\right).$$

Hence condition iv) Theorem 6.3.1 holds if and only if $t \in [-1, 1]$. We thus deduce that $M(t)$ is the mean first passage matrix for some Markov chain if and only if $-1 \leq t \leq 1$.

Our next sequence of results is directed towards establishing that an $n \times n$ positive matrix B is the inverse of a row and column diagonally dominant M–matrix if and only if B can be constructed as a fairly simple function of the mean first passage matrix for a Markov chain on $n+1$ states.

In order to establish our results, we require the following notation. Let $n \in \mathbb{N}$, with $n \geq 2$; for each $k = 1, \ldots, n$, we define the following $(n-1) \times n$ matrix:

$$P^{(k)} = \begin{bmatrix} e_1 & \ldots & e_{k-1} & -\mathbf{1} & e_k & \ldots & e_{n-1} \end{bmatrix}.$$

That is, $P^{(k)}$ is formed from the identity matrix of order $n-1$ by inserting $-\mathbf{1}$ between columns $k-1$ and k, with the obvious modification being made if k is 1 or n. (While the notation for $P^{(k)}$ does not explicitly reference it, the parameter n will always be clear from the context.) Note that $P^{(k)}\mathbf{1} = 0, k = 1, \ldots, n$.

Suppose now that we have an $n \times n$ matrix B with all positive entries. It is not difficult to determine that for each $k = 1, \ldots, n$, the $(n-1) \times (n-1)$ matrix $H^{(k)} = \left[h_{i,j}^{(k)}\right]_{i,j=1,\ldots,n-1}$ defined as $H^{(k)} = -P^{(k)}(B -$

$B_{dg})P^{(k)^t}$ has entries given by

$$h_{i,j}^{(k)} = \begin{cases} b_{i,k} + b_{k,i} & \text{if } i = j, \\ b_{i,k} + b_{k,j} - b_{i,j} & \text{if } i \neq j, \end{cases} \quad \text{for } i,j = 1,\ldots,n-1. \tag{6.24}$$

We note in passing that if B happens to be the mean first passage matrix for a Markov chain, then from Theorem 6.2.1 we see that $H^{(k)} \geq 0$ for each $k = 1,\ldots,n$. Our next results describes $H^{(k)}$ in more detail in the case that B is a mean first passage matrix.

PROPOSITION 6.3.5 ([102]) *Suppose that $T \in \mathcal{IS}_n$ with stationary vector w and mean first passage matrix M. Let $W = \text{diag}(w)$. For each $k = 1,\ldots,n$, let $H^{(k)} = -P^{(k)}(M - M_{dg})P^{(k)^t}$. Then*

$$W_{(k)}(I - T_{(k)})H^{(k)} = I.$$

Proof: Without loss of generality, we take the index k to be n. Set $Q = I-T$, and recall from Theorem 6.1.1 that $M = (I - Q^{\#} + JQ^{\#}_{dg})W^{-1}$. Consequently $M - M_{dg} = (-Q^{\#} + JQ^{\#}_{dg})W^{-1}$, and since $P^{(n)}\mathbf{1} = 0$, we find that $H^{(n)} = P^{(n)}Q^{\#}W^{-1}P^{(n)^t}$.

Write T as

$$T = \left[\begin{array}{c|c} T_{(n)} & \mathbf{1} - T_{(n)}\mathbf{1} \\ \hline u^t & 1 - u^t\mathbf{1} \end{array}\right].$$

From the eigen-equation $w^tT = w^t$ and the fact that $w^t\mathbf{1} = 1$, it follows that necessarily

$$\frac{1}{1 + u^t(I - T_{(n)})^{-1}\mathbf{1}} = w_n,$$

while

$$\frac{1}{1 + u^t(I - T_{(n)})^{-1}\mathbf{1}} u^t(I - T_{(n)})^{-1} = \mathbf{1}^tW_{(n)}.$$

Appealing to Theorem 2.5.1, and again using the fact that $P^{(n)}\mathbf{1} = 0$, we find that

$$P^{(n)}Q^{\#} = \left[\ (I - T_{(n)})^{-1} - (I - T_{(n)})^{-1}JW_{(n)} \mid -w_n(I - T_{(n)})^{-1}\mathbf{1}\ \right].$$

Hence we have

$$H^{(n)} = P^{(n)}Q^{\#}W^{-1}P^{(n)^t} =$$

$$\left[\ (I - T_{(n)})^{-1} - (I - T_{(n)})^{-1}JW_{(n)} \mid -w_n(I - T_{(n)})^{-1}\mathbf{1}\ \right] \left[\begin{array}{c} W_{(n)}^{-1} \\ \hline -\frac{1}{w_n}\mathbf{1}^t \end{array}\right] =$$

$$(I - T_{(n)})^{-1}W_{(n)}^{-1}.$$

The conclusion now follows. □

REMARK 6.3.6 It is straightforward to see that the converse to Theorem 6.3.5 holds. That is, maintaining the notation of that theorem, if H is an $(n-1)\times(n-1)$ matrix that can be written as $(W_{(k)}(I-T_{(k)}))^{-1}$ for some $T \in \mathcal{IS}_n$ with stationary vector w, and some index k between 1 and n, then necessarily $H = -P^{(k)}(M - M_{dg})P^{(k)^t}$, where M is the mean first passage matrix for the Markov chain with transition matrix T.

Our next result, which appears in [102], establishes a link between a special case of the inverse M-matrix problem and the existence of a particular mean first passage matrix.

THEOREM 6.3.7 *Suppose that A is an $n\times n$ real matrix. The following are equivalent:*
a) A is invertible and A^{-1} is an M-matrix that is both row diagonally dominant and column diagonally dominant;
b) there is a scalar $c > 0$ and a Markov chain on $n+1$ states, whose transition matrix is irreducible, such that for the corresponding mean first passage matrix M, we have, for each $i, j = 1, \ldots, n$,

$$a_{i,j} = \begin{cases} c(m_{i,n+1} + m_{n+1,i}) & \text{if } i = j, \\ c(m_{i,n+1} + m_{n+1,j} - m_{i,j}) & \text{if } i \neq j. \end{cases} \tag{6.25}$$

<u>**Proof**</u>: Suppose that b) holds. Observe that the matrix A satisfies a) if and only if the matrix $\frac{1}{c}A$ does, so we will assume for simplicity of exposition here that $c = 1$. Let T be the $(n+1)\times(n+1)$ transition matrix for the appropriate Markov chain, and denote its stationary vector by w. As usual, let $W = \text{diag}(w)$.

From (6.24), we see that A can be written as

$$A = -P^{(n+1)}(M - M_{dg})P^{(n+1)^t},$$

where $P^{(n+1)}$ is the $n \times (n+1)$ matrix $\left[\begin{array}{c|c} I & -\mathbf{1} \end{array}\right]$. From Proposition 6.3.5, we find that A is the inverse of the matrix $W_{(n+1)}(I - T_{(n+1)})$. Since T is irreducible and stochastic, $T_{(n+1)}\mathbf{1} \le \mathbf{1}$, so that $W_{(n+1)}(I - T_{(n+1)})\mathbf{1} \ge 0$. Thus A is row diagonally dominant. Also, denoting the submatrix of T on its last row and first n columns by the vector u^t, we find from the equation $w^t(I-T) = 0^t$ that $\mathbf{1}^t W_{(n+1)}(I - T_{(n+1)}) = w_{n+1}u^t \ge 0^t$. Consequently A is column diagonally dominant. Thus we see that a) holds.

Now suppose that a) holds. Observe that A satisfies b) if and only if sA satisfies b) for any scalar $s > 0$. Consequently, we will assume without loss of generality that $\text{trace}(A^{-1}) + \mathbf{1}^t A^{-1}\mathbf{1} \le 1$. For each $i = 1, \ldots, n$, denote the i–th diagonal entry of A^{-1} by d_i, and note that each $d_i > 0$ (otherwise, from row diagonal dominance of A^{-1}, it would follow that A^{-1} has an all zero row). Note also from row diagonal dominance that $A^{-1}\mathbf{1}$ is a nonnegative and nonzero vector, so that $\mathbf{1}^t A^{-1}\mathbf{1} > 0$. Next, we select positive numbers $w_i, i = 1, \ldots, n+1$ such that $w_i \ge d_i, i = 1, \ldots, n, w_{n+1} \ge \mathbf{1}^t A^{-1}\mathbf{1}$, and in addition $\sum_{i=1}^{n+1} w_i = 1$. Let W be the diagonal matrix whose i–th diagonal entry is $w_i, i = 1, \ldots, n+1$.

Now consider the matrix T of order $n+1$ given by

$$T = I - W^{-1}\left[\begin{array}{c|c} A^{-1} & -A^{-1}\mathbf{1} \\ \hline -\mathbf{1}^t A^{-1} & \mathbf{1}^t A^{-1}\mathbf{1} \end{array}\right]. \tag{6.26}$$

We claim that T is both stochastic and irreducible. Since A^{-1} is an M–matrix, its off-diagonal entries are nonpositive; we have observed above that $A^{-1}\mathbf{1} \ge 0$, and from the fact that A^{-1} is column diagonally dominant, we also have that $\mathbf{1}^t A^{-1} \ge 0^t$. Hence we find that the offdiagonal entries of T are all nonnegative. Further, from our construction of the w_is from the diagonal entries of A^{-1} and from $\mathbf{1}^t A^{-1}\mathbf{1}$, it follows that the diagonal entries of T are also nonnegative. It is readily verified that $T\mathbf{1} = \mathbf{1}$, so we have established that T is a stochastic matrix.

In order to see that T is irreducible, suppose to the contrary that T is a reducible matrix. It follows then that we may simultaneously permute the first n rows and columns of T (via a permutation matrix R) so that the resulting matrix RTR^t has the form

$$\left[\begin{array}{c|c} \hat{T}_{1,1} & \hat{T}_{1,2} \\ \hline \hat{T}_{2,1} & \hat{T}_{2,2} \end{array}\right], \tag{6.27}$$

where one of $\hat{T}_{1,2}$ and $\hat{T}_{2,1}$ is an all zero block. Without loss of generality we assume that T itself has the form given by (6.27). Partition W conformally with T as

$$\left[\begin{array}{c|c} W_1 & 0 \\ \hline 0 & W_2 \end{array}\right].$$

Observe then that A^{-1} can be partitioned as

$$\left[\begin{array}{c|c} (A^{-1})_{1,1} & (A^{-1})_{1,2} \\ \hline (A^{-1})_{2,1} & (A^{-1})_{2,2} \end{array}\right]$$

so that

$$T_{1,2} = -W_1^{-1}\left[\ (A^{-1})_{1,2}\ \middle|\ -((A^{-1})_{1,1}\mathbf{1} + (A^{-1})_{1,2}\mathbf{1})\ \right], \quad (6.28)$$

$$\text{and } T_{2,1} = -W_2^{-1}\left[\begin{array}{c}(A^{-1})_{2,1} \\ \hline -(\mathbf{1}^t(A^{-1})_{1,1} + \mathbf{1}^t(A^{-1})_{2,1})\end{array}\right].$$

Suppose that $T_{1,2} = 0$. Then from (6.28) we find that $(A^{-1})_{1,2} = 0$ and $(A^{-1})_{1,1}\mathbf{1} = 0$. But then A^{-1} is equal to

$$\left[\begin{array}{c|c}(A^{-1})_{1,1} & 0 \\ \hline (A^{-1})_{2,1} & (A^{-1})_{2,2}\end{array}\right],$$

where $(A^{-1})_{1,1}$ is singular, contradicting the invertibility of A^{-1}. A similar contradiction arises if $T_{2,1} = 0$. Consequently, it must be the case that T is irreducible.

From (6.26), we find that $A = \left(W_{(n+1)}(I - T_{(n+1)})\right)^{-1}$. Appealing to Remark 6.3.6, we find that A can be written as $A = -P^{(n+1)}(M - M_{dg})P^{(n+1)^t}$, where M is the mean first passage matrix for the Markov chain with transition matrix T. It now follows that A satisfies b). □

Suppose that A is an $n \times n$ matrix whose inverse is a row and column diagonally dominant M–matrix. Observe that if $\text{trace}(A^{-1}) + \mathbf{1}^t(A^{-1})\mathbf{1} < 1$, then the proof technique of Theorem 6.3.7 allows for the construction of a family of stochastic matrices such that if M is the mean first passage matrix of any member of that family, then (6.25) holds for all i and j, with $c = 1$. Indeed an inspection of the proof of Theorem 6.3.7 reveals that each admissible choice of the values $w_1, \ldots, w_{n+1}$ leads to a different stochastic matrix with the desired property. The following example illustrates.

EXAMPLE 6.3.8 Consider the matrix

$$A = \begin{bmatrix} 33 & 22 & 22 & 11 & 11 \\ 22 & 33 & 22 & 11 & 11 \\ 22 & 22 & 22 & 11 & 11 \\ 11 & 11 & 11 & 22 & 11 \\ 11 & 11 & 11 & 11 & 11 \end{bmatrix};$$

a computation shows that

$$A^{-1} = \frac{1}{11}\begin{bmatrix} 1 & 0 & -1 & 0 & 0 \\ 0 & 1 & -1 & 0 & 0 \\ -1 & -1 & 3 & 0 & -1 \\ 0 & 0 & 0 & 1 & -1 \\ 0 & 0 & -1 & -1 & 3 \end{bmatrix}.$$

Thus A^{-1} is a row and column diagonally dominant M–matrix, and note that trace$(A^{-1}) + \mathbf{1}^t(A^{-1})\mathbf{1} = \frac{10}{11}$.

Let w be the vector given by

$$w^t = \frac{1}{11}\left[\begin{array}{cccccc} 1 & 1 & 3 & 1 & 3 & 2 \end{array}\right].$$

Following the construction in the proof of Theorem 6.3.7, and using the values $w(1), \ldots, w(6)$, we produce the stochastic matrix

$$T = \left[\begin{array}{cccccc} 0 & 0 & 1 & 0 & 0 & 0 \\ 0 & 0 & 1 & 0 & 0 & 0 \\ \frac{1}{3} & \frac{1}{3} & 0 & 0 & \frac{1}{3} & 0 \\ 0 & 0 & 0 & 0 & 1 & 0 \\ 0 & 0 & \frac{1}{3} & \frac{1}{3} & 0 & \frac{1}{3} \\ 0 & 0 & 0 & 0 & \frac{1}{2} & \frac{1}{2} \end{array}\right].$$

It then follows that the mean first passage matrix for the Markov chain with transition matrix T is given by

$$M = \left[\begin{array}{cccccc} 11 & 11 & 1 & 16 & 6 & 15 \\ 11 & 11 & 1 & 16 & 6 & 15 \\ 10 & 10 & \frac{11}{3} & 15 & 5 & 14 \\ 17 & 17 & 7 & 11 & 1 & 10 \\ 16 & 16 & 6 & 10 & \frac{11}{3} & 9 \\ 18 & 18 & 8 & 12 & 2 & \frac{11}{2} \end{array}\right].$$

Next, we let $\tilde{w}$ be given by

$$\tilde{w}^t = \frac{1}{11}\left[\begin{array}{cccccc} 2 & 1 & 3 & 1 & 3 & 1 \end{array}\right].$$

Again using the construction in the proof of Theorem 6.3.7 with the values $\tilde{w}_1, \ldots, \tilde{w}_6$, we produce the following stochastic matrix:

$$\tilde{T} = \left[\begin{array}{cccccc} \frac{1}{2} & 0 & \frac{1}{2} & 0 & 0 & 0 \\ 0 & 0 & 1 & 0 & 0 & 0 \\ \frac{1}{3} & \frac{1}{3} & 0 & 0 & \frac{1}{3} & 0 \\ 0 & 0 & 0 & 0 & 1 & 0 \\ 0 & 0 & \frac{1}{3} & \frac{1}{3} & 0 & \frac{1}{3} \\ 0 & 0 & 0 & 0 & 1 & 0 \end{array}\right].$$

The mean first passage matrix for the Markov chain with transition matrix $\tilde{T}$ is given by

$$\tilde{M} = \begin{bmatrix} \frac{11}{2} & 12 & 2 & 18 & 8 & 18 \\ 10 & 11 & 1 & 17 & 7 & 17 \\ 9 & 10 & \frac{11}{3} & 16 & 6 & 16 \\ 15 & 16 & 6 & 11 & 1 & 11 \\ 14 & 15 & 5 & 10 & \frac{11}{3} & 10 \\ 15 & 16 & 6 & 11 & 1 & 11 \end{bmatrix}.$$

It is now straightforward to determine that $-P^{(6)}(M - M_{dg})P^{(6)^t} = A = P^{(6)}(\tilde{M} - \tilde{M}_{dg})P^{(6)^t}$.

REMARK 6.3.9 Suppose that the matrix A satisfies condition a) in Theorem 6.3.7, and consider stochastic matrix T arising in the proof of Theorem 6.3.7, namely

$$T = I - W^{-1}\left[\begin{array}{c|c} A^{-1} & -A^{-1}\mathbf{1} \\ \hline -\mathbf{1}^t A^{-1} & \mathbf{1}^t A^{-1}\mathbf{1} \end{array}\right].$$

Observe that the stationary vector for T is equal to $W\mathbf{1}$. From this it follows that

$$W^{\frac{1}{2}}TW^{-\frac{1}{2}} = I - W^{-\frac{1}{2}}\left[\begin{array}{c|c} A^{-1} & -A^{-1}\mathbf{1} \\ \hline -\mathbf{1}^t A^{-1} & \mathbf{1}^t A^{-1}\mathbf{1} \end{array}\right]W^{-\frac{1}{2}}.$$

In particular, if A happens to be symmetric, then so is A^{-1}, and we find readily that T is the transition matrix for a reversible Markov chain. Conversely, if the Markov chain associated with T is reversible, then A^{-1}, and hence A, must be symmetric.

6.4 A partitioned approach to the mean first passage matrix

Suppose that we have a matrix $T \in \mathcal{IS}_n$, and partition of $\{1, \ldots, n\}$ as $S \cup \overline{S}$, where $S = \{1, \ldots, k\}, \overline{S} = \{k+1, \ldots, n\}$. Write T in partitioned form as

$$T = \left[\begin{array}{c|c} T[S,S] & T[S,\overline{S}] \\ \hline T[\overline{S},S] & T[\overline{S},\overline{S}] \end{array}\right]. \tag{6.29}$$

(Here $T[S,S]$ denotes the submatrix of T on rows and columns indexed by S, $T[S,\overline{S}]$ denotes the submatrix of T on rows indexed by S and columns indexed by $\overline{S}$, and similarly for the other blocks.) Now consider the corresponding mean first passage matrix M, partitioned conformally with T as

$$M = \left[\begin{array}{c|c} M[S,S] & M[S,\overline{S}] \\ \hline M[\overline{S},S] & M[\overline{S},\overline{S}] \end{array}\right].$$

How might we compute the blocks of M without computing the entire matrix M? What interpretation might we attach to those blocks of M? In this section we consider these question, and provide formulas for the blocks of M in terms of mean first passage matrices for certain stochastic complements (introduced in section 5.1) associated with T.

We remark that our viewpoint here is philosophically aligned with that of [96]. As we outlined in section 5.1, it is shown in [96] that for an irreducible stochastic matrix T partitioned as in (6.29), then partitioning the stationary vector w for T conformally with (6.29), we have

$$w = \left[\begin{array}{c} a_1 u_1 \\ \hline a_2 u_2 \end{array}\right],$$

where u_1 and u_2 are the stationary vectors of the respective stochastic complements of T, and where a_1, a_2 are the respective entries in the stationary distribution of the 2×2 coupling matrix. As noted above, our results in this section describe the blocks of the mean first passage matrix in terms of mean first passage matrices corresponding to the respective stochastic complements, thus paralleling the results of [96] for the stationary distribution.

We will derive our results by considering (6.2) in partitioned form. The resulting formulas for the blocks of M will lead in section 8.3 to a "divide and conquer" strategy for computing the group inverse associated with the transition matrix. Our results in this section will necessitate the simultaneous use of several mean first passage matrices, and so in the interests of clarity, in this section we will adopt the following notation: for a given irreducible stochastic matrix T, its mean first passage matrix will be denoted M_T.

The following result from a paper of Kirkland, Neumann, and Xu [85] will be useful in the sequel.

LEMMA 6.4.1 *Suppose that $T \in \mathcal{IS}_n$, and fix an index k with $1 \leq k \leq n-1$. Let $S = \{1, \ldots, k\}$ and $\overline{S} = \{k+1, \ldots, n\}$, and partition T*

as

$$T = \left[\begin{array}{c|c} T[S,S] & T[S,\overline{S}] \\ \hline T[\overline{S},S] & T[\overline{S},\overline{S}] \end{array}\right]. \tag{6.30}$$

Partition M_T *conformally with* T *as*

$$M = \left[\begin{array}{c|c} M[S,S] & M[S,\overline{S}] \\ \hline M[\overline{S},S] & M[\overline{S},\overline{S}] \end{array}\right]. \tag{6.31}$$

Then

$$\begin{aligned}&(I - \mathcal{P}(T)_S)\, M_T[S,S] = \\ &J - \mathcal{P}(T)_S\,(M_T[S,S])_{dg} + T[S,\overline{S}]\left(I - T[\overline{S},\overline{S}]\right)^{-1} J.\end{aligned} \tag{6.32}$$

Proof: We proceed by considering the key equation (6.2) in block form. From the $(1,1)$ block on each side of (6.2), we find that

$$M_T[S,S] = T[S,S](M_T[S,S] - (M_T[S,S])_{dg}) + T[S,\overline{S}]M_T[\overline{S},S] + J. \tag{6.33}$$

Similarly, from the $(2,1)$ blocks on each side of (6.2), we have

$$M_T[\overline{S},S] = T[\overline{S},S](M_T[S,S] - (M_T[S,S])_{dg}) + T[\overline{S},\overline{S}]M_T[\overline{S},S] + J. \tag{6.34}$$

Solving (6.34) for $M_T[\overline{S},S]$ now yields

$$M_T[\overline{S},S]] = (I - T[\overline{S},\overline{S}])^{-1}\left(T[\overline{S},S]M_T[S,S] - T[\overline{S},S](M_T[S,S])_{dg}) + J\right). \tag{6.35}$$

Next, we substitute (6.35) into (6.33) to find that

$$\begin{aligned}&M_T[S,S] = \\ &(T[S,S] + T[S,\overline{S}](I - T[\overline{S},\overline{S}])^{-1}T[\overline{S},S])(M_T[S,S] - (M_T[S,S])_{dg})) \\ &\qquad + T[S,\overline{S}](I - T[\overline{S},\overline{S}])^{-1}J + J.\end{aligned} \tag{6.36}$$

Since $\mathcal{P}(T)_S = T[S,S] + T[S,\overline{S}](I - T[\overline{S},\overline{S}])^{-1}T[\overline{S},S]$, we see that (6.32) follows readily. □

We now make use of Lemma 6.4.1 in order to produce an expression for the blocks of M_T in (6.31). Our next result maintains the notation of Lemma 6.4.1.

THEOREM 6.4.2 ([85]) *Let* T, M_T, S *and* $\overline{S}$ *be as in Lemma 6.4.1,* *and partition T and M_T as in (6.30) and (6.31), respectively. Denote the stationary vector for T, partitioned conformally with T as*

$$w = \left[\frac{w[S]}{w[\overline{S}]} \right],$$

and set $\gamma_S = \frac{1}{\mathbf{1}^t w[S]}$ and $\gamma_{\overline{S}} = \frac{1}{\mathbf{1}^t w[\overline{S}]}$. Let V_S and $V_{\overline{S}}$ be the matrices of sizes $k \times k$ and $(n-k) \times (n-k)$, respectively, given by

$$\begin{aligned} V_S = (I - \mathcal{P}(T)_S)^{\#} T[S, \overline{S}](I - T[\overline{S}, \overline{S}])^{-1} J - \\ ((I - \mathcal{P}(T)_S)^{\#} T[S, \overline{S}](I - T[\overline{S}, \overline{S}])^{-1} J)^t, \\ V_{\overline{S}} = (I - \mathcal{P}(T)_{\overline{S}})^{\#} T[\overline{S}, S](I - T[S, S])^{-1} J - \\ ((I - \mathcal{P}(T)_{\overline{S}})^{\#} T[\overline{S}, S](I - T[S, S])^{-1} J)^t. \end{aligned} \tag{6.37}$$

Then the following expressions hold:

$$M_T[S, S] = \gamma_S M_{\mathcal{P}(T)_S} + V_S; \tag{6.38}$$

$$M_T[\overline{S}, \overline{S}] = \gamma_{\overline{S}} M_{\mathcal{P}(T)_{\overline{S}}} + V_{\overline{S}}; \tag{6.39}$$

$$\begin{aligned} M_T[\overline{S}, S]] = (I - T[\overline{S}, \overline{S}])^{-1}(T[\overline{S}, S] M_T[S, S] \\ - T[\overline{S}, S](M_T[S, S])_{dg}) + J); \end{aligned} \tag{6.40}$$

$$\begin{aligned} M_T[S, \overline{S}]] = (I - T[S, S])^{-1}(T[S, \overline{S}] M_T[\overline{S}, \overline{S}] \\ - T[S, \overline{S}](M_T[\overline{S}, \overline{S}])_{dg}) + J). \end{aligned} \tag{6.41}$$

Proof: Applying (6.2) to $\mathcal{P}(T)_S$, we find that

$$(I - \mathcal{P}(T)_S) M_{\mathcal{P}(T)_S} = J - \mathcal{P}(T)_S (M_{\mathcal{P}(T)_S})_{dg}. \tag{6.42}$$

Denote the matrix $M_T[S, S] - \gamma_S M_{\mathcal{P}(T)_S}$ by V_S; we will show that in fact V_S is given by (6.37). Recall that for each of M_T and $M_{\mathcal{P}(T)_S}$, the diagonal entries are given by the reciprocals of the corresponding entries in the associated stationary distribution. Since the stationary distribution for $\mathcal{P}(T)_S$ is given by $\gamma_S w[S]$, it now follows that V_S has zero diagonal.

From (6.32) and (6.42), it follows that

$$(I - \mathcal{P}(T)_S) V_S = (1 - \gamma_S) J + T[S, \overline{S}](I - T[\overline{S}, \overline{S}])^{-1} J. \tag{6.43}$$

Set $h = T[S, \overline{S}](I - T[\overline{S}, \overline{S}])^{-1}\mathbf{1}$. Multiplying (6.43) on the left by $(I - \mathcal{P}(T)_S)^{\#}$, we find that $(I - \gamma_S \mathbf{1} w[S]^t) V_S = (I - \mathcal{P}(T)_S)^{\#} h \mathbf{1}^t$, so that necessarily V_S has the form $\mathbf{1} z^t + (I - \mathcal{P}(T)_S)^{\#} h \mathbf{1}^t$ for some vector z. As V_S has zero diagonal, it now follows that V_S is given by (6.37). The expression for $M_T[S, S]$ in (6.38) now follows readily.

An similar argument establishes (6.39). Note that (6.40) is simply a restatement of (6.34), while (6.41) follows analogously. □

REMARK 6.4.3 It is straightforward to see that the quantities γ_S and $\gamma_{\overline{S}}$ in Theorem 6.4.2 can be determined from the coupling matrix associated with the partitioning of T in (6.30). As noted in the proof of Theorem 6.4.2, the stationary distribution for $\mathcal{P}(T)_S$ is given by $\gamma_S w[S]$, and similarly the stationary distribution for $\mathcal{P}(T)_{\overline{S}}$ is given by $\gamma_{\overline{S}} w[\overline{S}]$. Consequently, it follows that the vector

$$\begin{bmatrix} \frac{1}{\gamma_S} \\ \frac{1}{\gamma_{\overline{S}}} \end{bmatrix}$$

serves as the stationary distribution of the 2×2 coupling matrix corresponding to the partition in (6.30).

REMARK 6.4.4 Observe that our expression for V_S in (6.37) explicitly involves $(I - \mathcal{P}(T)_S)^{\#}$. Recall from Remark 6.1.2 that if we happen to know $M_{\mathcal{P}(T)_S}$, then we can find $(I - \mathcal{P}(T)_S)^{\#}$ from knowledge of its mean first passage matrix $M_{\mathcal{P}(T)_S}$. Thus (6.37) can be reformulated in terms of $M_{\mathcal{P}(T)_S}$.

REMARK 6.4.5 Suppose that we have an irreducible stochastic matrix with period $d \geq 2$, say

$$T = \left[\begin{array}{c|c|c|c|c|c} 0 & T_1 & 0 & 0 & \cdots & 0 \\ \hline 0 & 0 & T_2 & 0 & \cdots & 0 \\ \hline \vdots & & \ddots & \ddots & & \vdots \\ \hline 0 & \cdots & & 0 & T_{d-2} & 0 \\ \hline 0 & 0 & \ddots & & 0 & T_{d-1} \\ \hline T_d & 0 & 0 & \cdots & 0 & 0 \end{array}\right].$$

We can recover the expression given in (6.10) for the diagonal blocks of M_T by making use of Theorem 6.4.2 as follows. Let S denote the collection of indices corresponding to the first block in the partitioning of T above, and let $\overline{S}$ denote the remaining indices. It is readily determined that $\mathcal{P}(T)_S = T_1 T_2 \ldots T_d$ and that $\gamma_S = d$. Further, since

$$T[S,\overline{S}](I - T[\overline{S},\overline{S}])^{-1} =$$

$$\left[\begin{array}{c|c|c|c|c} T_1 & 0 & 0 & \ldots & 0 \end{array}\right] \left[\begin{array}{c|c|c|c|c} I & T_2 & T_2T_3 & \cdots & T_2 \ldots T_{d-1} \\ \hline 0 & I & T_3 & \cdots & T_3 \ldots T_{d-1} \\ \hline \vdots & & \ddots & & \vdots \\ \hline 0 & 0 & \cdots & 0 & I \end{array}\right],$$

we find that $T[S,\overline{S}](I-T[\overline{S},\overline{S}])^{-1}\mathbf{1}=(d-1)\mathbf{1}$. Hence $V_S=0$ by (6.37). The expression for the diagonal block $M_T[S,S]$ in (6.10) now follows immediately from (6.39).

In our next example, we again discuss stochastic Leslie matrices.

EXAMPLE 6.4.6 Suppose that we have an irreducible stochastic Leslie matrix L given by

$$L=\begin{bmatrix} a_1 & a_2 & \dots & a_{n-1} & a_n \\ 1 & 0 & 0 & \dots & 0 \\ 0 & 1 & 0 & \dots & 0 \\ & & \ddots & & \\ 0 & 0 & \dots & 1 & 0 \end{bmatrix}.$$

Let $S=\{1,\dots,k\}$ and $\overline{S}=\{k+1,\dots,n\}$. The stationary vector for L is given by

$$w=\frac{1}{\sum_{j=1}^n ja_j}\begin{bmatrix} \sum_{j=1}^n a_j \\ \sum_{j=2}^n a_j \\ \sum_{j=3}^n a_j \\ \vdots \\ a_n \end{bmatrix},$$

from which we find that $\gamma_S=\frac{\sum_{j=1}^n ja_j}{\sum_{j=1}^n \min\{j,k\}a_j}$.

Since $(I-L[\overline{S},\overline{S}])^{-1}$ is the lower triangular matrix of order $n-k$ with ones on and below the diagonal, it follows that $\mathcal{P}(L)_S$ is a stochastic Leslie matrix of order k, namely

$$\mathcal{P}(L)_S=\begin{bmatrix} a_1 & a_2 & \dots & a_{k-1} & \sum_{j=k}^n a_j \\ 1 & 0 & 0 & \dots & 0 \\ 0 & 1 & 0 & \dots & 0 \\ & & \ddots & & \\ 0 & 0 & \dots & 1 & 0 \end{bmatrix}.$$

Applying (6.37), we find that

$$V_S=k\left((I-\mathcal{P}(L)_S)^{\#}e_1\mathbf{1}^t-\mathbf{1}e_1^t((I-\mathcal{P}(L)_S)^{\#})^t\right).$$

In view of Remark 6.4.4, we see from (6.39) of Theorem 6.4.2 that $M_L[S,S]$ has the appealing property that it can be computed in terms of the mean first passage matrix for a stochastic Leslie matrix of lower order.

REMARK 6.4.7 From (6.37), we see that the matrix V_S in Theorem 6.4.2 is skew–symmetric. Consequently, we find that

$$M_T[S,S] + M_T[S,S]^t = \gamma_S(M_{\mathcal{P}(T)_S} + M^t_{\mathcal{P}(T)_S}).$$

Thus we find that for any pair of indices i and j between 1 and k, the sum of the mean first passage time from i to j and the mean first passage time from j to i in the Markov chain with transition matrix T is the same as the corresponding sum for the Markov chain with transition matrix $\mathcal{P}(T)_S$, scaled by γ_S.

The following example illustrates the conclusion of Remark 6.4.7.

EXAMPLE 6.4.8 Once again we revisit Markov chain in Example 5.2.4. Recall that the transition matrix for that example is given by

$$T = \left[\begin{array}{c|c|c} 0 & \frac{1}{2}I & \frac{1}{2}\mathbf{1} \\ \hline I & 0 & 0 \\ \hline \frac{1}{2k}\mathbf{1}^t & 0 & 0 \end{array}\right],$$

where the first two diagonal blocks are $k \times k$ and the third diagonal block is 1×1. The corresponding mean first passage matrix is given in Example 6.1.12 as

$$M = \left[\begin{array}{c|c|c} 4kJ - 2kI & (8k-1)J - 4kI & 3\mathbf{1} \\ \hline (4k+1)J - 4kI & 8kJ - 4kI & 4\mathbf{1} \\ \hline (4k-3)\mathbf{1}^t & (8k-4)\mathbf{1}^t & 4 \end{array}\right].$$

Let $S = \{1, \ldots, 2k\}$ and $\overline{S} = \{2k+1\}$. The stationary vector for T is

$$\left[\begin{array}{c} \frac{1}{2k}\mathbf{1} \\ \hline \frac{1}{4k}\mathbf{1} \\ \hline \frac{1}{4} \end{array}\right],$$

and so a straightforward computation shows that $\gamma_S = \frac{4}{3}$. Observe that

$$\mathcal{P}(T)_S = \left[\begin{array}{c|c} \frac{1}{2k}J & \frac{1}{2}I \\ \hline I & 0 \end{array}\right],$$

where the diagonal blocks are both $k \times k$. It now follows that

$$M_{\mathcal{P}(T)_S} = \left[\begin{array}{c|c} 3kJ - \frac{3k}{2}I & (6k-1)J - 3kI \\ \hline (3k+1)J - 3kI & 6kJ - 3kI \end{array}\right].$$

Set $S_1 = \{1, \ldots, k\}$ and $S_2 = \{k+1, \ldots, 2k\}$. From the formulas above, we see that $M_{\mathcal{P}(T)_S}[S_1, S_1] = \gamma_S M_T[S_1, S_1]$ and $M_{\mathcal{P}(T)_S}[S_2, S_2] = \gamma_S M_T[S_2, S_2]$; neither of these equations is anticipated by Remark 6.4.7, which merely assures us that $M_{\mathcal{P}(T)_S}[S_i, S_i] + (M_{\mathcal{P}(T)_S}[S_i, S_i])^t = \gamma_S(M_T[S_i, S_i] + (M_T[S_i, S_i])^t), i = 1, 2$. Note further that

$$M_{\mathcal{P}(T)_S}[S_1, S_2] + M_{\mathcal{P}(T)_S}[S_2, S_1] = 9kJ - 6kI = \gamma_S(M_T[S_1, S_2] + (M_T[S_2, S_1])),$$

as guaranteed by Remark 6.4.7.

REMARK 6.4.9 Observe that the matrix V_S of Theorem 6.4.2 is zero if and only if $(I - \mathcal{P}(T)_S)^{\#} T[S, \overline{S}](I - T[\overline{S}, \overline{S}])^{-1} J = cJ$ for some scalar c. Indeed it follows readily (multiplying on the left by $w[S]^t$) that c must be 0. Hence, $V_S = 0$ if and only if

$$(I - \mathcal{P}(T)_S)^{\#} T[S, \overline{S}](I - T[\overline{S}, \overline{S}])^{-1}\mathbf{1} = 0. \tag{6.44}$$

Since the null space of $(I - \mathcal{P}(T)_S)^{\#}$ is spanned by $\mathbf{1}$, we now find from (6.44) that $V_S = 0$ if and only if $T[S, \overline{S}](I - T[\overline{S}, \overline{S}])^{-1}\mathbf{1}$ is a scalar multiple of $\mathbf{1}$.

Our next sequence of results uses Theorem 6.4.2 in order to generalise some results in a paper of Kirkland and Neumann [83]. We begin with a result concerning mean first passage matrices for transition matrices having a special structure.

PROPOSITION 6.4.10 *Let T be an irreducible stochastic matrix of the following form*

$$T = \left[\begin{array}{c|c|c} T_{1,1} & T_{1,2} & 0 \\ \hline T_{2,1} & T_{2,2} & T_{2,3} \\ \hline 0 & T_{3,2} & T_{3,3} \end{array}\right], \tag{6.45}$$

where $T_{2,2}$ is a 1×1 block. Let S_1, S_2 and S_3 denote the index sets such that $T[S_i, S_i] = T_{i,i}, i = 1, 2, 3$. Let $r = \frac{1}{1+T_{2,3}(I-T_{3,3})^{-1}\mathbf{1}}$, and let $\tilde{T}$ be the stochastic matrix given by

$$\tilde{T} = \left[\begin{array}{c|c} T_{1,1} & T_{1,2} \\ \hline rT_{2,1} & 1 - rT_{2,1}\mathbf{1} \end{array}\right]. \tag{6.46}$$

Let M and $\tilde{M}$ denote the mean first passage matrices corresponding to T and $\tilde{T}$, respectively. Then

$$M[(S_1 \cup S_2), (S_1 \cup S_2)] - (M[(S_1 \cup S_2), (S_1 \cup S_2)])_{dg} = \tilde{M} - \tilde{M}_{dg}.$$

Proof: We begin by remarking that since T is irreducible, it must also be the case that $\tilde{T}$ is irreducible, so that $\tilde{M}$ is well–defined. We proceed by applying Theorem 6.4.2 to both T and $\tilde{T}$ with S taken as S_1 in each case. We also adopt the notation of Theorem 6.4.2 in our proof.

Let w and $\tilde{w}$ denote the stationary vectors of T and $\tilde{T}$, respectively. It is straightforward to determine that

$$w[S_2] = \frac{1}{1 + T_{2,1}(I - T_{1,1})^{-1}\mathbf{1} + T_{2,3}(I - T_{3,3})^{-1}\mathbf{1}};$$

the special form of T in (6.45) then yields

$$w[S_1]^t = \frac{1}{1 + T_{2,1}(I - T_{1,1})^{-1}\mathbf{1} + T_{2,3}(I - T_{3,3})^{-1}\mathbf{1}} T_{2,1}(I - T_{1,1})^{-1}.$$

Similarly we find that

$$\tilde{w}[S_1]^t = \frac{1}{1 + rT_{2,1}(I - T_{1,1})^{-1}\mathbf{1}} rT_{2,1}(I - T_{1,1})^{-1}.$$

Computing $w[S_1]^t\mathbf{1}$ and $\tilde{w}[S_1]^t\mathbf{1}$, and substituting in the value of r, it now follows that $w[S_1]^t\mathbf{1} = \tilde{w}[S_1]^t\mathbf{1}$. In particular (with the obvious notation) we find that $\gamma_{S_1} = \tilde{\gamma}_{S_1}$.

From (6.45) and (6.46), we see that $T[S_1, S_2] = T_{1,2} = \tilde{T}[S_1, S_2]$, while $T[S_2, S_1] = T_{2,1}$ and $\tilde{T}[S_2, S_1] = rT_{2,1}$. Further, we have

$$\mathcal{P}(T)_{S_1} = T_{1,1} + \frac{1}{T_{2,1}\mathbf{1}} T_{1,2}T_{2,1} = \mathcal{P}(\tilde{T})_{S_1}.$$

Finally, from the formula for the partitioned form of the inverse ([57, section 0.7.3]) and the structure of T, it follows that

$$e_1^t(I - T[S_2 \cup S_3, S_2 \cup S_3])^{-1} = \frac{1}{T_{2,1}\mathbf{1}} \left[\ 1 \mid T_{2,3}(I - T_{3,3})^{-1}\ \right]$$

and

$$(I - T[S_2 \cup S_3, S_2 \cup S_3])^{-1} e_1 = \frac{1}{T_{2,1}\mathbf{1}}\mathbf{1}.$$

Since $T[S_1, S_2 \cup S_3] = T_{1,2}e_1^t$, and $\mathcal{P}(T)_{S_1} = \mathcal{P}(\tilde{T})_{S_1}$, we have (again with the obvious notation)

$$V_{S_1} = \frac{1}{rT_{2,1}\mathbf{1}}((I - \mathcal{P}(T)_{S_1})^{\#}T_{1,2}\mathbf{1}^t - \mathbf{1}((I - \mathcal{P}(T)_{S_1})^{\#}T_{1,2})^t) = \tilde{V}_{S_1}.$$

Referring to (6.39) (for both M and $\tilde{M}$), we find that

$$M[S_1, S_1] = \gamma_{S_1} M_{\mathcal{P}(T)_{S_1}} + V_{S_1} = \tilde{\gamma}_{S_1} M_{\mathcal{P}(\tilde{T})_{S_1}} + \tilde{V}_{S_1} = \tilde{M}[S_1, S_1].$$

From (6.40), it follows that

$$M[S_1, S_2] = (I - T_{1,1})^{-1}\mathbf{1} = \tilde{M}[S_1, S_2].$$

Finally, applying (6.41) to M yields

$$M[S_2, S_1] = \frac{1}{T_{2,1}\mathbf{1}} \mathbf{1} T_{2,1} \left(M[S_1, S_1] - (M[S_1, S_1])_{dg} + J \right),$$

while applying (6.41) to $\tilde{M}$ yields

$$\tilde{M}[S_2, S_1] = \frac{1}{rT_{2,1}\mathbf{1}} \mathbf{1} r T_{2,1} \left(M[S_1, S_1] - (M[S_1, S_1])_{dg} + J \right) = M[S_2, S_1].$$

□

REMARK 6.4.11 Suppose that we have a matrix T having the form (6.45); label the index corresponding to S_2 by i. Observe that in $\mathcal{D}(T)$, every path from a vertex in S_1 to a vertex in S_3 must pass through vertex i, and conversely, every path from a vertex in S_3 to a vertex in S_1 must also pass through vertex i.

THEOREM 6.4.12 *Suppose that $T \in \mathcal{IS}_n$ has the following form*

$$T = \left[\begin{array}{c|c|c|c|c|c} T_{1,1} & 0 & 0 & \cdots & 0 & T_{1,k+1} \\ \hline 0 & T_{2,2} & 0 & \cdots & 0 & T_{2,k+1} \\ \hline \vdots & & \ddots & & \vdots & \vdots \\ \hline 0 & \cdots & 0 & T_{k-1,k-1} & 0 & T_{k-1,k+1} \\ \hline 0 & 0 & \cdots & 0 & T_{k,k} & T_{k,k+1} \\ \hline T_{k+1,1} & T_{k+1,2} & \cdots & T_{k+1,k-1} & T_{k+1,k} & T_{k+1,k+1} \end{array}\right], \quad (6.47)$$

where all of the diagonal blocks are square, and $T_{k+1,k+1}$ is 1×1. For each $j = 1, \ldots, k$, let

$$r_j = \frac{1}{1 + \sum_{l=1,\ldots,k, l \neq j} T_{k+1,l}(I - T_{l,l})^{-1}\mathbf{1}},$$

and let

$$\tilde{T}(j) = \left[\begin{array}{c|c} T_{j,j} & T_{j,k+1} \\ \hline r_j T_{k+1,j} & 1 - r_j T_{k+1,j}\mathbf{1} \end{array}\right].$$

Then partitioning M_T conformally with (6.47), $M_T - (M_T)_{dg}$ can be written as

$$M_T - (M_T)_{dg} = \left[\begin{array}{c|c|c|c|c} A_1 & a_1\mathbf{1}^t + \mathbf{1}b_2^t & \dots & a_1\mathbf{1}^t + \mathbf{1}b_k^t & a_1 \\ \hline a_2\mathbf{1}^t + \mathbf{1}b_1^t & A_2 & \dots & a_2\mathbf{1}^t + \mathbf{1}b_k^t & a_2 \\ \hline \vdots & & \ddots & & \vdots \\ \hline a_k\mathbf{1}^t + \mathbf{1}b_1^t & a_k\mathbf{1}^t + \mathbf{1}b_2^t & \dots & A_k & a_k \\ \hline b_1^t & b_2^t & \dots & b_k^t & 0 \end{array}\right], \tag{6.48}$$

where, for each $j = 1, \dots, k$,

$$\left[\begin{array}{c|c} A_j & a_j \\ \hline b_j^t & 0 \end{array}\right] = M_{\tilde{T}(j)} - \left(M_{\tilde{T}(j)}\right)_{dg}. \tag{6.49}$$

Proof: In view of Proposition 6.4.10, we see that (6.49) is immediate. Let $S_1, \dots, S_k, S_{k+1}$ denote the partitioning of $\{1, \dots, n\}$ corresponding to the partitioning of T in (6.47). Suppose that we have distinct indices i, j between 1 and k, and note that in $\mathcal{D}(T)$, any path from a vertex in S_i to a vertex in S_j must pass through vertex n. From Theorem 6.2.1, we see that if $p \in S_i$ and $q \in S_j$, then $(M_T)_{p,q} = (M_T)_{p,n} + (M_T)_{n,q}$. The expressions for the offdiagonal blocks in (6.48) now follow readily. □

We close the section with an example illustrating the conclusion of Theorem 6.4.12.

EXAMPLE 6.4.13 Consider the directed graph D on $n = \sum_{j=1}^{k} n_j + 1$ vertices that consists of k directed cycles of lengths $n_j + 1, j = 1, \dots, k$, such that each pair of cycles intersects in a unique common "central" vertex, say vertex n. (See Figure 6.2, which illustrates the general structure, but only explicitly depicts the central vertex n.) Let A denote the $(0, 1)$ adjacency matrix of D, and let T be the stochastic matrix formed from A be dividing each row by the corresponding row sum of A. Then with an appropriate labelling of the vertices of D, we find that T can be written as

$$T = \left[\begin{array}{c|c|c|c|c} N_{n_1} & 0 & \dots & 0 & e_{n_1} \\ \hline 0 & N_{n_2} & 0 & \dots & e_{n_2} \\ \hline \vdots & & \ddots & & \vdots \\ \hline 0 & \dots & 0 & N_{n_k} & e_{n_k} \\ \hline \frac{1}{k}e_1^t & \frac{1}{k}e_1^t & \dots & \frac{1}{k}e_1^t & 0 \end{array}\right], \tag{6.50}$$

where N_{n_j} is the matrix of order n_j with ones on the superdiagonal and zeros elsewhere.

Adopting the notation of Theorem 6.4.12, we find that

$$r_j = \frac{k}{k + \sum_{l=1,\ldots,k,l\neq j} n_l}, j = 1, \ldots, k.$$

Setting

$$t_j = \frac{1}{k + \sum_{l=1,\ldots,k,l\neq j} n_l}, j = 1, \ldots, k,$$

we find that for each $j = 1, \ldots, k$,

$$\tilde{T}(j) = \left[\begin{array}{c|c} N_{n_j} & e_{n_j} \\ \hline t_j e_1^t & 1 - t_j \end{array}\right].$$

It is straightforward to determine that

$$M_{\tilde{T}(j)} - (M_{\tilde{T}(j)})_{dg} =$$

$$\begin{bmatrix} 0 & 1 & 2 & \ldots & n_j - 1 & n_j \\ n_j - 1 + \frac{1}{t_j} & 0 & 1 & 2 & \ldots & n_j - 1 \\ n_j - 2 + \frac{1}{t_j} & n_j - 1 + \frac{1}{t_j} & 0 & 1 & \ldots & n_j - 2 \\ \vdots & & \ddots & & & \vdots \\ 1 + \frac{1}{t_j} & 2 + \frac{1}{t_j} & \ldots & n_j - 1 + \frac{1}{t_j} & 0 & 1 \\ \frac{1}{t_j} & 1 + \frac{1}{t_j} & 2 + \frac{1}{t_j} & \ldots & n_j - 1 + \frac{1}{t_j} & 0 \end{bmatrix}. \tag{6.51}$$

Using (6.51) and Theorem 6.4.12, we can readily determine the mean first passage time between any two distinct vertices on a common cycle in D, as well as the mean first passage time from n to any other vertex, and from any other vertex to n.

Finally, denote the partitioning of $\{1, \ldots, n\}$ induced by the partitioning of T by $S_1, \ldots, S_k, S_{k+1}$ (where S_{k+1} contains a single index, namely n). Suppose that $p \in S_i$ and $q \in S_j$ for distinct indices i, j between 1 and k. Denote the distance from p to n in D by d_p, and denote the distance from n to q in D by d_q. From Theorem 6.4.12 and (6.51), we find that the mean first passage time from p to q is given by

$$d_p + d_q - 1 + \frac{1}{t_j} = d_p + d_q - 1 + k + \sum_{l=1,\ldots,k,l\neq j} n_l.$$

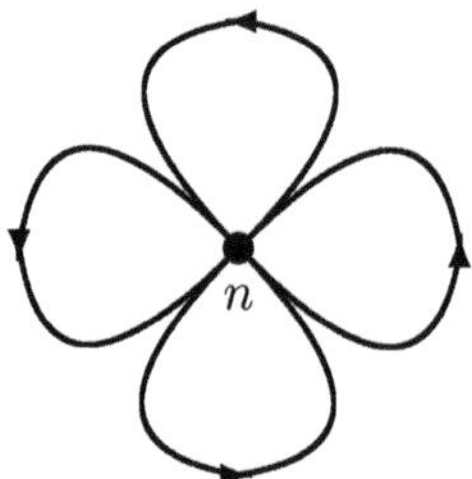

FIGURE 6.2
A directed graph of the type arising in Example 6.4.13

6.5 The Kemeny constant

Suppose that we have a matrix $T \in \mathcal{IS}_n$, with eigenvalues $1, \lambda_2, \lambda_2, \ldots, \lambda_n$. We denote the Kemeny constant associated with T by $\mathcal{K}(T)$, and recall that

$$\mathcal{K}(T) = \sum_{j=2}^{n} \frac{1}{1-\lambda_j}. \tag{6.52}$$

We have already encountered the Kemeny constant associated with T, once in section 5.3, where it arose in the context of the conditioning of the stationary distribution vector under perturbation of T, and once in section 6.1, where we saw its connection with mean first passage times. In this section we delve further into the properties of the Kemeny constant, with an emphasis on its interpretation in terms of mean first passage times and other quantities associated with the Markov chain in question.

Denote the mean first passage matrix associated with T by M, and let the stationary vector for T be w. From Theorem 6.1.9 we have

$$Mw = (\mathcal{K}(T) + 1)\mathbf{1},$$

or equivalently, recalling that $m_{i,i}w_i = 1, i = 1, \ldots, n$, we find that for each $i = 1, \ldots, n$,

$$\sum_{j=1,\ldots,n, j\neq i} m_{i,j}w_j = \mathcal{K}(T). \tag{6.53}$$

Observe that, remarkably, the right-hand side of (6.53) is independent of the choice of the index i.

In [60], Hunter provides the following interpretation of $\mathcal{K}(T)$, based on (6.53). Suppose that we have a random variable Y that takes on values in $\{1, \ldots, n\}$ and has the same distribution as w. Fix an index i between 1 and n, and consider the Markov chain u_k with transition matrix T, and initial state $u_0 = i$; we say that the Markov chain *achieves mixing* at time k_0 when k_0 is the first index $k \geq 1$ such that $u_k = Y$. The *time to mixing* for the Markov chain is then defined as the random variable $R = \min\{k \in \mathbb{N} | u_k = Y\}$. It is shown in [60] that the expected value of R is given by

$$E(R) = \sum_{j=1}^{n} m_{i,j} w_j.$$

(We note as an aside that the variance of the time to mixing depends on the initial state i in general; see Hunter's paper [61].) Thus we see that $E(R) = \mathcal{K}(T) + 1$, and so we may interpret the Kemeny constant associated with T in terms of the expected time to mixing for the corresponding Markov chain.

In [89], Levene and Loizou observe that

$$\mathcal{K}(T) + 1 = \sum_{i=1}^{n} w_i \sum_{j=1}^{n} m_{i,j} w_j,$$

and thus interpret $\mathcal{K}(T) + 1$ as the mean first passage time from an unknown initial state to an unknown destination state. This last observation has been applied to the modelling of vehicle traffic networks in a paper of Crisostomi, Kirkland, and Shorten [26], where $\mathcal{K}(T)$ is proposed as a measure of the overall efficiency of the network — i.e., if $\mathcal{K}(T)$ is small then the network is viewed as being efficient, since then the expected length of a random trip made by a vehicle is small.

Our first result on the Kemeny constant picks up on the theme of section 6.4.

THEOREM 6.5.1 *Suppose that $T \in \mathcal{IS}_n$. Let S be a nonempty proper subset of $\{1, \ldots, n\}$, and consider the corresponding stochastic complement $\mathcal{P}(T)_S$. Then*

$$\mathcal{K}(\mathcal{P}(T)_S) < \mathcal{K}(T). \tag{6.54}$$

Proof: Let w denote the stationary distribution for T, and note that the stationary vector for $\mathcal{P}(T)_S$ is given by $\frac{1}{\mathbf{1}^t w[S]} w[S]$. Referring to Theorem 6.4.2 (and keeping the notation of that theorem) we have

$$M_T[S, S] = \frac{1}{\mathbf{1}^t w[S]} M_{\mathcal{P}(T)_S} + V_S,$$

where V_S is given by (6.37). Since V_S is skew–symmetric, we find that

$$\frac{1}{\mathbf{1}^t w[S]} w[S]^t M_T[S,S]w[S] = \left(\frac{1}{\mathbf{1}^t w[S]} w[S]^t\right) M_{\mathcal{P}(T)_S} \left(\frac{1}{\mathbf{1}^t w[S]} w[S]\right).$$

Applying Theorem 6.1.9 to $\mathcal{P}(T)_S$, we have $M_{\mathcal{P}(T)_S}(\frac{1}{\mathbf{1}^t w[S]} w[S]) = (\mathcal{K}(\mathcal{P}(T)_S)+1)\mathbf{1}$, while an application of Theorem 6.1.9 to T yields that

$$\frac{1}{\mathbf{1}^t w[S]} w[S]^t M_T[S,S]w[S] = \frac{1}{\mathbf{1}^t w[S]} w[S]^t((\mathcal{K}(T)+1)\mathbf{1} - M[S,\overline{S}]w[\overline{S}]) =$$
$$(\mathcal{K}(T)+1) - \frac{1}{\mathbf{1}^t w[S]} w[S]^t M[S,\overline{S}]w[\overline{S}].$$

Consequently,

$$\mathcal{K}(\mathcal{P}(T)_S) = K(T) - \frac{1}{\mathbf{1}^t w[S]} w[S]^t M[S,\overline{S}]w[\overline{S}] < \mathcal{K}(T),$$

as desired. □

The conclusion of Theorem 6.5.1 yields the interpretation that the stochastic complement $\mathcal{P}(T)_S$ is "faster mixing" than T is, in the sense that the expected time to mixing of the former is strictly less than that of the latter. We note further that while the connection between the non–Perron eigenvalues of T and $\mathcal{P}(T)_S$ is not especially well-understood, Theorem 6.5.1 does provides some information on the relationship between those two sets of eigenvalues. The following example (which appears in the paper of Ipsen and Kirkland [62] in a different setting) illustrates.

EXAMPLE 6.5.2 Consider the 3×3 stochastic matrix T given by

$$T = \begin{bmatrix} \frac{5}{6} & 0 & \frac{1}{6} \\ \frac{3}{4} & \frac{1}{6} & \frac{1}{12} \\ \frac{2}{3} & \frac{1}{3} & 0 \end{bmatrix}.$$

The eigenvalues of T are 1 and 0, the latter with algebraic multiplicity two (and geometric multiplicity one). Hence $\mathcal{K}(T) = 2$. Letting $S = \{1,2\}$, we find that

$$\mathcal{P}(T)_S = \begin{bmatrix} \frac{17}{18} & \frac{1}{18} \\ \frac{29}{36} & \frac{7}{36} \end{bmatrix},$$

which has eigenvalues 1 and $\frac{5}{36}$. Hence $\mathcal{K}(\mathcal{P}(T)_S) = \frac{36}{31}$.

It is interesting to note here that the subdominant eigenvalue for $\mathcal{P}(T)_S$ is $\frac{5}{36}$, which exceeds the subdominant eigenvalue 0 of T. So, while the expected time to mixing for the Markov chain associated with $\mathcal{P}(T)_S$ is smaller than the expected time to mixing for the Markov chain corresponding to T, we see that the asymptotic rate of convergence to the stationary distribution of the latter chain is faster than that of the former.

Next, we consider the Kemeny constant for an irreducible periodic stochastic matrix.

THEOREM 6.5.3 *Suppose that $T \in \mathcal{IS}_n$ is periodic with period $d \geq 2$, and is written in periodic normal form as*

$$T = \left[\begin{array}{c|c|c|c|c|c} 0 & T_1 & 0 & 0 & \cdots & 0 \\ \hline 0 & 0 & T_2 & 0 & \cdots & 0 \\ \hline \vdots & & \ddots & \ddots & & \vdots \\ \hline 0 & \cdots & & 0 & T_{d-2} & 0 \\ \hline 0 & 0 & \cdots & & 0 & T_{d-1} \\ \hline T_d & 0 & 0 & \cdots & 0 & 0 \end{array}\right]. \tag{6.55}$$

Let $S_i = T_i T_{i+1} \dots T_d T_1 \dots T_{i-1}, i = 1, \dots, d$, and denote the order of S_i by n_i for $i = 1, \dots, d$. Then for each $i = 1, \dots, d, \mathcal{K}(T) = d(\mathcal{K}(S_i) - n_i) + n + \frac{d-1}{2}$.

<u>**Proof**</u>: Let w_1 denote the stationary distribution vector for S_1, and for each $j = 2, \dots, d$, let $w_j^t = w_1^t T_1 T_2 \dots T_{j-1}$. Since $\mathcal{K}(T) = \text{trace}(I - T)^{\#}$, we find from Theorem 5.2.2 that $K(T) = \sum_{i=1}^{d} \text{trace}((I - S_i)^{\#} + \frac{d-1}{2d} \mathbf{1} w_i^t)$. Fix an index i between 2 and d, let $U = T_i T_{i+1} \dots T_d$ and $V = T_1 \dots T_{i-1}$. Then $S_i = UV$ and $S_1 = VU$. From Lemma 5.2.1 we have $(I - UV)^{\#} = I - \mathbf{1} w_i^t + U(I - VU)^{\#} V$. Next, we note that

$$\begin{aligned} &\text{trace}(U(I - VU)^{\#} V) = \text{trace}((I - VU)^{\#} VU) = \\ &\text{trace}((I - VU)^{\#} - I + \mathbf{1} w_1^t) = \text{trace}((I - VU)^{\#} - (n_1 - 1). \end{aligned}$$

Consequently, we find that $\text{trace}((I - S_i)^{\#}) = n_i - n_1 + \text{trace}((I - S_1)^{\#})$ for each $i = 1, \dots, d$. Hence,

$$\begin{aligned} \sum_{i=1}^{d} \text{trace}(I - S_i)^{\#} &= \sum_{i=1}^{d} (\text{trace}((I - S_1)^{\#}) + n_i - n_1) \\ &= d(\text{trace}((I - S_1)^{\#}) - n_1) + n. \end{aligned}$$

It now follows readily that for each $j = 1, \ldots, d$,

$$\mathcal{K}(T) = d(\mathcal{K}(S_j) - n_j) + n + \frac{d-1}{2}$$

□

It is shown in [60] that for a matrix $T \in \mathcal{IS}_n$, $\mathcal{K}(T) \geq \frac{n-1}{2}$, with equality holding in the case that T is the adjacency matrix of a directed cycle of length n. Our next sequence of results considers the following related problem. Given a strongly connected directed graph Δ on n vertices, let

$$\mathcal{S}(\Delta) = \{T \in \mathbb{R}^{n \times n} | T \text{ is stochastic and } i \to j \in \mathcal{D}(T) \text{ only if } i \to j \in \Delta\}.$$

We want to find

$$\mu(\Delta) = \inf\{\mathcal{K}(T) | T \in \mathcal{S}(\Delta) \cap \mathcal{AS}_n\}.$$

Observe here that we have slightly expanded the class of matrices for with the Kemeny constant is defined; we now include the stochastic matrices having 1 as a simple eigenvalue. For such a matrix T with eigenvalues $1, \lambda_2, \ldots, \lambda_n$, we define $\mathcal{K}(T)$ via (6.52), but note that the interpretation of $\mathcal{K}(T)$ in terms of mean first passage times only holds when T is irreducible. We note in passing that since the irreducible members of $\mathcal{S}(\Delta)$ are dense in that set, it follows that $\mu(\Delta)$ is also equal to $\inf\{\mathcal{K}(T) | T \in \mathcal{S}(\Delta) \text{ is irreducible}\}$.

Following the development presented in Kirkland's paper [74], we will find $\mu(\Delta)$ via a sequence of technical results.

LEMMA 6.5.4 *Consider a stochastic matrix T of the form*

$$T = \left[\begin{array}{c|c} C & 0 \\ \hline X & N \end{array}\right], \tag{6.56}$$

where C is the adjacency matrix of a directed cycle of length ℓ and N is a nilpotent matrix. Then $\mathcal{K}(T) = \frac{2n-\ell-1}{2}$.

Proof: It is readily seen that

$$(I-T)^{\#} = \left[\begin{array}{c|c} (I-C)^{\#} & 0 \\ \hline (I-N)^{-1}X(I-C)^{\#} - \frac{1}{\ell}(I-N)^{-1}J & (I-N)^{-1} \end{array}\right],$$

where J is an all ones matrix of the appropriate order. Hence $K(T) = K(C) + n - \ell$. From Example 5.2.3, we find that $\text{trace}((I-C)^{\#}) = \frac{\ell-1}{2}$. The expression for $\mathcal{K}(T)$ now follows. □

COROLLARY 6.5.5 ([74]) *Let Δ be a strongly connected directed graph on n vertices, and denote the length of the longest cycle in Δ by k. Then $\mu(\Delta) \le \frac{2n-k-1}{2}$.*

<u>**Proof**</u>: We construct a spanning subgraph of Δ as follows. Begin with a cycle of length k in Δ, let V_0 denote the collection of vertices on that cycle, and let A_0 denote the collection of arcs on that cycle. Next, for each $l \ge 0$ such that $| \cup_{p=0}^{l} V_p| < n$, we let V_{l+1} be the set of vertices in Δ having an out–arc to some vertex in V_l. For each $i \in V_{l+1}$, select a single vertex $j_i \in V_l$ such that $i \to j_i$ is an arc in Δ, and let A_{l+1} denote a collection of arcs $i \to j_i, i \in V_{l+1}$. There is a minimum index m such that $| \cup_{p=0}^{m} V_p| = n$, and now we let $\hat{\Delta}$ be the spanning subgraph of Δ with arc set $\cup_{p=0}^{m} A_p$. Observe that $\hat{\Delta}$ contains a single directed cycle of length k, that every vertex in $\hat{\Delta}$ has outdegree 1, and that for any vertex j of $\hat{\Delta}$ not on the cycle, there is a directed path from j to the cycle.

It now follows that the $(0,1)$ adjacency matrix of $\hat{\Delta}$ (which is an element of $\mathcal{S}(\Delta)$ having 1 as a simple eigenvalue) can be written in the form (6.56), with C as the adjacency matrix of the directed cycle of length k. The conclusion now follows readily from Lemma 6.5.4. $\square$

We now show that there is a matrix $T \in \mathcal{S}(\Delta)$ yielding the minimum value of $\mathcal{K}(T)$.

LEMMA 6.5.6 ([74]) *Let Δ be a strongly connected directed graph. There is a matrix $T \in \mathcal{S}(\Delta)$ having 1 as a simple eigenvalue such that $\mathcal{K}(T) = \mu(\Delta)$.*

<u>**Proof**</u>: We first claim that if $S \in \mathcal{S}(\Delta)$ has 1 as a simple eigenvalue and $\mathcal{K}(S) \le n$, then for any eigenvalue $\lambda \ne 1$ of S, $|1-\lambda| \ge \frac{1-\cos\left(\frac{2\pi}{n}\right)}{n}$. To see the claim, first note that if $\lambda \in \mathbb{R}$, then $\frac{1}{|1-\lambda|} = \frac{1}{1-\lambda} \le \mathcal{K}(A) \le n$, and hence $|1-\lambda| \ge \frac{1}{n}$. On the other hand, if λ is complex, then writing $\lambda = x+iy$, we have $n \ge K(A) \ge \frac{1}{1-\lambda} + \frac{1}{1-\bar{\lambda}} = \frac{2(1-x)}{(1-x)^2+y^2}$. By a result of Dmitriev and Dynkin in [32], $|y| \le (1-x)\frac{\sin\left(\frac{2\pi}{n}\right)}{1-\cos\left(\frac{2\pi}{n}\right)}$. Hence $y^2 \le (1-x)^2\frac{\sin^2\left(\frac{2\pi}{n}\right)}{(1-\cos\left(\frac{2\pi}{n}\right))^2}$. Consequently, $\frac{2(1-x)}{(1-x)^2+y^2} \ge \frac{1-\cos\left(\frac{2\pi}{n}\right)}{1-x}$. We now find that $|1-\lambda| \ge 1-x \ge \frac{1-\cos\left(\frac{2\pi}{n}\right)}{n}$, as claimed.

Let T_m be a sequence of matrices in $\mathcal{S}(\Delta)$, each having 1 as a simple eigenvalue, such that $\mathcal{K}(T_m) \to \mu(\Delta)$ as $m \to \infty$. Since $\mathcal{S}(\Delta)$ is a compact set, there is a subsequence T_{m_j} of T_m such that T_{m_j} converges in

$\mathcal{S}(\Delta)$ as $j \to \infty$. Let $T = \lim_{j\to\infty} T_{m_j}$. From Corollary 6.5.5, it follows that $\mathcal{K}(T_{m_j}) \le n$ for all sufficiently large j. Hence, from our claim above, for all such j, and any eigenvalue $\lambda \ne 1$ of T_{m_j}, $|1-\lambda| \ge \frac{1-\cos\left(\frac{2\pi}{n}\right)}{n}$. Consequently the matrix T has 1 as a simple eigenvalue. From Theorem 5.3.30 it follows that $(I-T_{m_j})^{\#} \to (I-T)^{\#}$ as $j \to \infty$, and hence

$$\mu(\Delta) = \lim_{j\to\infty} \mathcal{K}(T_{m_j}) = \lim_{j\to\infty} \text{trace}\left((I-T_{m_j})^{\#}\right) = \text{trace}\left((I-T)^{\#}\right) = \mathcal{K}(T).$$

□

Our next technical result describes the behaviour of the Kemeny constant when a single row of the transition matrix is perturbed.

LEMMA 6.5.7 ([74]) *Suppose that $T, \tilde{T} \in \mathcal{AS}_n$ and that $\tilde{T} = T + e_i u^t$ for some vector u such that $u^t\mathbf{1} = 0$. Letting $Q = I - T$, we have*

$$\mathcal{K}(\tilde{T}) = \mathcal{K}(T) + \frac{u^t(Q^{\#})^2 e_i}{1 - u^t Q^{\#} e_i}. \tag{6.57}$$

Proof: Let w denote the stationary vector for T. From Corollary 5.3.31, we find that

$$(I-\tilde{T})^{\#} = Q^{\#} + \frac{1}{1-u^tQ^{\#}e_i} Q^{\#} e_i u^t Q^{\#} - \frac{w_i}{1-u^tQ^{\#}e_i} \mathbf{1}\left(u^t(Q^{\#})^2 + \frac{u^t(Q^{\#})^2 e_i}{1-u^tQ^{\#}e_i} u^t Q^{\#}\right).$$

Note that for any square rank one matrix ab^t (where a and b are vectors in $\mathbb{R}^n$) we have $\text{trace}(ab^t) = b^t a$. Thus we find that $\text{trace}(Q^{\#} e_i u^t Q^{\#}) = u^t(Q^{\#})^2 e_i$, while

$$\text{trace}\left(\mathbf{1}\left(u^t(Q^{\#})^2 + \frac{u^t(Q^{\#})^2 e_i}{1-u^tQ^{\#}e_i} u^t Q^{\#}\right)\right) = u^t(Q^{\#})^2\mathbf{1} + \frac{u^t(Q^{\#})^2 e_i}{1-u^tQ^{\#}e_i} u^t Q^{\#}\mathbf{1} = 0.$$

The conclusion now follows readily. □

The following result determines $\mu(\Delta)$.

THEOREM 6.5.8 ([74]) *Let Δ be a strongly connected directed graph on n vertices. Let k denote the length of a longest cycle in Δ. Then*

$$\mu(\Delta) = \frac{2n-k-1}{2}. \tag{6.58}$$

Proof: We begin with a claim that there is a $(0,1)$ matrix $A \in \mathcal{S}(\Delta)$ having 1 as a simple eigenvalue such that $\mathcal{K}(T) = \mu(\Delta)$. To see the claim, first let T be a matrix in $\mathcal{S}(\Delta)$ with 1 as a simple eigenvalue, such that $\mathcal{K}(T) = \mu(\Delta)$ (such a T is guaranteed to exist by Lemma 6.5.6). If T is a $(0,1)$ matrix, we have nothing to show, so suppose that some row of T, say the i-th, contains an entry that is strictly between 0 and 1. Then there are indices p, q such that $t_{i,p}, t_{i,q} > 0$. Let $Q = I - T$, and observe that for each $t \in (-s_{ip}, s_{iq})$, $T + te_i(e_p - e_q)^t \in \mathcal{S}(\Delta)$ for all such t. Further, $\mathcal{D}(T + te_i(e_p - e_q)^t) = \mathcal{D}(T)$ for each $t \in (-s_{ip}, s_{iq})$, from which we deduce that 1 is a simple eigenvalue of $T + te_i(e_p - e_q)^t$ for each such t. From Lemma 6.5.7, it follows that

$$\mathcal{K}(T + te_i(e_p - e_q)^t) = \mathcal{K}(T) + \frac{t(e_p - e_q)^t (Q^{\#})^2 e_i}{1 - t(e_p - e_q)^t Q^{\#} e_i}$$

for each $t \in (-s_{ip}, s_{iq})$. Since $\mathcal{K}(T) = \mu(\Delta)$, it must be the case that $(e_p - e_q)^t (Q^{\#})^2 e_i = 0$, otherwise we could find a small (positive or negative) value of t so that $\mathcal{K}(T + te_i(e_p - e_q)^t) < \mathcal{K}(T)$, a contradiction. Thus we have $\mu(\Delta) = \mathcal{K}(T) = \mathcal{K}(T + te_i(e_p - e_q)^t)$ for all $t \in (-s_{ip}, s_{iq})$. As in the proof of Lemma 6.5.6, the matrix $T - s_{ip} e_i(e_p - e_q)^t$ has 1 as a simple eigenvalue, and $\mathcal{K}(T - s_{ip} e_i(e_p - e_q)^t) = \lim_{t \to -s_{ip}^+} \mathcal{K}(T + te_i(e_p - e_q)^t)$. We deduce that $\mathcal{K}(T - s_{ip} e_i(e_p - e_q)^t) = \mu(\Delta)$. Note that $T - s_{ip} e_i(e_p - e_q)^t$ has fewer positive entries than T does, and so iterating the argument if necessary, we find that there is a $(0,1)$ matrix $A \in \mathcal{S}(\Delta)$ having 1 as a simple eigenvalue such that $\mathcal{K}(A) = \mu(\Delta)$, as claimed.

Since A is $(0,1)$ with 1 as a simple eigenvalue, we find that A can be written in the form (6.56), where C is the adjacency matrix of a directed cycle, say of length l, and where N is nilpotent. Applying Lemma 6.5.4 yields that $\mu(\Delta) = \mathcal{K}(A) = \frac{2n-l-1}{2} \geq \frac{2n-k-1}{2}$. From Corollary 6.5.5, we also have $\mu(\Delta) \leq \frac{2n-k-1}{2}$, so that necessarily $l = k$ and $\mu(\Delta) = \frac{2n-k-1}{2}$.
□

The following is immediate.

COROLLARY 6.5.9 *Suppose that* $T \in \mathcal{IS}_n$, *and that the longest cycle in* $\mathcal{D}(T)$ *has length* l. *Then*

$$\mathcal{K}(T) \geq \frac{2n - l - 1}{2}. \tag{6.59}$$

REMARK 6.5.10 Observe that the proof of Theorem 6.5.8 shows that there is a $(0,1)$ matrix A in $\mathcal{S}(\Delta)$ whose Kemeny constant coincides with

$\mu(\Delta)$; however note that such a $(0,1)$ matrix may be reducible. While we will not pursue the matter here, in [74] there is a characterisation of all of the matrices in $\mathcal{S}(\Delta)$ having minimum Kemeny constant $\mu(\Delta)$. In Example 6.5.12 below, we exhibit one class of irreducible matrices yielding equality in Corollary 6.5.9.

Theorem 6.5.8 leads to the following result, which gives an inequality relating mean first passage times, stationary vector entries, and the length of a longest cycle.

COROLLARY 6.5.11 *Suppose that Δ is a strongly connected directed graph on n vertices, and let $T \in \mathcal{S}(\Delta)$ be irreducible with stationary distribution vector w and mean first passage matrix M. Let k be the length of a longest cycle in Δ. For each index $i = 1, \ldots, n$, there is an index $j \neq i$ such that $m_{ij} \geq \frac{2n-k-1}{2(1-w_i)}$.*

Proof: From Theorem 6.5.8, we find that $\mathcal{K}(T) \geq \frac{2n-k-1}{2}$. Since $\mathcal{K}(T) = \sum_{l \neq i} m_{il} w_l$, we see that $\frac{\mathcal{K}(T)}{1-w_i}$ is a weighted average of the quantities $m_{il}, l = 1, \ldots, n, l \neq i$. The conclusion now follows readily. □

We close this chapter by finding the Kemeny constant for a Markov chain whose transition matrix is of the form in Example 6.4.13.

EXAMPLE 6.5.12 Suppose that for parameters $n_1, \ldots, n_k \in \mathbb{N}, T$ is the stochastic matrix given by (6.50). Set $n = \sum_{j=1}^{k} n_j + 1$, so that T is $n \times n$. It is straightforward to determine that the stationary distribution vector for T is given by

$$w = \frac{1}{n-1+k} \left[\begin{array}{c} \mathbf{1}_{n-1} \\ \hline k \end{array} \right],$$

where the subscript on the all ones vector denotes its order. Let M denote the mean first passage matrix for T. From Theorem 6.1.9 we find that $\mathcal{K}(T) = e_n^t (M - M_{dg}) w$.

For each $j = 1, \ldots, k$, let $t_j = \frac{1}{k + \sum_{l \neq j} n_j}$, and let b_j be the vector of order n_j given by

$$b_j^t = \left[\begin{array}{ccccc} \frac{1}{t_j} & (\frac{1}{t_j} + 1) & (\frac{1}{t_j} + 2) & \ldots & (\frac{1}{t_j} + n_j - 1) \end{array} \right].$$

From Theorem 6.4.12, (6.51), and the fact that $\mathcal{K}(T) = e_n^t (M - M_{dg}) w$, it now follows that $\mathcal{K}(T) = \frac{1}{n-1+k}\sum_{j=1}^{k} b_j^t \mathbf{1}$. For each $j = 1, \ldots, k$, we find that

$$b_j^t \mathbf{1} = \frac{n_j}{t_j} + \frac{n_j(n_j-1)}{2} = n_j\left(n-1+k-\frac{n_j+1}{2}\right).$$

Consequently, we have

$$\mathcal{K}(T) = \frac{(n-1)(2n+2k-3) - \sum_{j=1}^{k} n_j^2}{2(n-1+k)}.$$

In the special case that all of the n_js are equal, say $n_j = m, j = 1, \ldots, k$, we see that every cycle in $\mathcal{D}(T)$ has length $m+1$ (see Figure 6.2). When this occurs, the expression for $\mathcal{K}(T)$ simplifies to $\frac{2n-(m+1)-1}{2}$, so that equality holds in (6.59).

7

Applications of the Group Inverse to Laplacian Matrices

In this chapter we focus on Laplacian matrices for weighted graphs, paying special attention to the interplay between the group inverse of the Laplacian matrix and the underlying structure of the weighted graph. In section 7.2, we consider the class of weighted trees, and derive formulas expressing the entries of the group inverse of the corresponding Laplacian matrices in terms of certain weighted distances between vertices. In particular, we characterise those weighted trees for which the group inverse of the Laplacian matrix is an M-matrix. Section 7.3 derives bounds on the algebraic connectivity of a weighted graph in terms of the group inverse of the corresponding Laplacian matrix, and discusses classes of graphs for which equality holds. In section 7.4, we deal with the resistance distance between vertices of a weighted graph, develop some of its main properties, and discuss the Kirchhoff index for weighted graphs. Finally, in section 7.5 we discuss the use of the group inverse of the Laplacian matrix in the analysis of electrical networks.

Throughout this chapter we will assume a basic knowledge of graph theory, and we will employ some common graph-theoretic terminology and notation. The interested reader may refer to the books of Bondy and Murty [13] or Harary [51] for the necessary background material.

7.1 Introduction to the Laplacian matrix

Suppose that we have an undirected graph $\mathcal{G}$ on vertices labelled $1, \ldots, n$. There are a number of different matrices that can be associated with $\mathcal{G}$, including the vertex-edge incidence matrix, the adjacency matrix, and the Seidel matrix; the study of graph spectra considers the algebraic and analytic properties of these and other matrices, in order to obtain information about the structure and properties of $\mathcal{G}$. We refer the reader to the books of Cvetkovic, Doob, and Sachs [28] and Cvetkovic, Doob, Gutman, and Torgasev [27], as well as the references contained therein, for an overview of key results in spectral graph theory.

For each index $i = 1, \ldots, n$, we let d_i be the *degree* of vertex i in $\mathcal{G}$ – that is, the number of edges in $\mathcal{G}$ that are incident with i. Given distinct indices i, j we write $i \sim j$ to denote the fact that i and j are adjacent in $\mathcal{G}$, and $i \not\sim j$ if they are not adjacent. The *Laplacian matrix* for $\mathcal{G}$, denoted by $L(\mathcal{G})$ (or simply by L when the graph in question is clear from the context) is given by $L(\mathcal{G}) = [l_{i,j}]_{i,j=1,\ldots,n}$, where

$$l_{i,j} = \begin{cases} -1 & \text{if } i \sim j, \\ 0 & \text{if } i \neq j \text{ and } i \not\sim j, \\ d_i & \text{if } i = j. \end{cases} \tag{7.1}$$

Evidently $L(\mathcal{G})$ has nonpositive off-diagonal entries and nonnegative diagonal entries, and observing that $L(\mathcal{G})\mathbf{1} = 0$, it follows readily that in fact $L(\mathcal{G})$ is an M-matrix (see Exercise 4.1.4 of [12]). In the case that $\mathcal{G}$ is connected, we find that $L(\mathcal{G})$ is irreducible, so that 0 is a simple eigenvalue of $L(\mathcal{G})$. More generally, it is not difficult to show that the multiplicity of 0 as an eigenvalue of $L(\mathcal{G})$ coincides with the number of connected components in $\mathcal{G}$.

It turns out that some information about the nature of $\mathcal{G}$ can be determined from the corresponding Laplacian matrix. For instance, suppose that $\mathcal{G}$ has connected components $C_1, \ldots, C_k$, and for each $i = 1, \ldots, k$ define the $(0,1)$ vector $z(C_i)$ by

$$z(C_i)_j = \begin{cases} 1 & \text{if } j \in C_i, \\ 0 & \text{if } j \notin C_i \end{cases}$$

(evidently $z(C_i)$ is just an indicator vector for C_i). It is readily determined that the vectors $z(C_i), i = 1, \ldots, k$ form a basis for the null space of $L(\mathcal{G})$. Thus, knowledge of the structure of the null space of the Laplacian yields information on the connected components of the underlying graph. In a different direction, the matrix–tree theorem, attributed to Kirchhoff, states that the determinant of any $(n-1) \times (n-1)$ principal submatrix of $L(\mathcal{G})$ coincides with the number of *spanning trees* of $\mathcal{G}$, i.e., the number of spanning subgraphs that are connected and have no cycles; see [15, section 2.5] for a proof.

A *weighted graph* $(\mathcal{G}, w)$ consists of a graph $\mathcal{G}$ (on vertices $1, \ldots, n$, say) together with a function w on the edges of $\mathcal{G}$ such that whenever vertices i and j are adjacent in $\mathcal{G}, w_{i,j} = w_{j,i} > 0$. The function w is known as a *weighting* of $\mathcal{G}$ and the value $w_{i,j}$ is known as the *weight* of the edge between vertices i and j. If e is an edge of $\mathcal{G}$ with end points i, j, we will occasionally use the alternate notation $w(e)$ to denote the weight of the edge e, so that $w(e) = w_{i,j}$. Given such a weighted graph, we may form the corresponding Laplacian matrix $L(\mathcal{G}, w) = [l_{i,j}]_{i,j=1,\ldots,n}$, where

$$l_{i,j} = \begin{cases} -w_{i,j} & \text{if } i \sim j, \\ 0 & \text{if } i \neq j \text{ and } i \not\sim j, \\ \sum_{k=1,\ldots,n,k\sim i} w_{i,k} & \text{if } i = j. \end{cases}.$$

In the special case that all of the edge weights are equal to one, we recover the usual Laplacian matrix $L(\mathcal{G})$ from (7.1); in that case we say that the graph $\mathcal{G}$ is *unweighted.*

As above, we see that $L(\mathcal{G}, w)$ is a symmetric, singular M-matrix. Further, the multiplicity of zero as an eigenvalue of $L(\mathcal{G}, w)$ is the same as the number of connected components of $\mathcal{G}$, and those connected components correspond to a basis of the null space of $L(\mathcal{G}, w)$, as in the case of the usual Laplacian matrix. It is not difficult to determine that any symmetric singular M-matrix B having $\mathbf{1}$ as a null vector can be thought of as a weighted graph, where vertices i and j are adjacent whenever $i \neq j$ and $b_{i,j} < 0$, and where the weight of the corresponding edge is $|b_{i,j}|$.

Our goal in this chapter is to consider Laplacian matrices for connected graphs (either weighted or unweighted), to analyse the corresponding group inverses, and to use the group inverse as a tool in generating graph-theoretic insight. Since the Laplacian matrix for a weighted graph is symmetric, then as noted in section 2.4, its group inverse coincides with its Moore–Penrose inverse. In order to be consistent with the preceding chapters, we shall phrase our results in this chapter in terms of the group inverse, rather than in terms of the Moore–Penrose inverse.

7.2 Distances in weighted trees

In this section we focus on weighted trees, and develop various expressions for the entries in the group inverses of the corresponding Laplacian matrices. These expressions will be given in terms of both the edge weights and various combinatorial quantities associated with the tree. Much of our development will follow the approach of Kirkland, Neumann, and Shader in [78].

We now introduce some of the notation and terminology that will be used throughout the remainder of this chapter. Let $\mathcal{T}$ be a tree on vertices $1, \ldots, n$. For each edge e of $\mathcal{T}$, we let $\mathcal{T} \setminus e$ denote the graph formed from $\mathcal{T}$ by deleting the edge e; evidently $\mathcal{T} \setminus e$ has two connected components, each of which is a tree. For each edge e and vertex i of $\mathcal{T}$, we let $\beta_i(e)$ denote the set of vertices in the connected component of $\mathcal{T} \setminus e$ that does not contain vertex i. Given any pair of distinct indices i, j, we let $P_{i,j}$ denote the (unique) path in $\mathcal{T}$ from vertex i to vertex j; it is straightforward to see that e is an edge of $P_{i,j}$ if and only if $j \in \beta_i(e)$. A number of our arguments in the sequel involve expressions of the form $|\beta_i(e)|(n - |\beta_i(e)|)$ for some edge e and index i; we remark here that in fact the quantity $|\beta_i(e)|(n - |\beta_i(e)|)$ is independent of the choice of i, and coincides with the number of paths in $\mathcal{T}$ that contain the edge e. Evidently the notations $\beta_i(e)$ and $P_{i,j}$ suppress the explicit dependence on the underlying tree $\mathcal{T}$; however, the tree under consideration will always be clear from the context.

We begin with a basic result that holds for general connected weighted graphs.

OBSERVATION 7.2.1 Suppose that $\mathcal{G}$ is a connected graph on n vertices, and let w be a weighting of $\mathcal{G}$. The following is immediate from Observation 2.3.4. Consider the Laplacian matrix $L(\mathcal{G}, w)$, and let B denote the inverse of the $(n-1) \times (n-1)$ leading principal submatrix of $L(\mathcal{G}, w)$. Then

$$L(\mathcal{G}, w)^{\#} = \frac{\mathbf{1}^t B \mathbf{1}}{n^2} J + \left[\begin{array}{c|c} B - \frac{1}{n} BJ - \frac{1}{n} JB & -\frac{1}{n} B\mathbf{1} \\ \hline -\frac{1}{n} \mathbf{1}^t B & 0 \end{array} \right].$$

Referring to Observation 7.2.1, we see that the matrix B facilitates the computation of $L(\mathcal{G}, w)^{\#}$. We now turn our attention to the problem of providing a combinatorial description of the entries in the matrix B. In order to do so, we require some notation and terminology. For any subgraph $\mathcal{H}$ of $\mathcal{G}$, we define the *weight* of $\mathcal{H}$, $w(\mathcal{H})$, to be the product of the weights of the edges in $\mathcal{H}$, where we take the convention that a graph with no edges has weight 1. Recall that an undirected graph is a *forest* if it does not contain any cycles. Let $\mathcal{S}$ denote the set of spanning trees of $\mathcal{G}$, and for each triple of indices $i, j, k = 1, \ldots, n$, we let $\mathcal{S}_k^{\{i,j\}}$ denote the set of spanning forests of $\mathcal{G}$ with the property that each forest contains exactly two connected components, one of which contains vertex k and the other of which contains both vertex i and vertex j.

Fix indices i and j, let $\tilde{L}_{\{n\},\{n\}}$ denote the principal submatrix of L formed by deleting row and column n, and let $\tilde{L}_{\{j,n\},\{i,n\}}$ denote the submatrix of L formed by deleting rows j, n and columns i, n. From the adjoint formula for the inverse ([57, section 0.8.2]), we find that for each $i, j = 1, \ldots, n$,

$$b_{i,j} = (-1)^{i+j} \frac{\det(\tilde{L}_{\{j,n\},\{i,n\}})}{\det(\tilde{L}_{\{n\},\{n\}})}. \tag{7.2}$$

Evidently combinatorial descriptions of $\det(\tilde{L}_{\{j,n\},\{i,n\}})$ and $\det(\tilde{L}_{\{n\},\{n\}})$ will in turn furnish a combinatorial interpretation of $b_{i,j}$. Fortunately, the following result, which is a consequence of the all-minors matrix tree theorem (see Chaiken's paper [19]), yields the desired descriptions.

PROPOSITION 7.2.2 *Let $\mathcal{G}$ be a connected graph on n vertices, and let w be a weighting of its edges. Denote the Laplacian matrix for the weighted graph $(\mathcal{G}, w)$ by L. Fix indices i and j between 1 and $n-1$, let $\tilde{L}_{\{n\},\{n\}}$ be the submatrix of L formed by deleting row and column n, and let $\tilde{L}_{\{j,n\},\{i,n\}}$ be the submatrix of L formed by deleting rows j, n and columns i, n. Then*

$$\det(\tilde{L}_{\{n\},\{n\}}) = \sum_{\mathcal{T} \in \mathcal{S}} w(\mathcal{T})$$

and

$$\det(\tilde{L}_{\{j,n\},\{i,n\}}) = (-1)^{i+j} \sum_{\mathcal{F} \in \mathcal{S}_n^{\{i,j\}}} w(\mathcal{F}).$$

The matrix B above is known as the *bottleneck matrix based at vertex n* for the weighted graph $(\mathcal{G}, w)$. The motivation for that name stems from the following result, which is an easy consequence of Proposition 7.2.2.

COROLLARY 7.2.3 ([78]) *Let $(\mathcal{T}, w)$ be a weighted tree on vertices $1, \ldots, n$, and let B denote the corresponding bottleneck matrix based at vertex n for T. Then for each $i, j = 1, \ldots, n-1$,*

$$b_{i,j} = \sum_{e \in P_{i,n} \cap P_{j,n}} \frac{1}{w(e)},$$

where the sum is taken over all edges e on both $P_{i,n}$ and $P_{j,n}$, and is *interpreted as zero in the case that $P_{i,n}$ and $P_{j,n}$ do not have any edges in common.*

Proof: Fix a pair of (not necessarily distinct) indices i, j between 1 and $n-1$. Observe that any spanning forest with exactly two connected components can be constructed by deleting precisely one edge e from $\mathcal{T}$. We thus find that a spanning forest $\mathcal{F}$ is in $\mathcal{S}_n^{\{i,j\}}$ if and only if it is constructed by deleting an edge, say $e(\mathcal{F})$ that is on both the path from i to n and on the path from j to n in $\mathcal{T}$. Further, for such a spanning forest, $w(\mathcal{F}) = \frac{w(\mathcal{T})}{w(e(\mathcal{F}))}$. The conclusion now follows from (7.2) and Proposition 7.2.2. □

In the case that $\mathcal{T}$ is an unweighted tree, we see from Corollary 7.2.3 that $b_{i,j}$ is given by the number of edges on both $P_{i,n}$ and $P_{j,n}$; thus, in colloquial terms, $b_{i,j}$ can be thought of as the number of "bottlenecks" (i.e., edges in common) on the routes from vertices i and j to vertex n. Hence the term "bottleneck matrix".

Corollary 7.2.3 assists us in finding an expression for the diagonal entries of $L(\mathcal{T}, w)^{\#}$ for any tree $\mathcal{T}$.

COROLLARY 7.2.4 ([78]) *Let $(\mathcal{T}, w)$ be a weighted tree on vertices $1, \ldots, n$, and let L denote the corresponding Laplacian matrix. Then for each $k = 1, \ldots, n$ we have*

$$l_{k,k}^{\#} = \frac{1}{n^2} \sum_{e \in \mathcal{T}} \frac{|\beta_k(e)|^2}{w(e)},$$

where the sum is taken over all edges e in $\mathcal{T}$.

Proof: Without loss of generality we take $k = n$. Let B denote the bottleneck matrix for $(\mathcal{T}, w)$ based at vertex n. From Observation 7.2.1, we find that the n-th diagonal entry of $L^{\#}$ is equal to $\frac{1}{n^2}\mathbf{1}^t B\mathbf{1}$. From Corollary 7.2.3, it follows that

$$\mathbf{1}^t B \mathbf{1} = \sum_{i=1}^{n-1} \sum_{j=1}^{n-1} \sum_{e \in P_{i,n} \cap P_{j,n}} \frac{1}{w(e)}.$$

Given indices $i, j = 1, \ldots, n-1$, and an edge e of $\mathcal{T}$, we have $e \in P_{i,n} \cap P_{j,n}$ if and only if $i, j \in \beta_n(e)$. Consequently we see that

$$\sum_{i=1}^{n-1} \sum_{j=1}^{n-1} \sum_{e \in P_{i,n} \cap P_{j,n}} \frac{1}{w(e)} = \sum_{e \in \mathcal{T}} \sum_{i,j \in \beta_n(e)} \frac{1}{w(e)} = \sum_{e \in \mathcal{T}} \frac{|\beta_n(e)|^2}{w(e)}.$$

The conclusion now follows. □

The expression in Corollary 7.2.4 allows us to discuss the maximum and minimum diagonal entries in $L^{\#}$ in terms of the structure of the underlying tree. Recall that a vertex i of a tree $\mathcal{T}$ on n vertices is a *centroid* if it has the property that for any edge e of $\mathcal{T}$ that is incident with i, $|\beta_i(e)| \leq \frac{n}{2}$. It is known that for any tree, there is either a unique centroid, or there are exactly two centroids, which are necessarily adjacent ([51, Chapter 4]).

PROPOSITION 7.2.5 ([78]) *Let $(\mathcal{T}, w)$ be a weighted tree on n vertices with Laplacian matrix L. The largest diagonal entry of $L^{\#}$ occurs at an index corresponding to a pendent vertex of $\mathcal{T}$. The smallest diagonal entry of $L^{\#}$ occurs at an index corresponding to a centroid of $\mathcal{T}$.*

Proof: Suppose that vertex k is not pendent, say with k adjacent to vertices j and m; let e and $\hat{e}$ denote the edges between k and j, and between k and m, respectively. From Corollary 7.2.4, we find that $l^{\#}_{k,k} \geq l^{\#}_{j,j}$ if and only if $|\beta_k(e)| \geq |\beta_j(e)|$. Since $|\beta_k(e)|+|\beta_j(e)| = n$, we see that $l^{\#}_{k,k} \geq l^{\#}_{j,j}$ if and only if $|\beta_k(e)| \geq \frac{n}{2}$. Similarly, we find that $l^{\#}_{k,k} \geq l^{\#}_{m,m}$ if and only if $|\beta_k(\hat{e})| \geq \frac{n}{2}$. Since the sets $\beta_k(e)$ and $\beta_k(\hat{e})$ are disjoint and neither contains vertex k, it cannot be the case that both have cardinalities at least $\frac{n}{2}$. We conclude that either $l^{\#}_{k,k} < l^{\#}_{j,j}$ or $l^{\#}_{k,k} < l^{\#}_{m,m}$. Hence the maximum diagonal entry of $L^{\#}$ must correspond to a pendent vertex of $\mathcal{T}$.

Next, consider a vertex i of $\mathcal{T}$ such that $l^{\#}_{i,i}$ is minimum over all choices of i. Then in particular, for a vertex j adjacent to i, we find as above that $|\beta_i(e)| \leq \frac{n}{2}$, where e denotes the edge between vertices i and j. Thus, if $l^{\#}_{i,i}$ is minimum, then for each edge f incident with vertex i, $|\beta_i(f)| \leq \frac{n}{2}$. We conclude that j must be a centroid of $\mathcal{T}$. □

COROLLARY 7.2.6 ([78]) *Let $(\mathcal{T}, w)$ be a weighted tree on vertices $1, \ldots, n$ with Laplacian matrix L. For each $i, j = 1, \ldots, n$ with $i \neq j$, we have*

$$l^{\#}_{i,j} = \frac{1}{n^2} \sum_{e \in \mathcal{T}} \frac{|\beta_j(e)|^2}{w(e)} - \frac{1}{n} \sum_{e \in P_{i,j}} \frac{|\beta_j(e)|}{w(e)}.$$

Proof: Without loss of generality, suppose that $j = n$. From Observation 7.2.1, we have $l^{\#}_{i,n} = \frac{\mathbf{1}^t B \mathbf{1}}{n^2} - \frac{1}{n} \sum_{k=1}^{n-1} b_{i,k}$, where B denotes the bottleneck matrix for $(\mathcal{T}, w)$ based at vertex n. As in Corollary 7.2.4, we find that

$\mathbf{1}^t B \mathbf{1} = \sum_{e \in \mathcal{T}} \frac{|\beta_n(e)|^2}{w(e)}$. From Corollary 7.2.3, we find that $\sum_{k=1}^{n-1} b_{i,k} = \sum_{k=1}^{n-1} \sum_{e \in P_{i,n} \cap P_{k,n}} \frac{1}{w(e)}$. Since $e \in P_{i,n} \cap P_{k,n}$ if and only if $i, k \in \beta_n(e)$, we see that

$$\sum_{k=1}^{n-1} \sum_{e \in P_{i,n} \cap P_{k,n}} \frac{1}{w(e)} = \sum_{e \in \mathcal{T} \ni i \in \beta_n(e)} \sum_{k \in \beta_n(e)} \frac{1}{w(e)} = \sum_{e \in \mathcal{T} \ni i \in \beta_n(e)} \frac{|\beta_n(e)|}{w(e)} = \sum_{e \in P_{i,n}} \frac{|\beta_n(e)|}{w(e)}.$$

The conclusion follows readily. □

REMARK 7.2.7 From Corollary 7.2.6, we find that for indices i, j, k between 1 and n, $l^{\#}_{i,j} - l^{\#}_{k,j} = \frac{1}{n}\left(\sum_{e \in P_{k,j}} \frac{|\beta_j(e)|}{w(e)} - \sum_{e \in P_{i,j}} \frac{|\beta_j(e)|}{w(e)}\right)$. In particular, if vertex i is on the path between vertex k and vertex j, then $l^{\#}_{i,j} > l^{\#}_{k,j}$. Thus the entries in the j-th column of $L^{\#}$ are strictly decreasing as we move away from vertex j along any path in the tree.

EXAMPLE 7.2.8 Here we consider the Laplacian matrix L for an unweighted path P_n on n vertices, where the vertices are labelled so that vertices 1 and n are pendent, and for each $k = 2, \ldots, n-1$, vertex k is adjacent to vertices $k-1$ and $k+1$. Suppose that for some p between 1 and $n-1$, e is the edge between vertices p and $p+1$. Then for each $k = 1, \ldots, n$, we find readily that $|\beta_k(e)| = \begin{cases} n-p & \text{if } k \le p, \\ p & \text{if } k \ge p+1. \end{cases}$

Applying Corollaries 7.2.4 and 7.2.6, we now find that the entries of $L^{\#}$ are given as follows:

$$l^{\#}_{i,j} = \begin{cases} \frac{j(j-1)(2j-1)+(n-j)(n-j+1)(2n-2j+1)}{6n^2} - \frac{|j-i|(j+i-1)}{2n} & \text{if } i \le j, \\ \frac{i(i-1)(2i-1)+(n-i)(n-i+1)(2n-2i+1)}{6n^2} - \frac{|j-i|(j+i-1)}{2n} & \text{if } i \ge j+1. \end{cases}$$

Our next result provides another way of writing the group inverse of the Laplacian matrix for a weighted tree.

PROPOSITION 7.2.9 *Let $(\mathcal{T}, w)$ be a weighted tree on vertices $1, \ldots, n$, and let L denote the corresponding Laplacian matrix. For each edge e of $\mathcal{T}$, denote the connected components of $\mathcal{T} \setminus e$ by $\mathcal{T}_1(e), \mathcal{T}_2(e)$, and let $v(e)$ be the vector in $\mathbb{R}^n$ whose entries are given by*

$$v_i(e) = \begin{cases} \frac{|\beta_i(e)|}{n} & \text{if } i \in \mathcal{T}_1(e), \\ \frac{-|\beta_i(e)|}{n} & \text{if } i \in \mathcal{T}_2(e). \end{cases} \tag{7.3}$$

Then $L^{\#}$ can be written as

$$L^{\#} = \sum_{e \in \mathcal{T}} \frac{1}{w(e)} v(e)v(e)^t. \tag{7.4}$$

Proof: For each $i, j = 1, \ldots, n$, we have

$$l^{\#}_{i,j} = \frac{1}{n^2} \sum_{e \in \mathcal{T}} \frac{|\beta_j(e)|^2}{w(e)} - \frac{1}{n} \sum_{e \in P_{i,j}} \frac{|\beta_j(e)|}{w(e)},$$

where we interpret the second summation as zero if $i = j$. Note that for an edge e and vertices i, j, we have $e \in P_{i,j}$ if and only if $i \in \mathcal{T}_1(e), j \in \mathcal{T}_2(e)$ or $j \in \mathcal{T}_1(e), i \in \mathcal{T}_2(e)$. Fix an edge e of $\mathcal{T}$, and consider its contribution to the expression for $l^{\#}_{i,j}$ above. If $e \notin P_{i,j}$, then the edge e contributes

$$\frac{|\beta_j(e)|^2}{w(e)n^2} = \frac{|\beta_i(e)||\beta_j(e)|}{w(e)n^2},$$

(here we use the fact that $|\beta_i(n)| = |\beta_j(n)|$ whenever $e \notin P_{i,j}$). On the other hand, if $e \in P_{i,j}$, then the edge e contributes

$$\frac{|\beta_j(e)|^2}{w(e)n^2} - \frac{|\beta_j(e)|}{w(e)n} = -\frac{|\beta_j(e)|(n - |\beta_j(e)|)}{w(e)n^2} = -\frac{|\beta_i(e)||\beta_j(e)|}{w(e)n^2},$$

the last equality following from the fact that $|\beta_i(n)| + |\beta_j(n)| = n$ whenever $e \in P_{i,j}$. It is now readily determined that each edge e contributes $\frac{v_i(e)v_j(e)}{w(e)}$ to the expression for $l^{\#}_{i,j}$. The conclusion now follows. □

We note here that there is a little bit of flexibility in the definition of $v(e)$ in Proposition 7.2.9, since the entries of $v(e)$ depend in part on which of the connected components of $\mathcal{T} \setminus e$ is labelled $\mathcal{T}_1(e)$ and which is labelled $\mathcal{T}_2(e)$. Observe that exchanging labellings of those connected components has the effect of multiplying $v(e)$ by -1, but leaves the expression for $L^{\#}$ in (7.4) unchanged. In the sequel we will occasionally take advantage of this flexibility by choosing the labelling of the connected components of $\mathcal{T} \setminus e$ so that a particular entry of $v(e)$ is positive.

REMARK 7.2.10 For any connected graph on n vertices, the corresponding Laplacian matrix L is a symmetric positive semidefinite matrix of rank $n - 1$, and hence so is $L^{\#}$. In the special case that our graph is a tree, Proposition 7.2.9 explicitly expresses $L^{\#}$ as a sum of $n - 1$ symmetric positive semidefinite matrices, each of rank 1.

Our next example illustrates the conclusion of Proposition 7.2.9.

EXAMPLE 7.2.11 Let $(\mathcal{T}, w)$ be the weighted star on n vertices such that vertices $1, \ldots, n-1$ are pendent. For each $j = 1, \ldots, n-1$, let w_j denote the weight of the edge incident with vertex j. Referring to Proposition 7.2.9, we find that for each $j = 1, \ldots, n-1$, if e is the edge incident with vertex j, then we may take $v(e)$ to be given by $v(e) = e_j - \frac{1}{n}\mathbf{1}$. Consequently, we find from Proposition 7.2.9 that $L(\mathcal{T}, w)^\# = \sum_{j=1}^{n-1} \frac{1}{w_j}(e_j - \frac{1}{n}\mathbf{1})(e_j - \frac{1}{n}\mathbf{1})^t$.

In our next sequence of results, we develop some alternative expressions for the group inverse of the Laplacian matrix for a weighted tree. To that end, we give two definitions. Let $(\mathcal{T}, w)$ be a weighted tree on vertices $1, \ldots, n$. For each pair of distinct vertices i, j, we define the *inverse weighted distance* from i to j by $\tilde{d}(i,j) = \sum_{e \in P_{i,j}} \frac{1}{w(e)}$. By convention we take $\tilde{d}(i,i) = 0$. For any index i, we then define the *inverse weighted status* of vertex i as $\tilde{d}_i = \sum_{j=1}^n \tilde{d}(i,j)$ (here we have adapted a term for directed graphs due to Harary in [50]). Observe that for an unweighted tree, $\tilde{d}(i,j)$ coincides with the usual graph-theoretic distance from vertex i to vertex j, i.e., the number of edges on $P_{i,j}$. Our goal in the next couple of results is to express the group inverse of the Laplacian matrix of a weighted tree in terms of inverse weighted distances and inverse weighted statuses. We begin with the diagonal entries.

PROPOSITION 7.2.12 ([78]) *Let $(\mathcal{T}, w)$ denote a weighted tree on vertices $1, \ldots, n$ with Laplacian matrix L. For each $k = 1, \ldots, n$,*

$$l^\#_{k,k} = \frac{\tilde{d}_k}{n} - \frac{1}{n^2} \sum_{e \in \mathcal{T}} \frac{|\beta_k(e)|(n - |\beta_k(e)|)}{w(e)}.$$

<u>**Proof**</u>: Fix an index k and recall that for each edge $e \in \mathcal{T}$, the expression $|\beta_k(e)|(n - |\beta_k(e)|)$ is independent of the choice of the vertex k. From Corollary 7.2.4, we have $l^\#_{k,k} = \frac{1}{n^2} \sum_{e \in \mathcal{T}} \frac{|\beta_k(e)|^2}{w(e)}$, which we rewrite as

$$l^\#_{k,k} = \frac{1}{n} \sum_{e \in \mathcal{T}} \frac{|\beta_k(e)|}{w(e)} - \frac{1}{n^2} \sum_{e \in \mathcal{T}} \frac{|\beta_k(e)|(n - |\beta_k(e)|)}{w(e)}.$$

Thus, it suffices to show that $\tilde{d}_k = \sum_{e \in \mathcal{T}} \frac{|\beta_k(e)|}{w(e)}$.

Note that $\tilde{d}_k = \sum_{j=1}^n \sum_{e \in P_{k,j}} \frac{1}{w(e)}$, and observe that $e \in P_{k,j}$ if and only if $j \in \beta_k(e)$. Thus we find that $\tilde{d}_k$ can be written as

$\tilde{d}_k = \sum_{e\in\mathcal{T}} \sum_{j\in\beta_k(e)} \frac{1}{w(e)}$. It now follows that $\tilde{d}_k = \sum_{e\in\mathcal{T}} \frac{|\beta_k(e)|}{w(e)}$, as desired. □

COROLLARY 7.2.13 ([78]) *Let $(\mathcal{T}, w)$ denote a weighted tree on vertices $1, \ldots, n$ with Laplacian matrix L. Then for any index k between 1 and n we have*

$$\text{trace}(L^{\#}) = \frac{1}{n}\sum_{e\in\mathcal{T}} \frac{|\beta_k(e)|(n - |\beta_k(e)|)}{w(e)}.$$

Equivalently, we have

$$\sum_{1\le i<j\le n} \tilde{d}(i,j) = n\text{trace}(L^{\#}).$$

__Proof__: For each edge e of $\mathcal{T}$, let $v(e)$ be the vector defined in Proposition 7.2.9. We find from Proposition 7.2.9 that $\text{trace}(L^{\#}) = \sum_{e\in\mathcal{T}} \frac{1}{w(e)}\text{trace}(v(e)v(e)^t) = \sum_{e\in\mathcal{T}} \frac{v(e)^t v(e)}{w(e)}$. A straightforward computation shows that for any k between 1 and n, $v(e)^t v(e) = |\beta_k(e)|(n - |\beta_k(e)|)$, and the first conclusion now follows.

For the second conclusion, we note that

$$\sum_{1\le i<j\le n} \tilde{d}(i,j) = \sum_{1\le i<j\le n} \sum_{e\in P_{i,j}} \frac{1}{w(e)}.$$

Since $e \in P_{i,j}$ if and only if $j \in \beta_i(e)$, we find that $\sum_{1\le i<j\le n} \sum_{e\in P_{i,j}} \frac{1}{w(e)} = \sum_{e\in\mathcal{T}} \frac{|\beta_k(e)|(n-|\beta_k(e)|)}{w(e)}$. □

Next, we produce an alternate expression for the off-diagonal entries in the group inverse of the Laplacian matrix for a weighted tree.

PROPOSITION 7.2.14 ([78]) *Let $(\mathcal{T}, w)$ be a weighted tree on vertices $1, \ldots, n$ with Laplacian matrix L. For each pair of distinct indices i, j between 1 and n, we have*

$$l^{\#}_{i,j} = \frac{\tilde{d}_i + \tilde{d}_j}{2n} - \frac{\tilde{d}(i,j)}{2} - \frac{1}{2n^2}\sum_{k=1}^{n} \tilde{d}_k.$$

__Proof__: First we claim that $l^{\#}_{i,i} + l^{\#}_{j,j} - 2l^{\#}_{i,j} = \tilde{d}(i,j)$. To see the claim, observe from Corollaries 7.2.4 and 7.2.6 that

$$l^{\#}_{i,i} + l^{\#}_{j,j} - 2l^{\#}_{i,j} =$$
$$\frac{1}{n^2}\sum_{e\in\mathcal{T}}\frac{|\beta_i(e)|^2}{w(e)} + \frac{1}{n^2}\sum_{e\in\mathcal{T}}\frac{|\beta_j(e)|^2}{w(e)} - \frac{2}{n^2}\sum_{e\in\mathcal{T}}\frac{|\beta_j(e)|^2}{w(e)} + \frac{2}{n}\sum_{e\in P_{i,j}}\frac{|\beta_j(e)|}{w(e)} =$$
$$\frac{1}{n^2}\sum_{e\in\mathcal{T}}\frac{|\beta_i(e)|^2 - |\beta_j(e)|^2}{w(e)} + \frac{2}{n}\sum_{e\in P_{i,j}}\frac{|\beta_j(e)|}{w(e)}.$$

Observe that $\beta_i(e) = \beta_j(e)$ whenever $e \notin P_{i,j}$, while if $e \in P_{i,j}$, we have $|\beta_i(e)| = n - |\beta_j(e)|$. We thus find that

$$l^{\#}_{i,i} + l^{\#}_{j,j} - 2l^{\#}_{i,j} = \frac{1}{n^2}\sum_{e\in P_{i,j}}\frac{(n-|\beta_j(e)|)^2 - |\beta_j(e)|^2 + 2n|\beta_j(e)|}{w(e)} =$$
$$\sum_{e\in P_{i,j}}\frac{1}{w(e)} = \tilde{d}(i,j),$$

as claimed.

Applying our claim and Proposition 7.2.12, we thus find that

$$l^{\#}_{i,j} = \frac{l^{\#}_{i,i} + l^{\#}_{j,j}}{2} - \frac{\tilde{d}(i,j)}{2} = \frac{\tilde{d}_i + \tilde{d}_j}{2n} - \frac{\tilde{d}(i,j)}{2} - \frac{1}{n^2}\sum_{e\in\mathcal{T}}\frac{|\beta_i(e)|(n-|\beta_i(e)|)}{w(e)}.$$

By Corollary 7.2.13,

$$\sum_{e\in\mathcal{T}}\frac{|\beta_i(e)|(n-|\beta_i(e)|)}{w(e)} = \sum_{1\le i<j\le n}\tilde{d}(i,j) = \frac{1}{2}\sum_{i=1}^{n}\sum_{j=1}^{n}\tilde{d}(i,j) = \frac{1}{2}\sum_{k=1}^{n}\tilde{d}_k.$$

The conclusion now follows. □

Our final goal in this section is to address Question 3.3.8 (which seems to be quite difficult, in general) in the special case that the M-matrix in question is the Laplacian matrix of a weighted tree. Our approach here will follow that in Kirkland and Neumann's paper [82]. We begin with the following preliminary results.

LEMMA 7.2.15 *Let $\mathcal{T}$ be a tree on vertices $1,\ldots,n$. Label the edges of $\mathcal{T}$ as $\mathrm{e}(1),\ldots,\mathrm{e}(n-1)$, where for each $i = 1,\ldots,n-1$, the edge $\mathrm{e}(i)$ is incident with vertex i. Consider the $(n-1)\times(n-1)$ matrix A whose entries are given by*

$$a_{i,j} = \begin{cases} -|\beta_i(e_j)|^2 & \text{if } i \neq j, \\ |\beta_i(e_i)|(n-|\beta_i(e_i)|) & \text{if } i = j. \end{cases} \tag{7.5}$$

The matrix A is an M-matrix if and only if there is a weighting of $\mathcal{T}$ such that for the corresponding Laplacian matrix L, we have that $L^{\#}$ is an M-matrix. Further, whenever such a weighting w exists, it has the form $w(e(i)) = \frac{1}{u_i}, i = 1, \ldots, n-1$, where u is a positive vector such that $Au \geq 0$.

Proof: Suppose that w is some weighting of $\mathcal{T}$, and let L be the corresponding Laplacian matrix. From Remark 7.2.7, it follows that $L^{\#}$ is an M-matrix if and only if $l^{\#}_{i,j} \leq 0$ for any pair of distinct vertices i, j that are adjacent in $\mathcal{T}$. Suppose that i and j are adjacent in $\mathcal{T}$, and let e denote the edge between them. Referring to Corollary 7.2.6, we find that $l^{\#}_{i,j} \leq 0$ if and only if

$$\sum_{f \in \mathcal{T}} \frac{|\beta_i(f)|^2}{w(f)} \leq n \frac{|\beta_i(e)|}{w(e)}. \tag{7.6}$$

From the considerations above, and using our labelling of the edges of $\mathcal{T}$, we see that $L^{\#}$ is an M-matrix if and only if for each $m = 1, \ldots, n-1$,

$$n \frac{|\beta_m(e(m))|}{w(e_m)} - \sum_{k=1}^{n-1} \frac{|\beta_m(e(k))|}{w(e(k))} \geq 0. \tag{7.7}$$

Letting $u_k = \frac{1}{w(e(k))}, k = 1, \ldots, n-1$, we find that (7.7) holds for each $m = 1, \ldots, n-1$ if and only if $Au \geq 0$.

Observe that A has positive diagonal entries and negative off-diagonal entries. Appealing to [12, Theorem 2.3], we have that A is an M-matrix if and only if there is a positive vector v such that $Av \geq 0$. We now deduce that $\mathcal{T}$ admits a weighting w such that $L(\mathcal{T}, w)^{\#}$ is an M-matrix if and only if A is an M-matrix. Moreover, an examination of the argument above shows that if $\mathcal{T}$ admits such a weighting w, then there is a positive vector u such that $Au \geq 0$ and $w(e(k)) = \frac{1}{u_k}, k = 1, \ldots, n-1$. $\square$

We illustrate the technique suggested by Lemma 7.2.15 in the following example.

EXAMPLE 7.2.16 Consider the path on four vertices, P_4, where we label the vertices so that vertices 1 and 4 are pendent, and where vertex i is adjacent to vertices $i-1$ and $i+1, i = 2, 3$. Label the edges $e(1), e(2), e(3)$, with edge $e(i)$ incident with vertex i for each $i = 1, 2, 3$. Appealing to Lemma 7.2.15, in order to determine whether there is a weighting w of P_4 such that $L(P_4, w)^{\#}$ is an M-matrix, we need to consider the matrix

$$A = \begin{bmatrix} 3 & -4 & -1 \\ -1 & 4 & -1 \\ -1 & -4 & 3 \end{bmatrix}.$$

It is straightforward to show that A is a singular M-matrix, and that its null space is spanned by the vector $x = \begin{bmatrix} 2 \\ 1 \\ 2 \end{bmatrix}$. Consequently, we find from Lemma 7.2.15, that there is a weighting w such that $L(P_4, w)^{\#}$ is an M-matrix, and that any such weighting has the form $w(e(1)) = \theta, w(e(2)) = 2\theta, w(e(3)) = \theta$ for some $\theta > 0$.

Finally, we note that if we do have $w(e(1)) = \theta, w(e(2)) = 2\theta, w(e(3)) = \theta$ for some positive θ, then from Proposition 7.2.9, it follows readily that

$$L(P_4, w)^{\#} = \frac{1}{\theta}\begin{bmatrix} \frac{3}{4} & 0 & -\frac{1}{4} & -\frac{1}{2} \\ 0 & \frac{1}{4} & 0 & -\frac{1}{4} \\ -\frac{1}{4} & 0 & \frac{1}{4} & 0 \\ -\frac{1}{2} & -\frac{1}{4} & 0 & \frac{3}{4} \end{bmatrix}.$$

We now use Lemma 7.2.15 in conjunction with some graph-theoretic considerations to gain further insight into the trees such that the matrix A of Lemma 7.2.15 is an M-matrix.

LEMMA 7.2.17 ([82]) *Let $\mathcal{T}$ be a tree on $n \geq 4$ vertices. Suppose that $\mathcal{T}$ contains a subpath on vertices v_1, v_2, v_3, v_4, where vertex v_1 is pendent, and vertex v_i is adjacent to vertex v_{i+1} for $i = 1, 2, 3$. Let S be the set of vertices consisting of vertex v_2 and all of the branches at v_2 not containing vertices v_1 and v_3. If the matrix A given by (7.5) is an M-matrix, then $|S| \geq \frac{n-2}{2}$.*

Proof: Denote the degree of vertex v_3 by $t + 1$, where $t \geq 1$. We relabel the vertices and edges of $\mathcal{T}$ as follows: vertices v_1, v_2, v_3 are labelled $t+3, t+2, t+1$, respectively, and the neighbours of v_3 distinct from v_2 are labelled $1, \ldots, t$; for each $j = 1, \ldots, t$, the edge between vertex j and vertex $t + 1$ is $e(j)$, while the edge between vertices $t + 1$ and $t + 2$ is $e(t + 1)$, and the edge between vertices $t + 2$ and $t + 3$ is $e(t + 2)$; the remaining vertices and edges can then be relabelled in some fashion with the labels $t + 4, \ldots, n$ and $e(t + 3), \ldots, e(n - 1)$, respectively. Figure 7.1 illustrates the situation; in that figure, the circle enclosing vertex $t + 2$

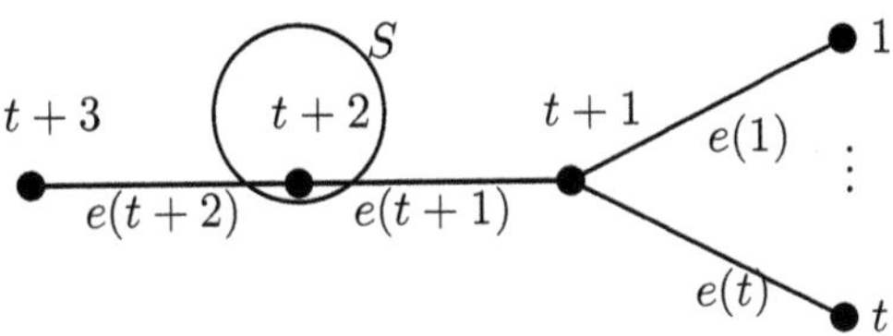

FIGURE 7.1
The labelled tree $\mathcal{T}$ of Lemma 7.2.17

represents the subgraph induced by the vertices of S, and the branches at vertices $1, \ldots, t$ not containing vertex $t+1$ have been suppressed.

For each $j = 1, \ldots, t$, let $k_j = |\beta_{t+1}(e(j))|$. Denoting the cardinality of S by s, we see that $\sum_{j=1}^{t} k_j = n - s - 2$. We also have the following observations:
$|\beta_j(e(j))| = n - k_j, j = 1, \ldots, t$;
$|\beta_i(e(j))| = k_j$ for any pair of distinct indices i, j between 1 and t;
$|\beta_{t+1}(e(j))| = k_j, s+1, 1$ according as $1 \le j \le t, j = t+1, j = t+2$, respectively; and
$|\beta_{t+2}(e(j))| = k_j, n-1-s, n-1$ according as $1 \le j \le t, j = t+1, j = t+2$, respectively.

Let D be the $t \times t$ diagonal matrix whose j-th diagonal entry is $k_j, j = 1, \ldots, t$. It follows that the principal submatrix of A on rows and columns $1, \ldots, t+2$ is given by

$$C = \left[\begin{array}{c|cc} nD - JD^2 & -(s+1)^2\mathbf{1} & -\mathbf{1} \\ \hline -\mathbf{1}^t D^2 & (s+1)(n-s-1) & -1 \\ -\mathbf{1}^t D^2 & -(n-s-1)^2 & n-1 \end{array}\right].$$

Since A is an M-matrix, so is C; in particular, $\det(C) \ge 0$. From the formula for the determinant of a partitioned matrix [57, section 0.8.5], it follows that $\det(C) \ge 0$ if and only if the determinant of the matrix

$$nD - JD^2 - \left(\left[\begin{array}{cc} (s+1)^2 & 1 \end{array} \right] \left[\begin{array}{cc} (s+1)(n-s-1) & -1 \\ -(n-s-1)^2 & n-1 \end{array} \right]^{-1} \left[\begin{array}{c} 1 \\ 1 \end{array} \right] \right) JD^2$$

is nonnegative. This last is readily seen to be equivalent to the condition

$$\det\left(I - \frac{s+1}{s(n-s-1)} JD \right) \ge 0.$$

Since

$$\det\left(I - \frac{s+1}{s(n-s-1)}JD\right) = 1 + \left(\frac{s+1}{s(n-s-1)}\right)\mathbf{1}^t D\mathbf{1},$$

and since the sum of the entries in D is $n-2-s$, we thus find that $\det(C) \geq 0$ if and only if $s \geq \frac{n-2}{2}$. The conclusion now follows. □

We are now sufficiently prepared to prove the following result.

THEOREM 7.2.18 ([82]) *Let $\mathcal{T}$ be a tree on n vertices. Then $\mathcal{T}$ admits a weighting w such that $L(\mathcal{T}, w)^{\#}$ is an M-matrix if and only if either $\mathcal{T}$ is a star, or $n = 4$ and $\mathcal{T} = P_4$. In the case that $\mathcal{T}$ is a star (with edges $e(1), \ldots, e(n-1)$), then for any weighting w such that $L(\mathcal{T}, w)^{\#}$ is an M-matrix, there is a nonnegative nonzero vector $y \in \mathbb{R}^{n-1}$ such that*

$$\begin{bmatrix} \frac{1}{w(e(1))} \\ \vdots \\ \frac{1}{w(e(n-1))} \end{bmatrix} = \frac{1}{n}(I+J)y.$$

In the case that $\mathcal{T} = P_4$ (with edges $e(1), e(2), e(3)$ and $e(2)$ as the nonpendent edge), if w is a weighting such that $L(P_4, w)^{\#}$ is an M-matrix, then for some $\theta > 0$, we have $w(e(1)) = w(e(3)) = \theta$ and $w(e(2)) = 2\theta$.

Proof: Suppose that $n \geq 5$ and that $\mathcal{T}$ is not a star. We claim then that there is no weighting w so that $L(\mathcal{T}, w)^{\#}$ is an M-matrix. To see the claim, first observe that there is a subpath of $\mathcal{T}$ of length $l \geq 3$ that joins two pendent vertices, say u and v. Denote their neighbours by u_0, v_0, respectively. If $\mathcal{T}$ admits a weighting w so that $L(\mathcal{T}, w)^{\#}$ is an M-matrix, then by Lemmas 7.2.15 and 7.2.17, we find that the subset of vertices consisting of u_0 and its branches containing neither u nor v contains at least $\frac{n-2}{2}$ vertices. Further, the subset of vertices consisting of v_0 and its branches containing neither u nor v also contains at least $\frac{n-2}{2}$ vertices. In particular, if $l \geq 4$, we get a contradiction.

Thus we must have $l = 3$, and from the considerations above, it must be the case that $\mathcal{T}$ has just two nonpendent vertices, namely u_0 and v_0, and that each has degree $\frac{n}{2}$ (see Figure 7.2). With a suitable labelling of vertices and edges (so that v_0 takes the label n), the matrix A of Lemma 7.2.15 is given by

$$A = \left[\begin{array}{c|c} nI - J & -\frac{n^2}{4}\mathbf{1} \\ \hline -\frac{n^2}{4}\mathbf{1}^t & \frac{n^2}{4} \end{array}\right].$$

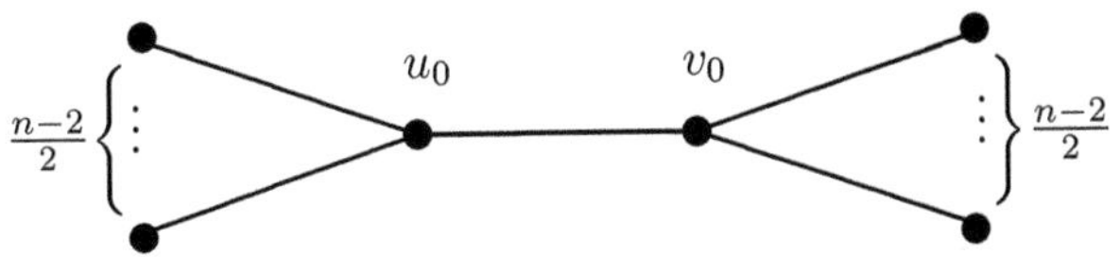

FIGURE 7.2
Tree for the case that $l = 3$, and u_0, v_0 each have degree $\frac{n}{2}$

It now follows that $\det(A) = \frac{n^2}{4}(1 - \mathbf{1}^t(nI - J)\mathbf{1})\det(nI - J) = \frac{n^2}{4}(1 - \frac{n-2}{2})(2(n^{n-3})) < 0$, a contradiction. We conclude that if $n \geq 5$ and $\mathcal{T}$ admits a weighting w such that $L(\mathcal{T}, w)^\#$ is an M-matrix, then $\mathcal{T}$ must be a star, as claimed.

Next suppose that $\mathcal{T}$ is a star on $n \geq 2$ vertices; we seek to determine the weightings w so that $L(\mathcal{T}, w)^\#$ is an M-matrix. Label the centre vertex of $\mathcal{T}$ as vertex n. From Lemma 7.2.15, we find that if a weighting w has that property that $L(\mathcal{T}, w)^\#$ is an M-matrix, then for some nonzero vector $y \geq 0$,

$$(nI - J)\begin{bmatrix} \frac{1}{w(e(1))} \\ \vdots \\ \frac{1}{w(e(n-1))} \end{bmatrix} = y.$$

Solving this equation then yields

$$\begin{bmatrix} \frac{1}{w(e(1))} \\ \vdots \\ \frac{1}{w(e(n-1))} \end{bmatrix} = \frac{1}{n}(I + J)y,$$

as desired.

The only remaining case to consider is the case that $n = 4$ and $\mathcal{T} = P_4$. So, suppose that $\mathcal{T} = P_4$, with edges $e(1), e(2), e(3)$ and $e(2)$ as the nonpendent edge. By Example 7.2.16, we find that P_4 admits a weighting w such that $L(P_4, w)^\#$ is an M-matrix if and only if there is a $\theta > 0$ such that $w(e(1)) = w(e(3)) = \theta$ and $w(e(2)) = 2\theta$. □

We close this section by noting that in the paper of Bendito, Carmona, Encinas, and Mitjana [11], Question 3.3.8 is addressed for the class of Laplacian matrices of distance-regular graphs.

7.3 Bounds on algebraic connectivity via the group inverse

Suppose that we have a weighted graph $(\mathcal{G}, w)$ on n vertices with Laplacian matrix L. As already noted above, L is a singular, symmetric positive definite matrix. Denote the eigenvalues of L by $0 = \mu_1 \leq \mu_2 \leq \ldots \leq \mu_n$. We saw in section 7.1 that the multiplicity of the eigenvalue zero of L coincides with number of connected components of the graph $\mathcal{G}$ (independently of the weighting function w). In particular, $\mathcal{G}$ is connected if and only if $\mu_2 > 0$. The quantity μ_2 is known as the *algebraic connectivity* of the weighted graph $(\mathcal{G}, w)$, and we denote it by $\alpha(\mathcal{G}, w)$; in the case that $\mathcal{G}$ is unweighted, we use the simpler notation $\alpha(\mathcal{G})$. An application of the variational characterisation of the eigenvalues of a symmetric matrix [57, section 4.2] readily shows that for any weighted graph $(\mathcal{G}, w), \alpha(\mathcal{G}, w) = \min\{u^t L(\mathcal{G}, w)u | u^t \mathbf{1} = 0, u^t u = 1\}$. In a landmark sequence of papers, Fiedler [37], [38], [39], initiated the study of algebraic connectivity for weighted graphs, and since Fielder's original papers, a large body of work on algebraic connectivity and related notions has emerged. The interested reader is referred to Abreu's paper [1] and Kirkland's article [73] for overviews of that material.

One of the key themes of the work on algebraic connectivity is that certain graph-theoretic properties of the graph $\mathcal{G}$ can be estimated via knowledge of the algebraic connectivity. We briefly present a few sample results in order to give the reader the flavour of that body of work. Let $\mathcal{G}$ be a connected unweighted graph and suppose that $\mathcal{G}$ is not a complete graph. Recall that the *vertex connectivity* of $\mathcal{G}, v(\mathcal{G})$, is the minimum number of vertices whose deletion yields a disconnected graph. It is shown in [37] that $\alpha(\mathcal{G}) \leq v(\mathcal{G})$, while in [76], Kirkland, Molitierno, Neumann and Shader characterise the graphs yielding equality. The *diameter* of a connected graph $\mathcal{G}$ is the maximum distance between any pair of vertices (where the distance between vertices is the number of edges on a shortest path between them). In Mohar's paper [99] it is shown that for a connected unweighted graph on n vertices with maximum degree Δ and diameter d,

$$\frac{4}{n\alpha(\mathcal{G})} \leq d \leq 2\left\lceil \frac{\Delta + \alpha(\mathcal{G})}{4\alpha(\mathcal{G})} ln(n-1) \right\rceil.$$

Finally, suppose that $\mathcal{G}$ is an unweighted graph on n vertices, let X be a subset of its vertex set of cardinality k, and suppose that $\mathcal{G}$ contains m edges having one end point in X and the other end point in its complement. Then

$$\alpha(\mathcal{G}) \leq \frac{nm}{k(n-k)}$$

(see Mohar's survey [100]). Thus we see that the algebraic connectivity can be used to estimate certain parameters associated with a graph, and that observation has motivated much of the analysis of algebraic connectivity.

In this section, we estimate the algebraic connectivity of a weighted graph by making use of the group inverse of the corresponding Laplacian matrix. Our first result of this type focuses on weighted trees.

THEOREM 7.3.1 *Let $(\mathcal{T}, w)$ be a weighted tree on n vertices. Then we have the following inequalities:*

$$\alpha(G,w) \leq \min\left\{ \frac{n(n-1)}{\sum_{e\in\mathcal{T}} \frac{|\beta_i(e)|^2}{w(e)}} \middle| i \text{ is a pendent vertex of } \mathcal{T} \right\}; \tag{7.8}$$

$$\alpha(G,w) \geq \frac{n}{\max\{\sum_{e\in P_{i,j}} \frac{|\beta_i(e)|(n-|\beta_i(e)|)}{w(e)} | i,j \text{ are pendent vertices of } \mathcal{T}\}}. \tag{7.9}$$

<u>**Proof**</u>: First, we apply the techniques of section 5.4 in order to establish (7.8). Consider the Laplacian matrix L for $(\mathcal{T}, w)$. Denote its eigenvalues by $\mu_1 = 0 < \alpha(\mathcal{T}, w) = \mu_2 \leq \mu_3 \leq \ldots \leq \mu_n$, and let $x^{(1)}, \ldots, x^{(n)}$ denote a corresponding orthonormal set of eigenvectors. Note that we may take $x^{(1)} = \frac{1}{\sqrt{n}}\mathbf{1}$. Then $L^{\#} = \sum_{j=2}^{n} \frac{1}{\mu_j} x^{(j)} x^{(j)^t}$. Hence we find that for any index i,

$$l^{\#}_{i,i} = \sum_{j=2}^{n} \frac{(x^{(j)}_i)^2}{\mu_j} \leq \frac{1}{\alpha(\mathcal{T}, w)} \sum_{j=2}^{n} (x^{(j)}_i)^2 = \frac{n-1}{n\alpha(\mathcal{T}, w)}.$$

Consequently, it follows that

$$\alpha(\mathcal{T}, w) \leq \frac{n-1}{n \max_i l^{\#}_{i,i}}.$$

From Proposition 7.2.9, we find that

$$l^{\#}_{i,i} = \sum_{e\in\mathcal{T}} \frac{v_i(e)^2}{w(e)} = \sum_{e\in\mathcal{T}} \frac{|\beta_i(e)|^2}{n^2 w(e)},$$

so that for any vertex i, we have

$$\alpha(\mathcal{G}, w) \leq \frac{n(n-1)}{\sum_{e \in \mathcal{T}} \frac{|\beta_i(e)|^2}{w(e)}}.$$

Finally, from Proposition 7.2.5, we see that the maximum diagonal entry of $L^{\#}$ occurs at an index corresponding to a pendent vertex. The inequality (7.8) now follows readily.

Next, we address (7.9). Recall from (5.16) that $\tau(L^{\#}) = \frac{1}{2}\max_{i,j} \sum_{k=1}^{n} |l^{\#}_{i,k} - l^{\#}_{j,k}|$, and observe that since $L^{\#}$ is symmetric, $\tau(L^{\#}) = \frac{1}{2}\max_{i,j} ||L^{\#}(e_i - e_j)||_1$. From Proposition 5.4.4, we have that for each $k = 2, \ldots, n$, $\frac{1}{\mu_k} \leq \tau(L^{\#})$. Consequently, we find that $\alpha(\mathcal{T}, w) \geq \frac{1}{\tau(L^{\#})}$. We claim that

$$\tau(L^{\#}) \leq \max\left\{ \sum_{e \in P_{i,j}} \frac{|\beta_i(e)|(n - |\beta_i(e)|)}{w(e)} \middle| i, j \text{ are pendent vertices of } \mathcal{T} \right\},$$

which is then sufficient to establish (7.9). Fix vertices i and j of $\mathcal{T}$. Then from Proposition 7.2.9, we find that $L^{\#}(e_i - e_j) = \sum_{e \in \mathcal{T}} \frac{v(e)^t(e_i - e_j)}{w(e)} v(e)$, where $v(e)$ is given by (7.3). Observe that $v(e)^t(e_i - e_j) = 0$ if $e \notin P_{i,j}$ while $v(e)^t(e_i - e_j) = 1$ if $e \in P_{i,j}$, so that $L^{\#}(e_i - e_j) = \sum_{e \in P_{i,j}} \frac{1}{w(e)} v(e)$. Since $||v(e)||_1 = \frac{2}{n}|\beta_i(e)|(n - |\beta_i(e)|)$ for each edge $e \in P_{i,j}$, it now follows that $\tau(L^{\#}) \leq \frac{1}{n}\max_{i,j} \sum_{e \in P_{i,j}} \frac{|\beta_i(e)|(n - |\beta_i(e)|)}{w(e)}$. It is straightforward to see that this last quantity is maximised for some pair of pendent vertices i and j, which establishes (7.9). □

Our next two examples illustrate (7.8) and (7.9), respectively.

EXAMPLE 7.3.2 Suppose that $\mathcal{T}$ is the unweighted star on n vertices, $K_{1,n-1}$. For any pendent vertex i of $\mathcal{T}$, it is readily established that $\sum_{e \in \mathcal{T}} |\beta_i(e)|^2 = (n-1)^2 + n - 2 = n^2 - n - 1$, so that the upper bound of (7.8) is then equal to $\frac{n^2 - n}{n^2 - n - 1}$. The eigenvalues of $L(K_{1,n-1})$ are readily seen to be $0, n$ and 1, the latter with multiplicity $n - 2$. Consequently $\alpha(K_{1,n-1}) = 1$, and so we find that for large values of n, the upper bound of (7.8) provides a reasonably close estimate for the true value of the algebraic connectivity in this case.

EXAMPLE 7.3.3 Here we take $\mathcal{T}$ to be the unweighted path on n

vertices, P_n. This tree has just two pendent vertices, say 1 and n, without loss of generality, and it is straightforward to show that

$$\sum_{e \in P_{1,n}} |\beta_1(e)|(n - |\beta_1(e)|) = \sum_{k=1}^{n-1} k(n-k) = \frac{n(n^2-1)}{6}.$$

The lower bound of (7.9) is thus equal to $\frac{6}{n^2-1}$. It is known (see the book of Gregory and Karney [47, p. 138]) that $\alpha(P_n) = 2(1 - \cos(\frac{\pi}{n}))$, which, for large values of n, is approximately $\frac{\pi^2}{n^2}$. Thus, for large values of n, our lower bound in (7.9) for this class of graphs is of the same order of magnitude (in n) as the true value of the algebraic connectivity.

In the course of establishing (7.9), we used the following fact: for a connected weighted graph $(\mathcal{G}, w)$ with Laplacian matrix L, we have

$$\alpha(\mathcal{G}, w) \geq \frac{1}{\tau(L^{\#})}. \tag{7.10}$$

While there is no known characterisation of the equality case in (7.10) at the moment, there are a few classes of unweighted graphs for which equality is known to hold in (7.10). We discuss some of those classes now, loosely following the approach in Kirkland, Molitierno and Neumann's paper [75], and [76]. In the sequel we will rely on the readily established fact that for a connected graph $\mathcal{G}$ on n vertices, $L(\mathcal{G})^{\#} = (L(\mathcal{G}) + J)^{\#} - \frac{1}{n}J$. We will also use the observation made above that $\tau(L(\mathcal{G})^{\#}) = \frac{1}{2}\max_{i,j} ||L(\mathcal{G})^{\#}(e_i - e_j)||_1$. We begin our discussion with the following technical result, part b) of which appears in [76].

LEMMA 7.3.4 *a) Let $(\mathcal{G}, w)$ be a weighted graph, and suppose that $t > 0$. Then $\tau((L(\mathcal{G}, w)+tI)^{-1}) \leq \frac{1}{t}$. If $\mathcal{G}$ is not connected, then $\tau((L(\mathcal{G}, w)+ tI)^{-1}) = \frac{1}{t}$.*
b) Let $\mathcal{G}$ be an unweighted graph on m vertices, and suppose that $t > 0$. Then each diagonal entry of $(L(\mathcal{G}) + tI)^{-1}$ is bounded below by $\frac{t+1}{t(m+t)}$.

<u>**Proof**</u>: a) First, note that $L(\mathcal{G}, w) + tI$ is a nonsingular M-matrix, with constant row sums t. Hence, $(L(\mathcal{G}, w) + tI)^{-1}$ is a nonnegative matrix with constant row sums $\frac{1}{t}$, from which we find readily that $\tau((L(\mathcal{G}, w)+ tI)^{-1}) \leq \frac{1}{t}$. Suppose now that $\mathcal{G}$ is not connected, say with $\mathcal{G} = \mathcal{G}_1 \cup \mathcal{G}_2$, where $\mathcal{G}_1, \mathcal{G}_2$ are proper subgraphs of $\mathcal{G}$. Then L can be written as

$$L = \left[\begin{array}{c|c} L_1 & 0 \\ \hline 0 & L_2 \end{array}\right],$$

where L_1, L_2 are the Laplacian matrices for the appropriate weightings of $\mathcal{G}_1, \mathcal{G}_2$, respectively. Evidently

$$(L+tI)^{-1} = \left[\begin{array}{c|c} (L_1+tI)^{-1} & 0 \\ \hline 0 & (L_2+tI)^{-1} \end{array}\right];$$

letting i, j be indices corresponding to the first and second parts of the partitioning of L, respectively, we find that $(e_i - e_j)^t(L+tI)^{-1}$ can be written as $(e_i - e_j)^t L = \left[\begin{array}{c|c} x_1^t & -x_2^t \end{array}\right]$, where each of x_1 and x_2 is a nonnegative vector with entries summing to $\frac{1}{t}$. It now follows readily that $\tau((L+tI)^{-1}) = \frac{1}{t}$.

b) Suppose first that $\mathcal{G}$ is not the complete graph on m vertices, K_m. Then for some pair of indices i, j, the vertices i, j are not adjacent in $\mathcal{G}$. Form $\hat{\mathcal{G}}$ from $\mathcal{G}$ by adding the edge between vertices i and j. Then $L(\hat{\mathcal{G}}) = L(\mathcal{G}) + (e_i - e_j)(e_i - e_j)^t$, from which we find that

$$(L(\hat{\mathcal{G}})+tI)^{-1} =$$
$$(L(\mathcal{G})+tI)^{-1} - \frac{(L(\mathcal{G})+tI)^{-1}(e_i-e_j)(e_i-e_j)^t(L(\mathcal{G})+tI)^{-1}}{1+(e_i-e_j)^t(L(\mathcal{G})+tI)^{-1}(e_i-e_j)}.$$

It now follows readily that for each $k = 1, \ldots, m$, the k–th diagonal entry of $(L(\mathcal{G})+tI)^{-1}$ is bounded below by the corresponding diagonal entry of $(L(\hat{\mathcal{G}})+tI)^{-1}$. We thus deduce that the diagonal entries of $(L(\mathcal{G})+tI)^{-1}$ are bounded below by the corresponding diagonal entries of $(L(K_m)+tI)^{-1}$. Since $(L(K_m)+tI)^{-1} = \frac{1}{m+t}I + \frac{1}{t(m+t)}J$, the conclusion now follows. □

Given unweighted graphs $\mathcal{G}$ and $\mathcal{H}$ on k and m vertices, respectively, the *join* of $\mathcal{G}$ and $\mathcal{H}$, denoted by $\mathcal{G} \vee \mathcal{H}$ is the graph on $k+m$ vertices formed from $\mathcal{G} \cup \mathcal{H}$ by making every vertex in $\mathcal{G}$ adjacent to every vertex in $\mathcal{H}$. It is straightforward to determine that the Laplacian matrix for $\mathcal{G} \vee \mathcal{H}$ can be written as

$$L(\mathcal{G} \vee \mathcal{H}) = \left[\begin{array}{c|c} L(\mathcal{G})+mI & -J \\ \hline -J & L(\mathcal{H})+kI \end{array}\right].$$

Using the join operation, we now inductively describe the class of *threshold graphs* as follows. Set $\Gamma_1 = \{K_p | p \geq 1\}$, where K_p denotes the complete graph on p vertices. For each $j \in \mathbb{N}$, we define $\Gamma_{j+1} = \{\mathcal{G}^c \vee K_p | \mathcal{G} \in \Gamma_j, p \geq 1\}$. (Here $\mathcal{G}^c$ is the complement of the graph $\mathcal{G}$, that is, the graph on the same vertex set as the vertex set for $\mathcal{G}$, with vertices i and j adjacent in $\mathcal{G}^c$ if and only if $i \neq j$ and vertices i

and j are not adjacent in $\mathcal{G}$.) The collection of graphs $\Gamma = \cup_{j\in\mathbb{N}}\Gamma_j$ is the family of connected threshold graphs (also known as *maximal graphs*, see the paper of Merris [92]), while any disconnected threshold graph is the union of a connected threshold graph with a collection of isolated vertices. Much is known about threshold graphs (see Mahadev and Peled [90]). For instance, the Laplacian spectrum of any connected threshold graph consists entirely of integers (see [92]), and the class of connected threshold graphs is known to provide the equality cases in a certain collection of inequalities between the degree sequence of a graph and its Laplacian spectrum (see Bai's paper [7]). As we will see in our next result, the connected threshold graphs also yield equality in (7.10), a fact that was first established in [75].

THEOREM 7.3.5 *a) Let $\mathcal{G}$ be an unweighted graph of the form $\mathcal{H}\vee K_p$, where $\mathcal{H}$ is disconnected and $p \in \mathbb{N}$. Then $\mathcal{G}$ yields equality in* (7.10), *with $\alpha(\mathcal{G}) = \frac{1}{\tau(L(\mathcal{G})^\#)} = p$.*
b) For any connected threshold graph $\mathcal{G}$ on two or more vertices, we have equality in (7.10).

Proof: a) Suppose that $\mathcal{H}$ has q vertices. Then we may write $L(\mathcal{G})$ as

$$L(\mathcal{G}) = \left[\begin{array}{c|c} L(\mathcal{H}) + pI & -J \\ \hline -J & (p+q)I - J \end{array}\right].$$

Since $L(\mathcal{G})^\# = (L(\mathcal{G}) + J)^{-1} - \frac{1}{(p+q)^2}J$, we have $\tau(L(\mathcal{G})^\#) = \tau((L(\mathcal{G}) + J)^{-1})$.

Next, we note that

$$\begin{aligned}(L(\mathcal{G}) + J)^{-1} &= \left[\begin{array}{c|c} L(\mathcal{H}) + pI + J & 0 \\ \hline 0 & (p+q)I \end{array}\right]^{-1} \\ &= \left[\begin{array}{c|c} (L(\mathcal{H}) + pI + J)^{-1} & 0 \\ \hline 0 & \frac{1}{p+q}I \end{array}\right] \\ &= \left[\begin{array}{c|c} (L(\mathcal{H}) + pI)^{-1} - \frac{1}{p(p+q)}J & 0 \\ \hline 0 & \frac{1}{p+q}I \end{array}\right]. \qquad (7.11)\end{aligned}$$

We claim that $\tau((L(\mathcal{G}) + J)^{-1}) = \frac{1}{p}$. To see the claim, first suppose that we have indices i, j with $1 \le i \le q < j \le p+q$. From (7.11), we find that $||(L(\mathcal{G})+J)^{-1}(e_i - e_j)||_1 = ||((L(\mathcal{H})+pI)^{-1} - \frac{1}{p(p+q)}J)e_i||_1 + \frac{1}{p+q}$. Since $(L(\mathcal{H}) + pI)^{-1}$ is a nonnegative matrix with constant row sums $\frac{1}{p}$, it follows that $||(L(\mathcal{G}) + J)^{-1}(e_i - e_j)||_1 \le \frac{1}{p} + \frac{q}{p(p+q)} + \frac{1}{p+q} = \frac{2}{p}$. Similarly, for indices i, j with $q+1 \le i, j \le p+q$, $||(L(\mathcal{G})+J)^{-1}(e_i - e_j)||_1 \le \frac{2}{p+q}$. Finally, for indices i, j between 1 and q, we have $||(L(\mathcal{G}) + J)^{-1}(e_i - e_j)||_1 = ||((L(\mathcal{H}) + pI)^{-1}(e_i - e_j)||_1$. Appealing to Lemma 7.3.4 a), we find that for each $i, j, = 1, \ldots, q$, $||(L(\mathcal{H}) + pI)^{-1}(e_i - e_j)||_1 \le \frac{2}{p}$, with equality holding for at least one such pair of indices i, j, since $\mathcal{H}$ is not connected. Assembling the observations above, we find that

$$\frac{1}{2}||(L(\mathcal{G}) + J)^{-1}(e_i - e_j)||_1 \le \frac{1}{p}$$

for all $i, j = 1, \ldots, p+q$, with equality holding for at least one pair of distinct indices i, j. Hence $\tau((L(\mathcal{G}) + J)^{-1}) = \frac{1}{p}$, as claimed.

We thus have $\frac{1}{\tau((L(\mathcal{G})^{\#})} = p \le \alpha(\mathcal{G})$. Since $\mathcal{G} = \mathcal{H} \vee K_p$, we can find a collection of p vertices whose deletion from $\mathcal{G}$ yields $\mathcal{H}$. As $\mathcal{H}$ is not connected, it then follows that the vertex connectivity $v(\mathcal{G})$ is at most p. Since $\alpha(\mathcal{G}) \le v(\mathcal{G}) \le p$, we thus find that $\frac{1}{\tau((L(\mathcal{G})^{\#})} = p = \alpha(\mathcal{G})$, as desired.

b) If $\mathcal{G}$ is a connected threshold graph, then either $\mathcal{G}$ is of the form $\mathcal{H} \vee K_p$, where $\mathcal{H}$ is disconnected, or $\mathcal{G}$ is a complete graph. In the former case, we find from part a) that $\frac{1}{\tau((L(\mathcal{G})^{\#})} = p = \alpha(\mathcal{G})$. On the other hand, if $\mathcal{G}$ is complete, say with $\mathcal{G} = K_m$, it is readily established that $\alpha(\mathcal{G}) = m = \frac{1}{\tau((L(K_m)^{\#})}$. □

Observe that in Theorem 7.3.5, we encounter graphs for which the algebraic connectivity and the vertex connectivity both coincide with $\frac{1}{\tau(L^{\#})}$. The final result of this section, which appears in [76], pursues that theme.

THEOREM 7.3.6 *Let $\mathcal{G}$ be a non–complete connected unweighted graph on n vertices with vertex connectivity r. Suppose that $r^2 \le n$. Then $\alpha(\mathcal{G}) = r$ if and only if $\frac{1}{\tau(L(\mathcal{G})^{\#})} = r$.*

Proof: Suppose first that $\frac{1}{\tau(L(\mathcal{G})^{\#})} = r$. Since $\frac{1}{\tau(L(\mathcal{G})^{\#})} \le \alpha(\mathcal{G}) \le r$, it must be the case that $\alpha(\mathcal{G}) = r$.

For the converse implication, suppose that $\alpha(\mathcal{G}) = r$, so that there is a subset of r vertices of $\mathcal{G}$ whose deletion yields a disconnected graph, say $\mathcal{G}_0 \cup \mathcal{G}_1$. For concreteness, we suppose that $\mathcal{G}_0$ and $\mathcal{G}_1$ have p and q

vertices, respectively. It follows then that $L(\mathcal{G})$ can be written as

$$L(\mathcal{G}) = \left[\begin{array}{c|c|c} L(\mathcal{G}_0)+D_0 & 0 & -X \\ \hline 0 & L(\mathcal{G}_1)+D_1 & -Y \\ \hline -X^t & -Y^t & L(\mathcal{G}_2)+D_2 \end{array}\right],$$

where X and Y are $(0,1)$ matrices, D_0, D_1, D_2 are diagonal matrices satisfying $D_0\mathbf{1} = X\mathbf{1}, D_1\mathbf{1} = Y\mathbf{1}, D_2\mathbf{1} = X\mathbf{1} + Y\mathbf{1}$, and where $\mathcal{G}_2$ is an induced subgraph of $\mathcal{G}$ on r vertices. Note that the diagonal entries of D_0 and D_1 are bounded above by r.

We claim that necessarily X and Y are all ones matrices. To see the claim, consider the vector u given by

$$u = \left[\begin{array}{c} q\mathbf{1} \\ \hline -p\mathbf{1} \\ \hline 0 \end{array}\right],$$

and note that $u^t\mathbf{1} = 0$. Note that $u^tL(\mathcal{G})u = q^2\mathbf{1}^tD_0\mathbf{1} + p^2\mathbf{1}^tD_1\mathbf{1} \leq q^2rp + p^2rq = ru^tu$. But since u is orthogonal to $\mathbf{1}$, we also have that $ru^tu = \alpha(\mathcal{G})u^tu \leq u^tL(\mathcal{G})u$, from which we conclude that $u^tL(\mathcal{G})u = ru^tu$. In particular, it must be the case that all diagonal entries in D_0 and D_1 are equal to r. Consequently, X and Y are all ones matrices, as claimed.

Thus we have

$$L(\mathcal{G}) = \left[\begin{array}{c|c|c} L(\mathcal{G}_0)+rI & 0 & -J \\ \hline 0 & L(\mathcal{G}_1)+rI & -J \\ \hline -J & -J & L(\mathcal{G}_2)+(p+q)I \end{array}\right].$$

As in the proof of Theorem 7.3.5, we find that

$$\tau(L(\mathcal{G})^{\#}) = \tau((L(\mathcal{G}) + J)^{-1}),$$

and it is straightforward to show that

$$(L(\mathcal{G}) + J)^{-1} = \left[\begin{array}{c|c|c} C_{1,1} & C_{1,2} & 0 \\ \hline C_{2,1} & C_{2,2} & 0 \\ \hline 0 & 0 & C_{3,3} \end{array}\right], \qquad (7.12)$$

where

$$C_{1,1} = (L(\mathcal{G}_0) + rI)^{-1} - \frac{1}{r(p+q+r)}J,$$

$$C_{1,2} = -\frac{1}{r(p+q+r)}J = C_{2,1}^t,$$

$$C_{2,2} = (L(\mathcal{G}_1) + rI)^{-1} - \frac{1}{r(p+q+r)}J, \text{ and}$$

$$C_{3,3} = (L(\mathcal{G}_2) + (p+q)I)^{-1} - \frac{1}{(p+q)(p+q+r)}J.$$

Suppose that we have indices i, j with $1 \le i, j \le p+q$. It follows readily from (7.12), (7.13), and the fact that each of $(L(\mathcal{G}_0) + rI)^{-1}, (L(\mathcal{G}_1) + rI)^{-1}$ is nonnegative with row sums $\frac{1}{r}$, that $||(L(\mathcal{G})+J)^{-1}(e_i-e_j)||_1 \le \frac{2}{r}$, with equality holding in the case that $1 \le i \le p < j \le p+q$. Similarly, for distinct indices i, j with $p+q+1 \le i, j \le p+q+r$, we find that $||(L(\mathcal{G}) + J)^{-1}(e_i - e_j)||_1 \le \frac{2}{p+q}$. Invoking our hypothesis that $n \ge r^2$, yields $p+q+r \ge r^2 \ge 2r$ (the last inequality follows from the assumption that there are two distinct indices between $p+q+1$ and $p+q+r$); it now now follows that $\frac{2}{p+q} \le \frac{2}{r}$, so that $||(L(\mathcal{G}) + J)^{-1}(e_i - e_j)||_1 \le \frac{2}{r}$.

Next, we consider an index i between 1 and p, and note that $||(L(\mathcal{G}) + J)^{-1}e_i||_1 = ||((L(\mathcal{G}_0) + rI)^{-1} - \frac{1}{r(p+q+r)}J)e_i||_1 + \frac{q}{r(p+q+r)}$. Letting d be the i–th diagonal entry of $(L(\mathcal{G}_0) + rI)^{-1}$, we find from Lemma 7.3.4 b) that $d \ge \frac{r+1}{r(p+r)}$, so that $d > \frac{1}{r(p+q+r)}$. We now estimate $||((L(\mathcal{G}_0)+rI)^{-1} - \frac{1}{r(p+q+r)}J)e_i||_1$ by separating out the i–th entry, and then applying the triangle inequality to the remaining entries. In view of the fact that the i–th row sum of $(L(\mathcal{G}_0) + rI)^{-1}$ is $\frac{1}{r}$, while the i–th row sum of the $p \times p$ matrix $\frac{1}{r(p+q+r)}J$ is $\frac{p}{r(p+q+r)}$, it now follows that $||((L(\mathcal{G}_0)+rI)^{-1} - \frac{1}{r(p+q+r)}J)e_i||_1 \le (d - \frac{1}{r(p+q+r)}) + (\frac{1}{r} - d) + \frac{p-1}{r(p+q+r)} = \frac{1}{r} + \frac{p-2}{r(p+q+r)}$. We thus find that $||(L(\mathcal{G}) + J)^{-1}e_i||_1 \le \frac{1}{r} + \frac{p+q-2}{r(p+q+r)}$. An analogous argument shows that if $p+q+1 \le j \le p+q+r$, then

$$||(L(\mathcal{G}) + J)^{-1}e_j||_1 =$$
$$\left|\left| \left(L(\mathcal{G}_2) + (p+q)I)^{-1} - \frac{1}{(p+q)(p+q+r)}J\right) e_{j-p-q}\right|\right|_1$$
$$\le \frac{1}{p+q} + \frac{r-2}{(p+q)(p+q+r)}.$$

Consequently, we find that if $1 \le i \le p$ and $p+q+1 \le j \le p+q+r$, then $||(L(\mathcal{G}) + J)^{-1}(e_i - e_j)||_1 \le \frac{1}{r} + \frac{p-2}{r(p+q+r)} + \frac{1}{p+q} + \frac{r-2}{(p+q)(p+q+r)} \le \frac{2}{r}$, the last inequality following from our hypothesis that $n \ge r^2$. Similarly,

$||(L(\mathcal{G})+J)^{-1}(e_i-e_j)||_1 \le \frac{2}{r}$ for $p+1 \le i \le p+q < j \le p+q+r$.

Assembling the observations above, it now follows that for any pair of indices i,j with $1 \le i,j \le n$,

$$\frac{1}{2}||(L(\mathcal{G})+J)^{-1}(e_i-e_j)||_1 \le \frac{1}{r},$$

with equality holding if $1 \le i \le p < j \le p+q$. Hence $\tau(L(\mathcal{G})^{\#}) = \tau((L(\mathcal{G})+J)^{-1}) = \frac{1}{r}$, as desired. □

The following example from [76] shows that in general, if the hypothesis $r^2 \le n$ is relaxed, then the conclusion of Theorem 7.3.6 can fail to hold.

EXAMPLE 7.3.7 Suppose that we have integers n,r such that $n-r \ge r \ge 2$ and $r^2 > n$. Consider the complete bipartite graph $K_{r,n-r}$, where the cardinalities of the partite sets are r and $n-r$. The corresponding Laplacian matrix is given by

$$L = \left[\begin{array}{c|c} (n-r)I & -J \\ \hline -J & rI \end{array}\right],$$

where the first diagonal block has order r and the second diagonal block has order $n-r$. It is readily determined that the eigenvalues of L are $0, r$ (with multiplicity $n-r-1$), $n-r$ (with multiplicity $r-1$) and n. In particular, $\alpha(K_{r,n-r}) = r$. Further, we find that the vertex connectivity of $K_{r,n-r}$ is also r, so that the algebraic and vertex connectivities of $K_{r,n-r}$ coincide.

We claim that $\tau(L^{\#}) > \frac{1}{r}$. To see the claim, first note that

$$(L+J)^{-1} = \left[\begin{array}{c|c} \frac{1}{n-r}I - \frac{1}{n(n-r)}J & 0 \\ \hline 0 & \frac{1}{r}I - \frac{1}{rn}J \end{array}\right].$$

Observe that

$$(L+J)^{-1}(e_1 - e_{r+1}) = \left[\begin{array}{c} \frac{1}{n-r}e_1 - \frac{1}{n(n-r)}\mathbf{1} \\ \hline -\frac{1}{r}e_1 + \frac{1}{rn}\mathbf{1} \end{array}\right].$$

A straightforward computation shows that $||(L+J)^{-1}(e_1-e_{r+1})||_1 = \frac{1}{n-r} + \frac{r-2}{n(n-r)} + \frac{1}{r} + \frac{n-r-2}{rn} = \frac{n^2+r(r-2)+(n-r)(n-r-2)}{r(n-r)n}$. From the hypothesis that $r^2 > n$, we find that $||(L+J)^{-1}(e_1-e_{r+1})||_1 > \frac{2}{r}$, which in turn yields $\tau((L+J)^{-1}) > \frac{1}{r}$. Thus we see that $r = \alpha(K_{r,n-r}) > \frac{1}{\tau(L(K_{r,n-r})^{\#})}$ when $r^2 > n$.

7.4 Resistance distance, the Weiner index and the Kirchhoff index

Suppose that we have a connected undirected graph $\mathcal{G}$ on vertices labelled $1, \ldots, n$. Recall that the *distance* between vertices i and j, denoted $d(i,j)$ is the number of edges on a shortest path from i to j in $\mathcal{G}$. The *Weiner index* of G is defined as $\mathcal{W}(\mathcal{G}) = \sum_{1 \le i < j \le n} d(i,j)$ (see the paper of Hosoya [58]). We saw in Corollary 7.2.13 that for a weighted tree $(\mathcal{T}, w)$ on n vertices with Laplacian matrix L, we have $n\text{trace}(L^{\#}) = \sum_{1 \le i < j \le n} \tilde{d}(i,j)$. Thus, in the special case that $\mathcal{T}$ is unweighted (so that $\tilde{d}(i,j) = d(i,j)$ for each i,j) we find that $\mathcal{W}(\mathcal{T}) = n\text{trace}(L(\mathcal{T})^{\#})$. As the following example shows, the corresponding relation between the Weiner index and trace$(L^{\#})$ may not hold for general unweighted graphs.

EXAMPLE 7.4.1 Consider the unweighted undirected cycle C_n on vertices $1, \ldots, n$, where $n \ge 3$. For indices $1 \le i < j \le n$, we find that $d(i,j) = \min\{j-i, n-(j-i)\}$. It now follows that

$$\mathcal{W}(C_n) = \begin{cases} \frac{n^3}{8} & \text{if } n \text{ is even,} \\ \frac{n(n^2-1)}{8} & \text{if } n \text{ is odd.} \end{cases}$$

It turns out that each diagonal entry of $L^{\#}$ is equal to $\frac{n^2-1}{12n}$ (in Example 7.4.8 below we outline a technique for establishing that fact); hence $n\text{trace}(L^{\#}) = \frac{n(n^2-1)}{12}$. Thus we find that for any $n \ge 3, \mathcal{W}(C_n) > n\text{trace}(L(C_n)^{\#})$.

Let $(\mathcal{G}, w)$ be a connected weighted graph on vertices $1, \ldots, n$ with corresponding Laplacian matrix L. In view of the connection between the Weiner index and the trace of $L^{\#}$ for the case of an unweighted tree, it is natural to seek an interpretation for the trace of $L^{\#}$ for general weighted graphs; much of the work in this section addresses that issue. To that end, we make the following definition (see Klein and Randic's paper [87]): for each pair of distinct indices i, j between 1 and n, the *resistance distance* from i to j, $r(i,j)$, is given by

$$r(i,j) = l^{\#}_{i,i} + l^{\#}_{j,j} - 2l^{\#}_{i,j}. \tag{7.13}$$

Equivalently, we have $r(i,j) = (e_i - e_j)^t L^{\#}(e_i - e_j)$. For consistency we define $r(i,i) = 0$ for each $i = 1, \ldots, n$.

In this section, we present some of the key properties of the resistance distance for weighted graphs. Most of the results appear in [87], though the proofs that we present in the sequel are somewhat different from those in [87].

REMARK 7.4.2 The resistance distance has a natural connection with the bottleneck matrices introduced in section 7.2. For concreteness, we take $j = n$, and let B denote the bottleneck matrix for the weighted graph $(\mathcal{G}, w)$ based at vertex n. It follows from Observation 7.2.1 that

$$(e_i - e_n)^t L^{\#}(e_i - e_n) =$$
$$(e_i - e_n)^t \left[\begin{array}{c|c} B - \frac{1}{n}BJ - \frac{1}{n}JB & -\frac{1}{n}B\mathbf{1} \\ \hline -\frac{1}{n}\mathbf{1}^t B & 0 \end{array}\right](e_i - e_n) = b_{i,i}.$$

Thus we find that for any pair of distinct indices i, j, the resistance distance from i to j is the diagonal entry of the bottleneck matrix based at vertex j that corresponds to vertex i. In particular, we see that $r(i,j) > 0$ whenever $i \neq j$.

If the resistance distance is to be interpretable as a distance, we should assure ourselves that it satisfies three natural properties for a distance function, namely:
a) $r(i,j) \geq 0$ for any i, j, with equality holding if and only if $i = j$;
b) $r(i,j) = r(j,i)$ for any i, j;
c) $r(i,k) + r(k,j) \geq r(i,j)$ for any i, j, k.
Remark 7.4.2 serves to establish a), while b) follows from (7.13). To see that property c) holds, observe that the desired inequality is

$$l^{\#}_{i,i} + l^{\#}_{k,k} - 2l^{\#}_{i,k} + l^{\#}_{j,j} + l^{\#}_{k,k} - 2l^{\#}_{j,k} \geq l^{\#}_{i,i} + l^{\#}_{j,j} - 2l^{\#}_{i,j}.$$

This last is readily seen to be equivalent to

$$l^{\#}_{k,k} - l^{\#}_{i,k} + l^{\#}_{i,j} - l^{\#}_{j,k} \geq 0,$$

which follows immediately from Proposition 6.2.4. Observe that Proposition 6.2.4 also shows that $r(i,k) + r(k,j) = r(i,j)$ if and only if either $k = i$, or $k = j$, or every path from vertex i to vertex j in $\mathcal{G}$ passes through vertex k. Thus we find that the resistance distance satisfies each of properties a), b), and c).

Part of the motivation for considering the resistance distance, instead of the ordinary graph–theoretic distance, is that for unweighted graphs, the former is influenced not only by the number of edges on a shortest path between vertices, but also by the number of paths between those vertices. The following example illustrates.

EXAMPLE 7.4.3 Fix an $n \geq 3$, and consider the unweighted complete bipartite graph $\mathcal{G} = K_{2,n-2}$, where the vertices are labelled so vertices $1, 2$ are in one subset of the bipartition, and vertices $3, \ldots, n$ are in the other subset of the bipartition. Observe that with this labelling of the vertices, the graph-theoretic distance from vertex 1 to vertex 2 is 2 (regardless of the value of n), and there are $n-2$ paths from 1 to 2. Our Laplacian matrix L is

$$L = \left[\begin{array}{c|c} (n-2)I & -J \\ \hline -J & 2I \end{array}\right].$$

It follows that the bottleneck matrix based at vertex 2 (which is $(n-1) \times (n-1)$) is

$$B = \left[\begin{array}{c|c} \frac{2}{n-2} & \frac{1}{n-2}\mathbf{1}^t \\ \hline \frac{1}{n-2}\mathbf{1} & \frac{1}{2}I + \frac{1}{2(n-2)}J \end{array}\right].$$

From Remark 7.4.2, we have $r(1,2) = b_{1,1} = \frac{2}{n-2}$, so we see that the resistance distance between vertices 1 and 2 is decreasing with n. Thus, for this family of examples, the resistance distance is sensitive to the number of paths between vertex 1 and vertex 2, but the graph-theoretic distance is not.

The following result from the paper of Gutman and Mohar [49] establishes a connection between resistance distances and Laplacian eigenvalues.

PROPOSITION 7.4.4 *Let $(\mathcal{G}, w)$ be a connected weighted graph on vertices $1, \ldots, n$ with Laplacian matrix L. Then*

$$\sum_{1 \leq i < j \leq n} r(i,j) = n\text{trace}L^{\#}.$$

Proof: We have $\sum_{1 \leq i < j \leq n} r(i,j) = \frac{1}{2}\sum_{i=1}^{n}\sum_{j=1}^{n} r(i,j)$. Next, note that

$$\sum_{i=1}^{n}\sum_{j=1}^{n} r(i,j) = \sum_{i=1}^{n}\sum_{j=1}^{n}(l^{\#}_{i,i} + l^{\#}_{j,j} - 2l^{\#}_{i,j}). \tag{7.14}$$

Since $L^{\#}$ has row sums zero, (7.14) now yields

$$\sum_{i=1}^{n}\sum_{j=1}^{n} r(i,j) = \sum_{i=1}^{n}(nl^{\#}_{i,i} + \text{trace}(L^{\#})) = 2n\text{trace}(L^{\#}),$$

and the conclusion follows. □

Given a connected weighted graph $(\mathcal{G}, w)$, the quantity $\text{Kf}(\mathcal{G}, w) = \sum_{1\le i<j\le n} r(i,j)$ is known as the *Kirchhoff index* for the weighted graph $(\mathcal{G}, w)$ (see [49]). In the case that $\mathcal{G}$ is unweighted, we use the simpler notation $\text{Kf}(\mathcal{G})$ for the Kirchhoff index. We note that in [44], Ghosh, Boyd and Saberi refer to the quantity $\sum_{1\le i<j\le n} r(i,j)$ as the *total effective resistance* for the weighted graph.

From Corollary 7.2.13 and Proposition 7.4.4, we find that for a weighted tree, the Kirchhoff and Weiner indices coincide. This suggests that for general weighted graphs, there may be a relationship between the resistance distances and some notion of a weighted distance between vertices. The next sequence of results explores that connection. Our first such result discusses the group inverse of the Laplacian matrix for a weighted graph when an edge weight is perturbed; it bears a resemblance to the Sherman–Morrison formula [57, section 0.7.4].

LEMMA 7.4.5 *Let $(\mathcal{G}, w)$ be a connected weighted graph on vertices $1, \ldots, n$ with Laplacian matrix L. Fix distinct indices i, j, let $\theta > 0$, and set $\hat{L} = L + \theta(e_i - e_j)(e_i - e_j)^t$. Then*

$$\hat{L}^{\#} = L^{\#} - \frac{\theta}{1+\theta(e_i - e_j)^t L^{\#}(e_i - e_j)} L^{\#}(e_i - e_j)(e_i - e_j)^t L^{\#}. \quad (7.15)$$

<u>**Proof**</u>: It is readily determined that the matrix

$$R = L^{\#} - \frac{\theta}{1+\theta(e_i - e_j)^t L^{\#}(e_i - e_j)} L^{\#}(e_i - e_j)(e_i - e_j)^t L^{\#}$$

commutes with $\hat{L}$, and that $R\hat{L} = \hat{L}R = I - \frac{1}{n}J$. Further, since $R\mathbf{1} = 0$ and $\mathbf{1}^t R = 0^t$, we deduce that $R = \hat{L}^{\#}$. □

REMARK 7.4.6 The proof of Lemma 7.4.5 also shows that (7.15) holds if $\theta < 0$, provided that $l_{i,j} - \theta < 0$ and $1+\theta(e_i - e_j)^t L^{\#}(e_i - e_j) \neq 0$.

Lemma 7.4.5 yields the following result, an equivalent version of which is found the book of Doyle and Snell [34, section 4.1] (see Observation 7.5.2 below).

COROLLARY 7.4.7 *Suppose that $(\mathcal{G}, w)$ is a connected weighted graph on vertices $1, \ldots, n$. Form the weighted graph $(\hat{\mathcal{G}}, \hat{w})$ by either adding a weighted edge to $(\mathcal{G}, w)$ or raising the weight of an existing edge in $(\mathcal{G}, w)$. Then for any pair of distinct indices i, j the resistance distance from i to j in $(\hat{\mathcal{G}}, \hat{w})$ is bounded above by the resistance distance from i to j in $(\mathcal{G}, w)$.*

Proof: Let L and $\hat{L}$ denote the Laplacian matrices for $(\mathcal{G}, w)$ and $(\hat{\mathcal{G}}, \hat{w})$, respectively. Let i_0, j_0 denote the end points of the edge in $\mathcal{G}, w$ that is either added or whose weight is increased to produce $(\hat{\mathcal{G}}, \hat{w})$. Then for some $\theta_0 > 0$, we have from Lemma 7.4.5 that

$$\hat{L}^{\#} = L^{\#} - \frac{\theta_0}{1 + \theta_0 (e_{i_0} - e_{j_0})^t L^{\#} (e_{i_0} - e_{j_0})} L^{\#} (e_{i_0} - e_{j_0})(e_{i_0} - e_{j_0})^t L^{\#}.$$

Letting $r(i, j)$ and $\hat{r}(i, j)$ denote the resistance distances from i to j in $(\mathcal{G}, w)$ and $(\hat{\mathcal{G}}, \hat{w})$, respectively, it now follows that

$$\hat{r}(i, j) = r(i, j) - \frac{\theta_0 \left((e_i - e_j)^t L^{\#} (e_{i_0} - e_{j_0}) \right)^2}{1 + \theta_0 (e_{i_0} - e_{j_0})^t L^{\#} (e_{i_0} - e_{j_0})}.$$

□

Our next example serves to justify one of the assertions made in Example 7.4.1.

EXAMPLE 7.4.8 Consider the unweighted undirected cycle C_n on vertices $1, \ldots, n$, where the labelling is such that for each $i = 1, \ldots, n$, vertex i is adjacent to vertices $i-1$ and $i+1$ (with indices taken modulo n). Observe that the Laplacian matrix for C_n, say $\hat{L}$, can be written as $L + (e_1 - e_n)(e_1 - e_n)^t$, where L is the Laplacian matrix for the path on n vertices (with vertices 1 and n pendent). From Lemma 7.4.5, it follows that

$$\hat{l}^{\#}_{1,1} = l^{\#}_{1,1} - \frac{(l^{\#}_{1,1} - l^{\#}_{1,n})^2}{1 + l^{\#}_{1,1} + l^{\#}_{n,n} - 2 l^{\#}_{1,n}}. \tag{7.16}$$

Referring to Example 7.2.8, we find that $l^{\#}_{1,1} = l^{\#}_{n,n} = \frac{(n-1)(2n-1)}{6n}$, while $l^{\#}_{1,n} = \frac{(n-1)(2n-1)}{6n} - \frac{n-1}{2}$. Substituting these values into (7.16), we find that $\hat{l}^{\#}_{1,1} = \frac{(n^2-1)}{12n}$. Indeed we find from an analogous argument (or from symmetry considerations) that $\hat{l}^{\#}_{i,i} = \frac{(n^2-1)}{12n}$ for each $i = 1, \ldots, n$.

Suppose that we have a connected weighted graph $(\mathcal{G}, w)$ on vertices $1, \ldots, n$. For each pair of distinct vertices i, j, let $\mathcal{P}(i, j)$ denote the set of paths in $\mathcal{G}$ from i to j. We define $\tilde{\delta}(i,j)$ as follows:

$$\tilde{\delta}(i,j) = \min\left\{\sum_{e\in P}\frac{1}{w(e)}\middle|P\in\mathcal{P}(i,j)\right\}.$$

Observe that in the special case that $(\mathcal{G},w)$ is a weighted tree, $\mathcal{P}(i,j)$ contains just one path, so that $\tilde{\delta}(i,j)$ coincides with the inverse weighted distance $\tilde{d}(i,j)$. The following result establishes a relationship between $r(i,j)$ and $\tilde{\delta}(i,j)$.

THEOREM 7.4.9 ([87]) *Let $(\mathcal{G},w)$ be a connected weighted graph on vertices $1,\ldots,n$. For each pair of distinct vertices i,j, we have*

$$r(i,j)\le\tilde{\delta}(i,j). \tag{7.17}$$

Equality holds in (7.17) *if and only if there is a unique path in $\mathcal{G}$ from i to j.*

<u>**Proof**</u>: First, suppose that there is a unique path P from vertex i to vertex j in $\mathcal{G}$. By Remark 7.4.2, $r(i,j)$ is the diagonal entry corresponding to vertex i in the bottleneck matrix based at vertex j. Thus, by (7.2) and Proposition 7.2.2, we find that

$$r(i,j)=\frac{\sum_{\mathcal{F}\in\mathcal{S}_j^{\{i,i\}}}w(\mathcal{F})}{\sum_{\mathcal{T}\in\mathcal{S}}w(\mathcal{T})}.$$

Consider the graph $\hat{\mathcal{G}}$ formed from $\mathcal{G}$ by deleting the intermediate vertices of P and all edges incident with them (if P has just one edge, then we just delete that edge). Since P is the unique path from i to j, it follows that $\hat{\mathcal{G}}$ is the disjoint union of two connected graphs $\mathcal{G}_1,\mathcal{G}_2$, one of which contains vertex i while the other contains vertex j. Let Σ_1,Σ_2 be the collections of spanning trees of $\mathcal{G}_1,\mathcal{G}_2$, respectively. It follows that any spanning tree of $\mathcal{S}$ can be decomposed as $\mathcal{T}_1\cup\mathcal{T}_2\cup P$, where $\mathcal{T}_1\in\Sigma_1$, and $T_2\in\Sigma_2$. Similarly, each forest in $\mathcal{S}_j^{\{i,i\}}$ can be decomposed as $T_1\cup T_2\cup(P\setminus e)$, for some $\mathcal{T}_1\in\Sigma_1$, $\mathcal{T}_2\in\Sigma_2$ and edge $e\in P$. It now follows that

$$\sum_{\mathcal{T}\in\mathcal{S}}w(\mathcal{T})=\sum_{\mathcal{T}_1\in\Sigma_1,\mathcal{T}_2\in\Sigma_2}w(\mathcal{T}_1)w(\mathcal{T}_2)w(P),$$

while

$$\sum_{\mathcal{F}\in\mathcal{S}_j^{\{i,i\}}}w(\mathcal{F})=\sum_{\mathcal{T}_1\in\Sigma_1,\mathcal{T}_2\in\Sigma_2}\sum_{e\in P}w(T_1)w(T_2)\frac{w(P)}{w(e)}.$$

We thus find that $r(i,j) = \sum_{e\in P} \frac{1}{w(e)} = \tilde{\delta}(i,j)$.

Now suppose that there are at least two paths in $\mathcal{G}$ from i to j. We claim that $r(i,j) < \tilde{\delta}(i,j)$. Let P denote a path from i to j in $\mathcal{G}$ such that $\sum_{e\in P} \frac{1}{w(e)} = \tilde{\delta}(i,j)$. By Corollary 7.4.7, deleting an edge in $\mathcal{G}$ cannot increase $r(i,j)$, so in order to establish the claim, it suffices to consider the case that $\mathcal{G}$ is a unicyclic graph (i.e., a connected graph containing exactly one cycle) with exactly two paths from i to j, one of which is P. Denote the other path from i to j by P_0, and let e be an edge of P_0 that is not on P, say with end points a, b. Suppose without loss of generality that when traversing P_0 from i to j, we pass through vertex a before vertex b.

Observe that the graph $\mathcal{T} = \mathcal{G} \setminus e$ is a tree. Consider the path $P_{i,a}$ in $\mathcal{T}$ from i to a, and let i_0 be the vertex on $P \cap P_{i,a}$ at maximum distance from i (possibly we have $i_0 = i$). Similarly, let $P_{j,b}$ be the path in $\mathcal{T}$ from j to b, and let j_0 be the vertex on $P \cap P_{j,b}$ at maximum distance from j (possibly $j_0 = j$). Finally, let P_{i_0,j_0} denote the path from i_0 to j_0 in $\mathcal{T}$. Figure 7.3 sketches the relative positions of i, j, a, b, i_0 and j_0 in $\mathcal{T}$.

Let L denote the Laplacian matrix for $\mathcal{T}$, where the edge weights for $\mathcal{T}$ are inherited from $\mathcal{G}$, and let $\hat{L}$ be the Laplacian matrix for $(\mathcal{G}, w)$. Setting $\theta = w(e)$ it follows from Lemma 7.4.5 that

$$r(i,j) = (e_i - e_j)^t L^{\#}(e_i - e_j) - \frac{\theta((e_i - e_j)^t L^{\#}(e_a - e_b))^2}{1 + \theta(e_i - e_j)^t L^{\#}(e_i - e_j)}.$$

Observe that since $\mathcal{T}$ is a (weighted) tree containing the (weighted) path P, we find from our argument above that $(e_i - e_j)^t L^{\#}(e_i - e_j) = \tilde{\delta}(i,j)$. Consequently, it suffices to show that $(e_i - e_j)^t L^{\#}(e_a - e_b) \neq 0$.

For each edge e of $\mathcal{T}$, let $v(e)$ be the vector defined in Proposition 7.2.9; without loss of generality, we assume that $v_i(e) > 0$ for each edge e of $\mathcal{T}$. Since $L^{\#} = \sum_{e\in\mathcal{T}} \frac{1}{w(e)} v(e)v(e)^t$, we find that $(e_i - e_j)^t L^{\#}(e_a - e_b) = \sum_{e\in\mathcal{T}} \frac{(v_i(e)-v_j(e))(v_a(e)-v_b(e))}{w(e)}$. Observe that $v_i(e) = v_j(e)$ whenever $e \notin P$, while if $e \in P$, we have $v_i(e) - v_j(e) = 1$. Hence we find that $(e_i - e_j)^t L^{\#}(e_a - e_b) = \sum_{e\in P} \frac{v_a(e)-v_b(e)}{w(e)}$. Next, consider an edge e on P. As above, if $e \notin P_{i_0,j_0}$, then $v_a(e) = v_b(e)$, while if $e \in P_{i_0,j_0}$, then $v_a(e) - v_b(e) = 1$. Thus we have $(e_i - e_j)^t L^{\#}(e_a - e_b) = \sum_{e\in P_{i_0,j_0}} \frac{1}{w(e)} > 0$, as claimed. □

COROLLARY 7.4.10 *Suppose that $\mathcal{G}$ is a connected unweighted graph on vertices $1, \ldots, n$. Then $\mathrm{Kf}(\mathcal{G}) \leq \mathcal{W}(\mathcal{G})$, with equality holding if and only if $\mathcal{G}$ is a tree.*

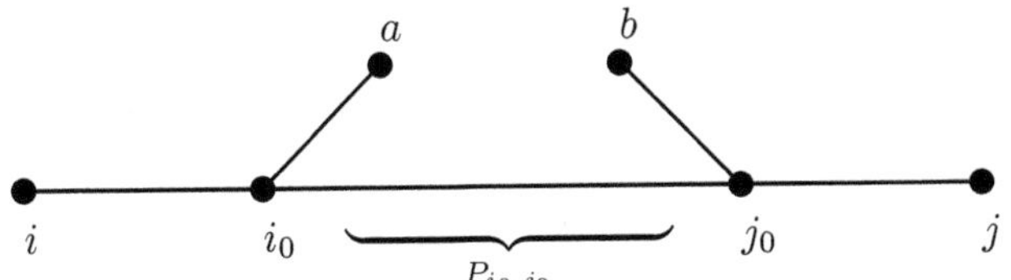

FIGURE 7.3
Sketch of the relevant portion of $\mathcal{T}$ in the proof of Theorem 7.4.9

Proof: For each $i, j = 1, \ldots, n$, let $d(i,j)$ denote the distance from i to j in $\mathcal{G}$. Applying Theorem 7.4.9 and the definitions of $\mathrm{Kf}(\mathcal{G})$ and $\mathcal{W}(\mathcal{G})$, we find that

$$\mathrm{Kf}(\mathcal{G}) = \sum_{1 \le i < j \le n} r(i,j) \le \sum_{1 \le i < j \le n} d(i,j) = \mathcal{W}(\mathcal{G}).$$

Further, appealing again to Theorem 7.4.9, we see that $\mathrm{Kf}(\mathcal{G}) = \mathcal{W}(\mathcal{G})$ if and only if for each pair of distinct vertices in G, there is a unique path between them. This last condition is known to hold if and only if $\mathcal{G}$ is a tree (see [51, Chapter 4]). □

Next, we present a lower bound on the Kirchhoff index of a weighted graph in terms of the diagonal entries of the corresponding Laplacian matrix. Before proving the result, we recall the notion of majorisation: given vectors $x, y \in \mathbb{R}^n$, we say that x *majorises* y if, when the entries in x and y are reordered as $\tilde{x}, \tilde{y}$, respectively, so that $\tilde{x}_1 \ge \tilde{x}_2 \ge \ldots \ge \tilde{x}_n, \tilde{y}_1 \ge \tilde{y}_2 \ge \ldots \ge \tilde{y}_n$, then we have $\sum_{j=1}^k \tilde{x}_j \ge \sum_{j=1}^k \tilde{y}_j, k = 1, \ldots, n-1$ and $\sum_{j=1}^n \tilde{x}_j = \sum_{j=1}^n \tilde{y}_j$. A standard result on majorisation asserts that if x majorises y and f is a convex function, then $\sum_{j=1}^n f(x_j) \ge \sum_{j=1}^n f(y_j)$. We refer the reader to Zhang [114, section 10.3] for details.

THEOREM 7.4.11 *Let $(\mathcal{G}, w)$ be a connected weighted graph on vertices $1, \ldots, n$. For each $j = 1, \ldots, n$, let d_j denote the sum of the weights of the edges of $\mathcal{G}$ that are incident with vertex j. Then for any $t > 0$,*

$$\mathrm{Kf}(\mathcal{G}, w) \ge n \sum_{j=1}^n \frac{1}{d_j + t} - \frac{1}{t}. \tag{7.18}$$

Proof: Let L be the Laplacian matrix for $(\mathcal{G}, w)$, and denote its eigenvalues by $\lambda_1 \ge \lambda_2 \ge \ldots \ge \lambda_{n-1} > \lambda_n = 0$. Observe that the diagonal entries of L are given by $d_1, \ldots, d_n$. Fix $t > 0$, and note that the matrix $L+tJ$ has eigenvalues $\lambda_1, \lambda_2, \ldots, \lambda_{n-1}, nt$, and diagonal entries $d_1+t, \ldots, d_n+t$. According to a theorem of Schur (see [114, section 10.4] for instance), the vector of eigenvalues of $L+tJ$ majorises the vector of diagonal entries of $L+tJ$.

Since the function $f(x)=\frac{1}{x}$ is convex for $x>0$, it follows that

$$\sum_{j=1}^{n-1} \frac{1}{\lambda_j} + \frac{1}{nt} \geq \sum_{j=1}^{n} \frac{1}{d_j+t}.$$

In view of Proposition 7.4.4, we thus find that $\mathrm{Kf}(\mathcal{G},w) = n\sum_{j=1}^{n-1}\frac{1}{\lambda_j} \geq n\sum_{j=1}^{n}\frac{1}{d_j+t} - \frac{1}{t}$. □

REMARK 7.4.12 By applying standard calculus techniques, it is readily established that, considered as a function of t, the right side of (7.18) attains its maximum value on $(0,\infty)$ at the unique $t_0>0$ such that $n\sum_{j=1}^{n}\left(\frac{t_0}{d_j+t_0}\right)^2 = 1$. In the special case that all of the d_js are equal to a common value d (for instance, if we are considering an unweighted graph whose vertices all have degree d), then $t_0=\frac{d}{n-1}$. In that case, the right side of (7.18) becomes $\frac{(n-1)^2}{d}$.

COROLLARY 7.4.13 *Let $(\mathcal{G},w)$ be a connected weighted graph on n vertices, and denote the sum of all edge weights by m. Then*

$$\mathrm{Kf}(\mathcal{G},w) \geq \frac{n(n-1)^2}{2m}. \tag{7.19}$$

Proof: We adopt the notation of Theorem 7.4.11, and observe that for any $t>0$,

$$\sum_{j=1}^{n}\frac{1}{d_j+t} \geq \frac{n^2}{\sum_{j=1}^{n}(d_j+t)}$$

(see the book of Marcus and Minc [91, Chapter II, section 3.5] for instance). Observing that $\sum_{j=1}^{n} d_j = 2m$, we now find from (7.18) that for any $t>0$,

$$\mathrm{Kf}(\mathcal{G},w) \geq \frac{n^3}{2m+nt} - \frac{1}{t}.$$

Substituting the choice $t=\frac{2m}{n(n-1)}$ now yields (7.19). □

The following example illustrates (7.19).

EXAMPLE 7.4.14 Let $p, q \in \mathbb{N}$ and consider the unweighted complete bipartite graph $K_{p,q}$. From Example 7.3.7, it follows that the eigenvalues of the corresponding Laplacian matrix are $0, q$ (with multiplicity $p-1$), p (with multiplicity $q-1$), and $p+q$. Consequently,

$$\mathrm{Kf}(K_{p,q}) = (p+q)\left(\frac{p-1}{q} + \frac{q-1}{p} + \frac{1}{p+q}\right).$$

Since $K_{p,q}$ has $p+q$ vertices and pq edges each of weight 1, it follows that for $K_{p,q}$, the right-hand side of (7.19) is equal to $\frac{(p+q)(p+q-1)^2}{2pq}$. A straightforward computation shows that

$$\mathrm{Kf}(K_{p,q}) - \frac{(p+q)(p+q-1)^2}{2pq} = 1 + ((p-q)^2 - 1)\left(\frac{p+q}{2pq}\right).$$

Observe that if p and q are not too far apart and both are large, then the difference between $\mathrm{Kf}(K_{p,q})$ and the lower bound afforded by (7.19) is close to 1.

In view of the interpretation of resistance distances as a notion of "closeness" between vertices in a weighted graph, we might ask how to assign weights to a given undirected graph so as to minimise the Kirchhoff index. It is natural to normalise the problem so that the sum of all weights coincides with the number of edges in the graph. To that end, we make the following definition. Let $\mathcal{G}$ be an undirected graph on n vertices with m edges. The *minimum Kirchhoff index* for $\mathcal{G}$, $\underline{\mathrm{Kf}}(\mathcal{G}) = \inf\{\mathrm{Kf}(\mathcal{G}, w) | w(e) \geq 0 \text{ for all } e \in \mathcal{G}, \sum_{e \in \mathcal{G}} w(e) = m\}$. In other words, the minimum Kirchhoff index for $\mathcal{G}$ is the infimum of the Kirchhoff indices for nonnegative weightings of $\mathcal{G}$ whose average edge weight is one. Observe that here we have relaxed the restriction that each edge weight be positive, so that zero edge weights are allowed. Our interpretation of $L(\mathcal{G}, w)$ in the case that some edges are weighted zero that $L(\mathcal{G}, w) = L(\hat{\mathcal{G}}, \hat{w})$, where $\hat{\mathcal{G}}$ is the subgraph of $\mathcal{G}$ induced by the edges e such that $w(e) > 0$, and where $\hat{w}(e) = w(e)$ for each $e \in \hat{G}$. We remark here that the definition above is inspired by Fielder's definition of the *absolute algebraic connectivity of a graph* $\mathcal{G}$, which is defined as the maximum $\alpha(\mathcal{G}, w)$, where the maximum is taken over all nonnegative weightings w of $\mathcal{G}$ such that the sum of the weights is equal to the number of edges in $\mathcal{G}$. See Fiedler's paper [40] for further details.

REMARK 7.4.15 Let $\mathcal{G}$ be a connected graph having m edges. While $\underline{\mathrm{Kf}}(\mathcal{G})$ is defined as an infimum, we claim that in fact $\underline{\mathrm{Kf}}(\mathcal{G}) = \min\{\mathrm{Kf}(\mathcal{G}, w) | w(e) \geq 0 \text{ for all } e \in \mathcal{G}, \sum_{e \in \mathcal{G}} w(e) = m\}$. To see the claim, let $w_k, k \in \mathbb{N}$, be a sequence of weightings of $\mathcal{G}$ such that for each such $k, \sum_{e \in \mathcal{G}} w_k(e) = m$, and $\text{Kf}(\mathcal{G}, w_k) \to \underline{\text{Kf}}(\mathcal{G})$ as $k \to \infty$. By passing to convergent subsequences if necessary, we may also assume that there is a nonnegative weighting $\tilde{w}$ of the edges of $\mathcal{G}$ such that for each edge $e \in \mathcal{G}$, $w_k(e) \to \tilde{w}(e)$. If the weighted graph $(\mathcal{G}, \tilde{w})$ is not connected, then the corresponding Laplacian matrix has zero as an eigenvalue of multiplicity two or more; since $\text{Kf}(\mathcal{G}, w_k)$ is bounded below by $\frac{1}{\alpha(\mathcal{G}, w_k)}$, it then follows that $\text{Kf}(\mathcal{G}, w_k)$ diverges to infinity, contrary to our hypothesis. Hence, it must be the case that $(\mathcal{G}, \tilde{w})$ is connected. But then, considered as a function of the edge weights, $L(\mathcal{G}, \tilde{w})^{\#}$ is continuous in a neighbourhood of $\tilde{w}$. Consequently, we find that $\underline{\text{Kf}}(\mathcal{G}) = \lim_{k \to \infty} \text{Kf}(\mathcal{G}, w_k) = \text{Kf}(\mathcal{G}, \tilde{w})$. The claim now follows.

Next we consider the effect of an edge addition on the minimum Kirchhoff index.

REMARK 7.4.16 Suppose that $\mathcal{G}$ is a connected graph with m edges, and form $\hat{\mathcal{G}}$ from $\mathcal{G}$ by adding an edge, say $\hat{e}$. We claim that $\underline{\text{Kf}}(\hat{\mathcal{G}}) \le \frac{m}{m+1}\underline{\text{Kf}}(\mathcal{G})$. To see the claim, let w be a weighting of $\mathcal{G}$ such that $\text{Kf}(\mathcal{G}, w) = \underline{\text{Kf}}(\mathcal{G})$. Now let $\hat{w}$ be the weighting of $\hat{\mathcal{G}}$ given by

$$\hat{w}(e) = \begin{cases} \frac{m+1}{m} w(e) & \text{if } e \in \mathcal{G}, \\ 0 & \text{if } e = \hat{e}. \end{cases}$$

Then $L(\hat{\mathcal{G}}, \hat{w}) = \frac{m+1}{m} L(\mathcal{G}, w)$, from which we find that $\text{Kf}(\hat{\mathcal{G}}, \hat{w}) = \frac{m}{m+1}\text{Kf}(\mathcal{G}, w)$. Hence,

$$\underline{\text{Kf}}(\hat{\mathcal{G}}) \le \text{Kf}(\hat{\mathcal{G}}, \hat{w}) = \frac{m}{m+1}\text{Kf}(\mathcal{G}, w) = \frac{m}{m+1}\underline{\text{Kf}}(\mathcal{G}),$$

as claimed. In particular, the minimum Kirchhoff index is strictly decreasing under edge addition.

We have the following intuitively appealing result.

THEOREM 7.4.17 *Let $\mathcal{G}$ be a connected graph on n vertices. Then*

$$\underline{\text{Kf}}(\mathcal{G}) \ge n - 1. \tag{7.20}$$

Equality holds in (7.20) if and only if $\mathcal{G} = K_n$.

Proof: Suppose that $\mathcal{G}$ has m edges. It follows from Corollary 7.4.13 that $\underline{\text{Kf}}(\mathcal{G}) \ge \frac{n(n-1)^2}{2m}$. Since $2m \le n(n-1)$, (7.20) now follows readily. Further, if $\underline{\text{Kf}}(\mathcal{G}) = n - 1$, then necessarily $2m = n(n-1)$, i.e., $\mathcal{G}$

is a complete graph. On the other hand, if $\mathcal{G} = K_n$, then considering the unweighted graph K_n we find easily that $\text{Kf}(K_n) = (n-1)$. Thus $n-1 \geq \underline{\text{Kf}}(K_n) \geq n-1$, so that $\underline{\text{Kf}}(K_n) = n-1$. □

REMARK 7.4.18 Our notion of the minimum Kirchhoff index is closely related to that of the *optimal total effective resistance* for a graph, which is studied in [44]. In that paper, for a given connected graph $\mathcal{G}$, the authors consider the minimum value of the Kirchhoff index for a weighted graph, where the minimum is taken over all nonnegative weightings such that the sum of the edge weights is 1; the notation $R^*_{tot}(\mathcal{G})$ is used to denote the optimal total effective resistance for $\mathcal{G}$. In [44], the optimal total effective resistance is viewed from the perspective of convex optimisation, and the authors discuss both the computational and the analytic aspects of the optimisation problem associated with finding the optimal total effective resistance.

As is noted in [44], for a connected graph $\mathcal{G}$ with m edges, we have $R^*_{tot}(\mathcal{G}) = m\underline{\text{Kf}}(\mathcal{G})$. Thus, for graphs having the same number of edges, it is clear that comparing their optimal total effective resistances is equivalent to comparing their minimum Kirchhoff indices. However, for graphs $\mathcal{G}_1, \mathcal{G}_2$ having different numbers of edges, comparisons between $R^*_{tot}(\mathcal{G}_1)$ and $R^*_{tot}(\mathcal{G}_2)$ may not parallel comparisons between $\underline{\text{Kf}}(\mathcal{G}_1)$ and $\underline{\text{Kf}}(\mathcal{G}_2)$ very closely.

As an example of that phenomenon, suppose that n is divisible by 6, and consider the complete bipartite graph $K_{\frac{n}{2},\frac{n}{2}}$ and the complete tripartite graph $K_{\frac{n}{3},\frac{n}{3},\frac{n}{3}}$. From symmetry considerations (with respect to the edges), it follows that for each of $K_{\frac{n}{2},\frac{n}{2}}$ and $K_{\frac{n}{3},\frac{n}{3},\frac{n}{3}}$, the minimum Kirchhoff index is attained when every edge has weight 1 (see [44] for a more careful discussion of symmetry in edges). The spectrum of $L(K_{\frac{n}{2},\frac{n}{2}})$ is readily seen to be $0, n$ and $\frac{n}{2}$, the last with multiplicity $n-2$. It now follows that $\underline{\text{Kf}}(K_{\frac{n}{2},\frac{n}{2}}) = 2n-3$, and hence that $R^*_{tot}(K_{\frac{n}{2},\frac{n}{2}}) = \frac{n^2(2n-3)}{4}$. Similarly, the spectrum of $L(K_{\frac{n}{3},\frac{n}{3},\frac{n}{3}})$ is given by $0, n$ (with multiplicity 2) and $\frac{2n}{3}$ (with multiplicity $n-3$). Hence we find that $\underline{\text{Kf}}(K_{\frac{n}{3},\frac{n}{3},\frac{n}{3}}) = \frac{3n-5}{2}$, while $R^*_{tot}(K_{\frac{n}{3},\frac{n}{3},\frac{n}{3}}) = \frac{n^2(3n-5)}{6}$. Consequently, we see that for large values of n (that are divisible by 6) $\frac{R^*_{tot}(K_{\frac{n}{3},\frac{n}{3},\frac{n}{3}})}{R^*_{tot}(K_{\frac{n}{2},\frac{n}{2}})}$ is close to 1, while $\frac{\underline{\text{Kf}}(K_{\frac{n}{3},\frac{n}{3},\frac{n}{3}})}{\underline{\text{Kf}}(K_{\frac{n}{2},\frac{n}{2}})}$ is close to $\frac{3}{4}$. Thus we see that the minimum Kirchhoff index makes a larger distinction between $K_{\frac{n}{3},\frac{n}{3},\frac{n}{3}}$ and $K_{\frac{n}{2},\frac{n}{2}}$ than the optimal total effective resistance does.

We round out this section with an application of some of the techniques of section 7.2 to find the minimum Kirchhoff index for a tree. We recall the fact that for a tree $\mathcal{T}$ on vertices $1,\ldots,n$, and any edge $e \in \mathcal{T}$, the quantity $|\beta_i(e)|(n-|\beta_i(e)|)$ is independent of the choice of i. The analogue of the following result for R^*_{tot} is given in [44].

THEOREM 7.4.19 *Let $\mathcal{T}$ be a tree on vertices $1,\ldots,n$. For each edge e of $\mathcal{T}$, let $x(e)=|\beta_i(e)|(n-|\beta_i(e)|)$, for some i between 1 and n. Then*

$$\underline{\mathrm{Kf}}(\mathcal{T}) = \frac{1}{n-1}\left(\sum_{e\in\mathcal{T}} \sqrt{x(e)}\right)^2.$$

In particular, for any tree $\mathcal{T}$ on n vertices, we have

$$\underline{\mathrm{Kf}}(\mathcal{T}) \ge (n-1)^2; \tag{7.21}$$

Equality holds in (7.21) *if and only if $\mathcal{T}$ is the star $K_{1,n-1}$.*

Proof: Suppose that w is a weighting of $\mathcal{T}$, normalised so that $\sum_{e\in\mathcal{T}} w(e) = n-1$, and let L be the corresponding Laplacian matrix. For each edge e of $\mathcal{T}$, let $v(e)$ be the vector given by (7.3). From Propositions 7.4.4 and Corollary 7.2.13, we find that

$$\mathrm{Kf}(\mathcal{T},w) = n\mathrm{trace}(L^{\#}) = \sum_{e\in\mathcal{T}} \frac{x(e)}{w(e)}.$$

Thus, in order to find the minimum Kirchhoff index for $\mathcal{T}$, we must find the minimum of the quantity $\sum_{e\in\mathcal{T}} \frac{x(e)}{w(e)}$ subject to the constraint that the edge weights are nonnegative and $\sum_{e\in\mathcal{T}} w(e) = n-1$. Evidently this minimum is attained for some weighting w_0 such that $w_0(e) > 0$ for all $e \in \mathcal{T}$.

Using Lagrange multipliers, it is not difficult to determine that since w_0 yields the minimising weighting for $\mathcal{T}$, it must be the case that $\frac{x(e)}{w_0(e)^2}$ takes on a common value for each edge e of $\mathcal{T}$. It now follows that for each $e \in \mathcal{T}$, $w_0(e) = \frac{(n-1)\sqrt{x(e)}}{\sum_{e\in\mathcal{T}}\sqrt{x(e)}}$. The formula for $\underline{\mathrm{Kf}}(\mathcal{T})$ now follows readily by computing $\mathrm{Kf}(\mathcal{T}, w_0)$.

In order to establish (7.21), we note that if $\mathcal{T}$ is a tree on n vertices, then for each edge e of $\mathcal{T}$ (and any index i between 1 and n),

$$|\beta_i(e)|(n-|\beta_i(e)|) \ge n-1.$$

It is now straightforward to show that $\underline{\mathrm{Kf}}(\mathcal{T}) \geq (n-1)^2$. Further, equality holds in (7.21) if and only if $|\beta_i(e)|(n - |\beta_i(e)|) = n - 1$ for every edge e of $\mathcal{T}$. The latter condition is easily seen to be equivalent to the statement that each edge e is incident with a pendent vertex of $\mathcal{T}$. Thus $\underline{\mathrm{Kf}}(\mathcal{T}) = (n-1)^2$ if and only if $\mathcal{T} = K_{1,n-1}$. □

The following example illustrates the conclusion of Theorem 7.4.19.

EXAMPLE 7.4.20 Consider the P_n, the path on n vertices; for concreteness, we take vertices 1 and n as pendent, with vertex i adjacent to vertices $i-1$ and $i+1, i = 2, \ldots, n-1$. It is readily determined that $\{|\beta_i(e)|(n - |\beta_i(e)|)|e \in \mathcal{T}\} = \{k(n-k)|k = 1, \ldots, n-1\}$. Applying Theorem 7.4.19, we find that

$$\underline{\mathrm{Kf}}(P_n) = \frac{1}{n-1}\left(\sum_{k=1}^{n-1} \sqrt{k(n-k)}\right)^2.$$

In order to get a sense of the order of magnitude of $\underline{\mathrm{Kf}}(P_n)$, observe that

$$\frac{1}{n^2}\sum_{k=1}^{n-1} \sqrt{k(n-k)} = \sum_{k=1}^{n-1} \frac{1}{n}\sqrt{\frac{k}{n}\left(1 - \frac{k}{n}\right)} \to \int_0^1 \sqrt{x(1-x)}dx$$

as $n \to \infty$. Using the trigonometric substitution $x = \sin^2\left(\frac{\theta}{4}\right)$ and the identity $1 - \cos(\theta) = 8\sin^2\left(\frac{\theta}{4}\right)\cos^2\left(\frac{\theta}{4}\right)$, it then follows that $\int_0^1 \sqrt{x(1-x)}dx = \frac{\pi}{8}$. Hence we find that

$$\lim_{n\to\infty} \frac{\underline{\mathrm{Kf}}(P_n)}{n^3} = \frac{\pi^2}{64},$$

so that for large values of n, $\underline{\mathrm{Kf}}(P_n)$ is approximately $\left(\frac{\pi^2}{64}\right)n^3$.

REMARK 7.4.21 From Remark 7.4.16 it follows that over all connected graphs on n vertices, the minimum Kirchhoff index must be maximised by some tree. We claim that over all trees on n vertices, the quantity $\sum_{e\in\mathcal{T}} \sqrt{x(e)}$ of Theorem 7.4.19 is uniquely maximised by P_n.

To see the claim, suppose that $\mathcal{T}$ is a tree on n vertices that is not a path. Then $\mathcal{T}$ has a vertex of degree $d \geq 3$; without loss of generality we suppose that it is vertex n. Let $e(1), \ldots, e(d)$ denote the edges incident with n, and suppose that the vertices have been labelled so that for each $j = 1, \ldots, d$, vertex j is the end point of $e(j)$ different from n. Observe that for each $j = 1, \ldots, d$, $|\beta_n(e(j))|$ is the number of vertices on the branch at vertex n containing vertex j. Again without loss of generality we may suppose that $|\beta_n(e(1))| \le |\beta_n(e(2))| \le \ldots \le |\beta_n(e(d))|$. Since $d \ge 3$ and $\sum_{j=1}^{d} |\beta_n(e(j))| = n - 1$, it follows that $n - 1 \ge |\beta_n(e(1))| + |\beta_n(e(2))| + |\beta_n(e(3))| \ge 2|\beta_n(e(1))| + |\beta_n(e(2))|$.

Construct a new tree $\tilde{\mathcal{T}}$ from $\mathcal{T}$ by deleting the edge $e(2)$ and adding a new edge $\tilde{e}$ between vertices 1 and 2. Figure 7.4 illustrates $\mathcal{T}$ and $\tilde{\mathcal{T}}$ in the case that $d = 3$. In that figure, the circles enclosing vertices $1, 2$ and 3 in both $\mathcal{T}$ and $\tilde{\mathcal{T}}$ represent the subgraphs of $\mathcal{T}$ induced by $\beta_n(e(1)), \beta_n(e(2))$ and $\beta_n(e(3))$, respectively.

Observe that in $\tilde{\mathcal{T}}$, vertex n has degree $d - 1$, the branches at n containing vertices $3, \ldots, d$ are the same as those in $\mathcal{T}$, and the branch in $\tilde{\mathcal{T}}$ at vertex n containing vertex 1 has $|\beta_n(e(1))| + |\beta_n(e(2))|$ vertices. By a slight abuse of notation, for each edge e of $\tilde{\mathcal{T}}$, we let $\tilde{\beta}_n(e)$ be the set of vertices in the connected component of $\tilde{\mathcal{T}} \setminus e$ that does not contain vertex n. From our construction of $\tilde{\mathcal{T}}$ we find that if e is an edge of $\mathcal{T}$ with $e \ne e(1), e(2)$, then e is also an edge of $\tilde{\mathcal{T}}$, and $|\beta_n(e)| = |\tilde{\beta}_n(e)|$. Further, we also have $|\beta_n(e(2))| = |\tilde{\beta}_n(\tilde{e})|$. Finally, note that

$$
\begin{aligned}
&|\tilde{\beta}_n(e(1))|(n - |\tilde{\beta}_n(e(1))|) = \\
&(|\beta_n(e(1))| + |\beta_n(e(2))|)(n - |\beta_n(e(1))| - |\beta_n(e(2))|) = \\
&|\beta_n(e(1))|(n - |\beta_n(e(1))|) + |\beta_n(e(2))|(n - 2|\beta_n(e(1))| - |\beta_n(e(2))|) > \\
&|\beta_n(e(1))|(n - |\beta_n(e(1))|).
\end{aligned}
$$

It now follows readily that $\underline{\mathrm{Kf}}(\tilde{\mathcal{T}}) > \underline{\mathrm{Kf}}(\mathcal{T})$, and we conclude that for trees on n vertices, the minimum Kirchhoff index is uniquely maximised by P_n. (We note in passing that a similar argument is given in [44] to establish the claim above.)

In particular, we find that for any graph $\mathcal{G}$,

$$
\underline{\mathrm{Kf}}(\mathcal{G}) \le \frac{1}{n-1}\left(\sum_{k=1}^{n-1} \sqrt{k(n-k)}\right)^2,
$$

with equality holding if and only if $G = P_n$.

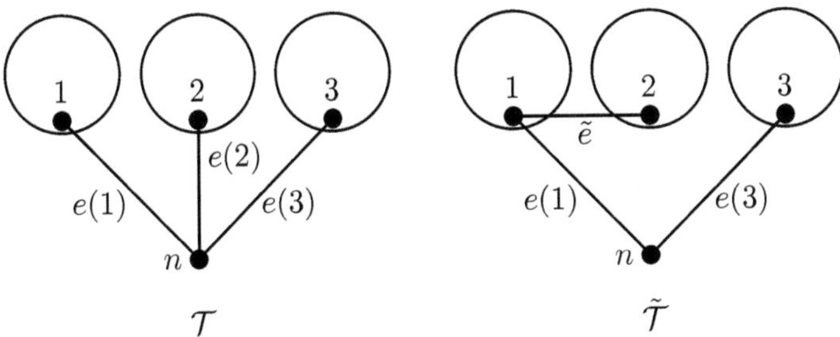

FIGURE 7.4
The trees $\mathcal{T}$ and $\tilde{\mathcal{T}}$ in Remark 7.4.21 when $d = 3$

7.5 Interpretations for electrical networks

In this section we explore some of the connections between Laplacian matrices for weighted graphs and resistive electrical networks. Not surprisingly, we will revisit some of the topics that appear in section 7.4. Our discussion below draws on the some of the material in Bapat's book [8, section 9.4]; here we rely on the machinery of group inverses for Laplacian matrices.

To fix ideas, suppose that $\mathcal{G}$ is a connected graph on vertices $1, \ldots, n$. We associate $\mathcal{G}$ with a network of resistors, so that for each each edge e of $\mathcal{G}$ we have a corresponding resistance $\rho(e)$. Fix a pair of distinct indices i_0, j_0 between 1 and n, and suppose that current is only allowed to enter our network at vertex i_0 and exit at vertex j_0. For each $k = 1, \ldots, n$, denote the voltage at vertex k by ν_k, and set $\nu_{i_0} = 1$ and $\nu_{j_0} = 0$. Our objective is to compute the voltages at the remaining vertices of $\mathcal{G}$, as well as, for each pair of adjacent vertices i, j, the current $\iota(i, j)$ flowing from vertex i to vertex j.

By Ohm's law, for each pair of adjacent vertices i, j we have $\iota(i, j) = \frac{\nu_i - \nu_j}{\rho(e)}$, where e denotes the edge between i and j. Further, for any index $k \neq i_0, j_0$, we have, according to Kirchhoff's current law, that

$$\sum_{j \sim k} \iota(k, j) = 0,$$

where the sum is taken over all vertices j adjacent to vertex k. Consequently, we find that for each index $k \neq i_0, j_0$,

$$\sum_{e=j\sim k} \frac{1}{\rho(e)}(\nu_k - \nu_j) = 0, \tag{7.22}$$

where the sum is taken over all vertices j adjacent to k, and where e denotes the corresponding edge. For each edge e of $\mathcal{G}$, let $w(e) = \frac{1}{\rho(e)}$ (this is known as the *conductance* of e), and let L denote the Laplacian matrix for the weighted graph $(\mathcal{G}, w)$. We see that for each $k \neq i_0, j_0$, (7.22) is equivalent to the equation

$$e_k^t L\nu = 0, \tag{7.23}$$

where ν is the vector of voltages at the vertices of $\mathcal{G}$.

For concreteness, we suppose that $i_0 = n-1$ and $j_0 = n$, and partition L as

$$L = \left[\begin{array}{c|c|c} L_{1,1} & L_{1,2} & L_{1,3} \\ \hline L_{2,1} & L_{2,2} & L_{2,3} \\ \hline L_{3,1} & L_{3,2} & L_{3,3} \end{array}\right],$$

where $L_{1,1}$ is $(n-2)\times(n-2)$, while $L_{2,2}$ and $L_{3,3}$ are both 1×1. From (7.23), we find that

$$L_{1,1}\hat{\nu} + \nu_{n-1}L_{1,2} = 0, \tag{7.24}$$

where $\hat{\nu}$ is the vector obtained from ν by deleting its last two entries. Recalling that $\nu_{n-1} = 1$, we find readily that

$$\hat{\nu} = -L_{1,1}^{-1}L_{1,2}. \tag{7.25}$$

Next, we consider the bottleneck matrix for $(\mathcal{G}, w)$ based at vertex n, say B. From the formula for the inverse of a partitioned matrix (see [57, section 0.7.3]), we have

$$B = \left[\begin{array}{c|c} L_{1,1} & L_{1,2} \\ \hline L_{2,1} & L_{2,2} \end{array}\right]^{-1} =$$

$$\left[\begin{array}{c|c} L_{1,1}^{-1} + \frac{1}{L_{2,2}-L_{2,1}L_{1,1}^{-1}L_{1,2}} L_{1,1}^{-1}L_{1,2}L_{2,1}L_{1,1}^{-1} & -\frac{1}{L_{2,2}-L_{2,1}L_{1,1}^{-1}L_{1,2}} L_{1,1}^{-1}L_{1,2} \\ \hline -\frac{1}{L_{2,2}-L_{2,1}L_{1,1}^{-1}L_{1,2}} L_{2,1}L_{1,1}^{-1} & \frac{1}{L_{2,2}-L_{2,1}L_{1,1}^{-1}L_{1,2}} \end{array}\right]. \tag{7.26}$$

In view of (7.24), we thus find that for each $k = 1, \ldots, n-2$,

$$\nu_k = \frac{b_{k,n-1}}{b_{n-1,n-1}}, \text{ while } \nu_{n-1} = 1 = \frac{b_{n-1,n-1}}{b_{n-1,n-1}}.$$

The current flowing into the network at vertex $n-1$ is equal to $\sum_{j\sim n-1} \iota(n-1,j) = \sum_{e=j\sim n-1} w(e)(\nu_{n-1}-\nu_j)$. Since $\nu_{n-1}=1$, and $\nu_n = 0$, we can thus rewrite this as $e_{n-1}^t L\nu$. Using the partitioning of L and (7.25), we then find that the current flowing in at vertex $n-1$ is

$$\left[\begin{array}{c|c} L_{2,1} & L_{2,2} \end{array}\right] \left[\begin{array}{c} -L_{1,1}^{-1}L_{1,2} \\ \hline 1 \end{array}\right] = L_{2,2} - L_{2,1}L_{1,1}^{-1}L_{1,2} = \frac{1}{b_{n-1,n-1}}, \quad (7.27)$$

the last equality following from (7.26). The *effective resistance* between $n-1$ and n is defined as the reciprocal of the current flowing in at vertex $n-1$ (see [34, section 3.4]), and so from (7.27), we see that the effective resistance between vertices $n-1$ and n is the $(n-1)$-st diagonal entry of the bottleneck matrix $(\mathcal{G}, w)$ based at vertex n. From Remark 7.4.2, we thus find that the effective resistance between vertices $n-1$ and n coincides with the resistance distance $r(n-1,n)$ of section 7.4. We were careful to explain in section 7.4 why the resistance distance r can be considered a "distance", and the derivation above now supports the application of the term "resistance" to r.

We can also derive alternative expressions for the various voltages and currents of interest using $L^{\#}$. From Observation 7.2.1 it follows that for each $k=1,\ldots,n-1, b_{k,n-1} = (e_k - e_n)^t L^{\#}(e_{n-1}-e_n)$, from which we deduce that

$$\nu_k = \frac{(e_k - e_n)^t L^{\#}(e_{n-1}-e_n)}{(e_{n-1} - e_n)^t L^{\#}(e_{n-1}-e_n)}, k = 1,\ldots,n. \quad (7.28)$$

For a pair of adjacent vertices i,j joined by edge e in $\mathcal{G}$, it follows that the current flowing from i to j is given by

$$\iota(i,j) = \frac{(e_i - e_j)^t L^{\#}(e_{n-1}-e_n)}{(e_{n-1} - e_n)^t L^{\#}(e_{n-1}-e_n)} w(e).$$

We note in passing that if instead we have the current entering the network at vertex n and exiting at vertex at vertex $n-1$, then denoting the corresponding voltages by $\tilde{\nu}_k, k=1,\ldots,n$, we find readily from (7.28) that $\nu_k + \tilde{\nu}_k = 1, k=1,\ldots,n$.

The following example illustrates (7.28).

EXAMPLE 7.5.1 Consider the path on 4 vertices, where vertices 1 and 4 are pendent, and vertex i is adjacent to vertices $i-1$ and $i+1$ for $i = 2, 3$. Suppose that the resistances for the two pendent edges are 1, while the resistance for the middle edge is $\frac{1}{2}$. The corresponding Laplacian matrix is given by

$$L = \begin{bmatrix} 1 & -1 & 0 & 0 \\ -1 & 3 & -2 & 0 \\ 0 & -2 & 3 & -1 \\ 0 & 0 & -1 & 1 \end{bmatrix}.$$

Observing that this is a weighting of P_4 of the type addressed in Example 7.2.16, it follows that

$$L^{\#} = \begin{bmatrix} \frac{3}{4} & 0 & -\frac{1}{4} & -\frac{1}{2} \\ 0 & \frac{1}{4} & 0 & -\frac{1}{4} \\ -\frac{1}{4} & 0 & \frac{1}{4} & 0 \\ -\frac{1}{2} & -\frac{1}{4} & 0 & \frac{3}{4} \end{bmatrix}. \tag{7.29}$$

Fix a pair of distinct indices i_0, j_0; allowing the current to enter the network only at vertex i_0 and exit at j_0, and setting $\nu_{i_0} = 1, \nu_{j_0} = 0$, we find from (7.28) that the corresponding voltages are given by

$$\nu_k = \frac{(e_k - e_{j_0})^t L^{\#}(e_{i_0} - e_{j_0})}{(e_{i_0} - e_{j_0})^t L^{\#}(e_{i_0} - e_{j_0})}, k = 1, \ldots, 4.$$

Using these expressions for the voltages in conjunction with (7.29) now yields the following voltage vectors:

$$\begin{bmatrix} 1 \\ 0 \\ 0 \\ 0 \end{bmatrix} \text{ for } (i_0, j_0) = (1, 2); \begin{bmatrix} 1 \\ \frac{1}{3} \\ 0 \\ 0 \end{bmatrix} \text{ for } (i_0, j_0) = (1, 3);$$

$$\begin{bmatrix} 1 \\ \frac{3}{5} \\ \frac{2}{5} \\ 0 \end{bmatrix} \text{ for } (i_0, j_0) = (1, 4); \text{ and } \begin{bmatrix} 1 \\ 1 \\ 0 \\ 0 \end{bmatrix} \text{ for } (i_0, j_0) = (2, 3).$$

We note that the voltage vectors corresponding to the (i_0, j_0) pairs not listed above can be deduced from the symmetry of the weighted graph (so that the voltage vectors corresponding to the pairs (i_0, j_0) and $(5-i_0, 5-j_0)$ are the same) and the fact that if ν and $\tilde{\nu}$ denote the voltage vectors corresponding to the pairs (i_0, j_0) and (j_0, i_0), respectively, then $\nu + \tilde{\nu} = \mathbf{1}$.

In [34, section 4.1], the following observation is identified as *Rayleigh's monotonicity law.*

OBSERVATION 7.5.2 If we increase one or more resistances in a circuit, then the effective resistance between any pair of points cannot decrease. Conversely, if we decrease one or more resistances in a circuit, then the effective resistance between any pair of points cannot increase.

In view of the fact that the effective resistance between two vertices in a weighted graph is the same as the resistance distance between those vertices, Rayleigh's monotonicity law can be deduced readily from Corollary 7.4.7 as follows. Suppose that we have a connected graph $\mathcal{G}$ on vertices $1, \ldots, n$ and two weightings of $\mathcal{G}$, w and $\tilde{w}$, such that w and $\tilde{w}$ agree on all but one edge of $\mathcal{G}$, say the edge $\hat{e}$. Suppose that $\tilde{w}(\hat{e}) > w(\hat{e})$, fix indices i, j between 1 and n, and denote the resistance distances between i and j in $(\mathcal{G}, w)$ and $(\mathcal{G}, \tilde{w})$ by $r(i,j)$ and $\tilde{r}(i,j)$, respectively. From Corollary 7.4.7 we find that $\tilde{r}(i,j) \le r(i,j)$. If $\tilde{w}(\hat{e}) < w(\hat{e})$, a similar argument (where we exchange the roles of $(\mathcal{G}, w)$ and $(\mathcal{G}, \tilde{w})$ when applying Corollary 7.4.7) shows that $\tilde{r}(i,j) \ge r(i,j)$. Recalling that the resistance on the edge $\hat{e}$ is the reciprocal of the corresponding edge weight, it now follows that the effective resistance between any two vertices is a nondecreasing function of the resistance on any edge of $\mathcal{G}$ — i.e., Rayleigh's monotonicity law holds.

It is natural to wonder whether the effective resistance between a pair of vertices is a strictly increasing function of the resistance on a particular edge, or whether there are circumstances under which the effective resistance is left unchanged when the resistance on a particular edge is perturbed. We take up this question now, and note that the topic is also explored in [87].

PROPOSITION 7.5.3 *Let $\mathcal{G}$ be a graph on vertices $1, \ldots, n$, and fix a particular edge $\hat{e}$ of $\mathcal{G}$. Consider two weightings of $\mathcal{G}, w$ and $\tilde{w}$, such that for each edge $e \ne \hat{e}$ of $\mathcal{G}, \tilde{w}(e) = w(e)$, while $\tilde{w}(\hat{e}) \ne w(\hat{e})$. Fix distinct indices i, j between 1 and n, and let $r(i,j)$ and $\tilde{r}(i,j)$ denote the resistance distances between i and j in $(\mathcal{G}, w)$ and $(\mathcal{G}, \tilde{w})$, respectively. If $\tilde{r}(i,j) \ne r(i,j)$, then there is a path from i to j in $\mathcal{G}$ that includes the edge $\hat{e}$.*

__Proof__: Without loss of generality, we take $i = n-1$ and $j = n$, and suppose that $\tilde{w}(\hat{e}) > w(\hat{e})$. Suppose that $\hat{e}$ is incident with vertices i_0 and j_0, where $i_0 < j_0$. Letting L and $\tilde{L}$ denote the Laplacian matrices for $(\mathcal{G}, w)$ and $(\mathcal{G}, \tilde{w})$, respectively, we find that $r(n-1,n)-\tilde{r}(n-1,n) = (e_{n-1}-e_n)^t L^{\#}(e_{n-1}-e_n)-(e_{n-1}-e_n)^t \tilde{L}^{\#}(e_{n-1}-e_n)$. Applying Lemma 7.4.5, we find that since $r(n-1,n)-\tilde{r}(n-1,n) \neq 0$, it must be the case that $(e_{n-1}-e_n)^t L^{\#}(e_{i_0}-e_{j_0}) \neq 0$.

Letting B denote the bottleneck matrix based at vertex n for $(\mathcal{G}, w)$, it follows from Observation 7.2.1 that

$$r(n-1,n)-\tilde{r}(n-1,n) = (e_{n-1}-e_n)^t L^{\#}(e_{i_0}-e_{j_0})$$
$$= \begin{cases} b_{n-1,i_0} - b_{n-1,j_0} & \text{if } j_0 \neq n, \\ b_{n-1,i_0} & \text{if } j_0 = n. \end{cases}$$

From Proposition 7.2.2 and the adjoint formula for the inverse, we find that for each index $k = 1, \ldots, n-1$,

$$b_{n-1,k} = \frac{\sum_{\mathcal{F} \in \mathcal{S}_n^{\{n-1,k\}}} w(\mathcal{F})}{\sum_{\mathcal{T} \in \mathcal{S}} w(\mathcal{T})}. \tag{7.30}$$

(Recall that $\mathcal{S}$ is the set of spanning trees in $\mathcal{G}$, while $\mathcal{S}_n^{\{n-1,k\}}$ is the set of spanning forests on $\mathcal{G}$ having two connected components, one containing vertices $n-1$ and k, and the other containing vertex n.)

Suppose first that $j_0 \neq n$. By hypothesis, $r(n-1,n) \neq \tilde{r}(n-1,n)$, and so we find from (7.30) that $\sum_{\mathcal{F} \in \mathcal{S}_n^{\{n-1,i_0\}}} w(\mathcal{F}) \neq \sum_{\mathcal{F} \in \mathcal{S}_n^{\{n-1,j_0\}}} w(\mathcal{F})$. In particular, it must be the case that $\mathcal{S}_n^{\{n-1,i_0\}} \neq \mathcal{S}_n^{\{n-1,j_0\}}$. Hence, we find without loss of generality that there is a spanning forest $\mathcal{F}$ that is in $\mathcal{S}_n^{\{n-1,i_0\}}$ but not in $\mathcal{S}_n^{\{n-1,j_0\}}$. Necessarily, $\mathcal{F}$ consists of two trees, one of which contains vertices $n-1$ and i_0, and the other of which contains vertices n and j_0. Consequently, there is a path P_1 in $\mathcal{F}$ from $n-1$ to i_0, and a path P_2 in $\mathcal{F}$ from j_0 to n. Note then that $P_1 \cup \{\hat{e}\} \cup P_2$ is a path in $\mathcal{G}$ from $n-1$ to n that includes the edge $\hat{e}$.

Finally, we suppose that $j_0 = n$. Since $r(n-1,n) \neq \tilde{r}(n-1,n)$, we have $b_{n-1,i_0} \neq 0$, so that necessarily $\mathcal{S}_n^{\{n-1,i_0\}} \neq \emptyset$. Thus, $\mathcal{G}$ contains a spanning forest $\mathcal{F}$ consisting of two trees, one containing vertices $n-1$ and i_0, the other containing vertex n. Consequently, there is a path P in $\mathcal{F}$ from vertex $n-1$ to vertex i_0, and since $\hat{e}$ is the edge between i_0 and n, it follows that $\mathcal{G}$ contains a path from vertex $n-1$ to vertex n that includes the edge $\hat{e}$. □

The following example shows that it is possible to have $r(i,j) = \tilde{r}(i,j)$ even though $\hat{e}$ is on a path from i to j in $\mathcal{G}$. Thus, the converse to

Proposition 7.5.3 fails.

EXAMPLE 7.5.4 Suppose that $k \in \mathbb{N}$, and consider the undirected cycle C_{4k} on vertices $1, \ldots, 4k$ where vertex i is adjacent to vertices $i-1$ and $i+1$ for $i = 2, \ldots, 4k-1$, and in addition vertices 1 and $4k$ are adjacent. Now construct a weighted graph $(\mathcal{G}, w)$ from C_{4k} by assigning weight 1 to each edge of C_{4k}, and adding an edge of weight $\theta > 0$ between vertices $k+1$ and $3k+1$. Figure 7.5 illustrates.

We claim that the effective resistance between vertices 1 and $2k+1$ is independent of the choice of θ. In order to establish the claim, we first introduce some notation. Let $\hat{L}$ denote the Laplacian matrix for C_{4k}, and let L denote the Laplacian matrix for the unweighted path formed from C_{4k} by deleting the edge between vertices 1 and $4k$. From Lemma 7.4.5, it suffices to show that $(e_1 - e_{2k+1})^t \hat{L}^{\#}(e_{k+1} - e_{3k+1}) = 0$ in order to prove the claim.

Since $\hat{L}$ is a circulant matrix, we find from Theorem 2.3.2 that $\hat{L}^{\#}$ is also a circulant matrix. From this it follows that $\hat{l}^{\#}_{2k+1,k+1} = \hat{l}^{\#}_{1,3k+1}$ and $\hat{l}^{\#}_{2k+1,3k+1} = \hat{l}^{\#}_{1,k+1}$. Observing that $(e_1 - e_{2k+1})^t \hat{L}^{\#}(e_{k+1} - e_{3k+1}) = \hat{l}^{\#}_{1,k+1} - \hat{l}^{\#}_{2k+1,k+1} - \hat{l}^{\#}_{1,3k+1} + \hat{l}^{\#}_{2k+1,3k+1} = 2(\hat{l}^{\#}_{1,k+1} - \hat{l}^{\#}_{1,3k+1})$, it suffices to prove that

$$\hat{l}^{\#}_{1,k+1} = \hat{l}^{\#}_{1,3k+1}. \tag{7.31}$$

Again appealing to Lemma 7.4.5, we have

$$\hat{L}^{\#} = L^{\#} - \frac{1}{1 + l^{\#}_{1,1} + l^{\#}_{4k,4k} - 2l^{\#}_{1,4k}} L^{\#}(e_1 - e_{4k})(e_1 - e_{4k})^t L^{\#}.$$

Consequently, for each $j = 1, \ldots, 4k$ we have

$$\hat{l}^{\#}_{1,j} = l^{\#}_{1,j} - \frac{l^{\#}_{1,1} - l^{\#}_{1,4k}}{1 + l^{\#}_{1,1} + l^{\#}_{4k,4k} - 2l^{\#}_{1,4k}} (l^{\#}_{1,j} - l^{\#}_{4k,j}).$$

Using the formulas for the entries in $L^{\#}$ from Example 7.2.8, we find that $\frac{l^{\#}_{1,1} - l^{\#}_{1,4k}}{1 + l^{\#}_{1,1} + l^{\#}_{4k,4k} - 2l^{\#}_{1,4k}} = \frac{4k-1}{4k}$, so that

$$\hat{l}^{\#}_{1,j} = \frac{4k+1}{4k} l^{\#}_{1,j} + \frac{4k-1}{4k} l^{\#}_{4k,j}, j = 1, \ldots, 4k.$$

It now follows that in order to establish (7.31), we need only show that

$$(4k+1)(l^{\#}_{1,k+1} - l^{\#}_{1,3k+1}) = (4k-1)(l^{\#}_{4k,k+1} - l^{\#}_{4k,3k+1}).$$

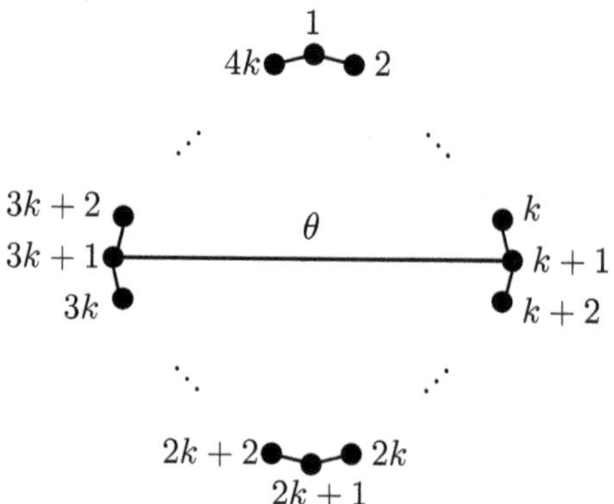

FIGURE 7.5
The weighted graph $(\mathcal{G}, w)$ for Example 7.5.4. Edge weights not shown explicitly are equal to 1

Again referring to Example 7.2.8, we find that $l^{\#}_{1,k+1} - l^{\#}_{1,3k+1} = \frac{4k-1}{4}$ and $l^{\#}_{4k,k+1} - l^{\#}_{4k,3k+1} = \frac{4k+1}{4}$. This then establishes (7.31), and so the effective resistance between vertices 1 and $2k+1$ is independent of the choice of θ, as claimed.

Note that the edge of weight θ between vertices $k+1$ and $3k+1$ lies on a path from vertex 1 to vertex $2k+1$, but nevertheless the effective resistance (equivalently, the resistance distance) between vertices 1 and $2k+1$ is independent of θ. The interested reader can compare this observation with the second assertion made in Lemma D of [87].

While Example 7.5.4 shows that the converse of Proposition 7.5.3 does not hold in general, our final result of this section establishes the converse for the class of weighted trees.

PROPOSITION 7.5.5 *Let $\mathcal{T}$ be a tree on vertices $1, \ldots, n$, and fix a particular edge $\hat{e}$ of $\mathcal{T}$. Consider two weightings of $\mathcal{T}, w$ and $\tilde{w}$ such that for each edge $e \neq \hat{e}$ of $\mathcal{T}, \tilde{w}(e) = w(e)$, while $\tilde{w}(\hat{e}) \neq w(\hat{e})$. Fix distinct indices i, j between 1 and n, and let $r(i,j)$ and $\tilde{r}(i,j)$ denote the resistance distances between i and j in $(\mathcal{T}, w)$ and $(\mathcal{T}, \tilde{w})$, respectively. If $\tilde{r}(i,j) = r(i,j)$, then the path from i to j in $\mathcal{T}$ does not include the edge $\hat{e}$.*

Proof: Without loss of generality we assume that $\tilde{w}(\hat{e}) > w(\hat{e})$. Let L denote the Laplacian matrix for $(\mathcal{T}, w)$, and denote the end points of the edge $\hat{e}$ by i_0, j_0. As in the proof of Proposition 7.5.3, we find from Lemma 7.4.5 that if $r(n-1,n) - \tilde{r}(n-1,n) = 0$, then $(e_i - e_j)^t L^{\#}(e_{i_0} - e_{j_0}) = 0$.

By Proposition 7.2.9, we have

$$L^{\#} = \sum_{e \in \mathcal{T}} \frac{1}{w(e)} v(e)v(e)^t,$$

where $v(e)$ is given by (7.3) for each edge e of $\mathcal{T}$. Without loss of generality, we can take each $v(e)$ to be normalised so that $v_{i_0}(e) > 0$. Thus we have

$$(e_i - e_j)^t L^{\#}(e_{i_0} - e_{j_0}) = \sum_{e \in \mathcal{T}} \frac{1}{w(e)} (e_i - e_j)^t v(e)v(e)^t(e_{i_0} - e_{j_0}) =$$

$$\frac{1}{w(\hat{e})}(e_i - e_j)^t v(\hat{e})v(\hat{e})^t(e_{i_0} - e_{j_0}),$$

the last equality following from the fact that for each edge $e \neq \hat{e}, v(e)^t(e_{i_0} - e_{j_0}) = 0$. From our normalisation of $v(\hat{e})$, we find that $v(\hat{e})^t(e_{i_0} - e_{j_0}) = 1$. Thus we have $(e_i - e_j)^t L^{\#}(e_{i_0} - e_{j_0}) = (e_i - e_j)^t v(\hat{e})$, so that necessarily $v_i(\hat{e}) = v_j(\hat{e})$. Referring to the definition of $v(\hat{e})$ in (7.3), we see that since $v_i(\hat{e}) = v_j(\hat{e})$, the path from i to j in $\mathcal{T}$ cannot include the edge $\hat{e}$. □

8

Computing the Group Inverse

In the preceding chapters, we have seen a number of situations in which group inverses of singular and irreducible M-matrices have facilitated the analysis of various phenomena. If these group inverses are to be useful in applied settings, then we need efficient and reliable methods to compute them. In this chapter, we consider some of the algorithmic and stability issues associated with computing the group inverse of irreducible singular M-matrices. Section 8.1 shows how the QR factorisation of an irreducible singular M-matrix yields a method for computing its group inverse, and discusses the use of the Cholesky factorisation for finding the group inverse of the Laplacian matrix of a connected weighted graph. In section 8.2, we consider two algorithms for computing the Drazin inverse of a singular matrix, and their specific implementation to the case of singular irreducible M-matrices. Next, in section 8.3 we pick up on a theme introduced in section 6.4, and present a method by which the group inverse of a matrix of the form $I - T$, where T is irreducible and stochastic, can be computed in a parallel fashion. Finally, in section 8.4, we present some results on the sensitivity and conditioning of the group inverse for various types of singular, irreducible M-matrices.

8.1 Introduction

At several points in Chapter 2, we encountered formulas or techniques for finding the group inverse of an irreducible singular M-matrix Q. For instance, if we have left and right null vectors for Q in hand, then we can readily factor Q as in (2.2), then use (2.3) to find $Q^{\#}$. Similarly, if a full rank factorisation of Q is known, then Proposition 2.3.1 yields $Q^{\#}$, while if the characteristic polynomial of Q is known, then $Q^{\#}$ can be found using Theorem 2.3.2. If Q happens to be of the form $I - T$ for an irreducible and stochastic matrix T, or if Q is the Laplacian matrix of a weighted graph, then Proposition 2.5.1 and Observation 7.2.1, respectively, give us expressions for the desired group inverse. While some of these expressions for the group inverse of Q have proven useful in our analysis of population models, Markov chains and electrical networks, they are not necessarily ideal means by which one might compute $Q^{\#}$ numerically.

We remark that the more general problem of computing the Drazin inverse of a square singular matrix B has received some attention. For example in the paper [113], Wilkinson discusses the use of a sequence of similarity transformations to bring B to the form

$$\left[\begin{array}{c|c} C & 0 \\ \hline 0 & N \end{array}\right],$$

where C is invertible and N is nilpotent, after which B^D can be readily computed. In a different direction, Greville's paper [48] shows how a certain iterative procedure, known as the Souriau–Frame algorithm, produces B^D as a polynomial in B. We note that one of the challenges in finding the Drazin inverse for a general square matrix B is that the index of B may not be known in advance, and so must be determined as part of the computation.

Fortunately for us, our goal is more modest. We assume that we are given an irreducible, singular M-matrix Q, and we simply want to compute $Q^{\#}$. In particular, the index of Q is known in advance, and the M-matrix structure helps to streamline some of our argumentation. In this section we show how some well–understood matrix factorisations can be used in order to compute $Q^{\#}$. We begin by considering some special cases for Q, namely those encountered in Chapters 5, 6, and 7.

Recall that a *floating point operation*, or *flop*, is a single arithmetic operation — either a single addition or a single multiplication. We refer the reader to the book of Golub and Van Loan [46, section 1.2.4] for a detailed discussion. We reiterate the point made in that book that estimating the flops required by a particular algorithm is useful in getting a general sense of the work required by the algorithm; however, flop counts do not take into consideration a number of other important factors that may affect the performance of a given algorithm. In the sequel, we will only keep track of the dominant terms when estimating flop counts for various algorithms.

OBSERVATION 8.1.1 Suppose that we have an M-matrix of the form $Q = I - T$, where $T \in \mathcal{IS}_n$. In [94], the implementation of the approach of Proposition 2.5.1 to computing $Q^{\#}$ is discussed. The bulk of the computational work consists of computing the inverse of a principal submatrix of Q of order $n-1$, with a corresponding cost of roughly $2n^3$ flops (note that [94] expresses the cost in terms of multiplications, rather than flops). It is also observed in [94] that once $Q^\#$ has been computed, the stationary vector w for T can be found at negligible expense from any row of $Q^\#$, as follows: since $Q^\# Q = I - \mathbf{1}w^t$, we see that for any index $i = 1, \ldots, n$, $w^t = e_i - (e_i^t Q^\#)Q$. Thus, computing $Q^\#$ is a somewhat simpler task than finding the fundamental matrix $Z = (I - T + \mathbf{1}w^t)^{-1}$ (discussed in [66, section 4.3]), since in order to find Z, we must first compute w, then compute the desired matrix inverse.

From a computational perspective, matrix inversion is seen typically as a task that should be avoided if possible, and approached with caution if inversion is necessary. Note that Observation 8.1.1 above involves the computation of a matrix inverse, as do many of the methods outlined in this section and sections 8.2 and 8.3. We refer the interested reader to Chapter 14 of Higham's book [55], which provides an extensive discussion of matrix inversion in the computational setting.

OBSERVATION 8.1.2 Consider the Laplacian matrix L of a connected weighted graph on n vertices. Recall that any symmetric positive definite matrix M has a *Cholesky factorisation* as $M = NN^t$, where N is lower triangular with positive entries on the diagonal (see [46, section 4.2]). In [25], Courrieu proposes the use of the Cholesky factorisation to compute Moore–Penrose inverses for positive semidefinite matrices. As the Moore–Penrose and group inverses for L coincide, we can apply the Cholesky factorisation idea of [25] to find $L^\#$ as follows. Let $\hat{L}$ denote a principal submatrix of L of order $n-1$; for concreteness we assume that $\hat{L}$ is the leading principal submatrix of order $n-1$. We now find a Cholesky factorisation of $\hat{L}$ as $\hat{L} = NN^t$, where N is lower triangular with positive entries on the diagonal. The bottleneck matrix for L based at vertex n is then given by $\hat{L}^{-1} = N^{-t}N^{-1}$, and applying Observation 7.2.1, we find that

$$L^\# = \frac{\mathbf{1}^t N^{-t} N^{-1} \mathbf{1}}{n^2} J + \left[\begin{array}{c|c} N^{-t}N^{-1} - \frac{1}{n} J N^{-t} N^{-1} - \frac{1}{n} N^{-t} N^{-1} J & -\frac{1}{n} N^{-t} N^{-1} \mathbf{1} \\ \hline -\frac{1}{n} \mathbf{1}^t N^{-t} N^{-1} & 0 \end{array}\right]. \quad (8.1)$$

We can compute a Cholesky factorisation of $\hat{L}$ in roughly $\frac{n^3}{3}$ flops ([46, section 4.2]). Since N is lower triangular, its inverse can be computed in $\frac{n^3}{3}$ flops ([46, section 3.1]), and from there the product $N^{-t}N^{-1}$ can be computed in a further $\frac{n^3}{3}$ flops. Referring to (8.1) for $L^{\#}$, we see that all other computations necessary for finding $L^{\#}$ are negligible. Thus, $L^{\#}$ can be computed in approximately n^3 flops.

REMARK 8.1.3 Suppose that we have a weighted tree $(\mathcal{T}, w)$ on n vertices, and let L denote the corresponding Laplacian matrix. We now describe a connection between the Cholesky factorisation method outlined in Observation 8.1.2 and the combinatorial information in $\mathcal{T}$.

We claim that there is a labelling of the vertices of $\mathcal{T}$ with the numbers $1, \ldots, n$, and the edges of $\mathcal{T}$ with the labels $e(1), \ldots, e(n-1)$, such that:
i) for each $j = 1, \ldots, n-1$, the edge $e(j)$ is incident with vertex j; and
ii) for each $j = 1, \ldots, n-1$, the label of the end point of $e(j)$ distinct from j, say k_j, has the property that $k_j > j$.

The claim is established by induction on n, as follows. Note that for a tree with two vertices, the result is immediate. Suppose now that the claim holds for any tree on n vertices and that $\mathcal{T}$ is a tree with $n+1$ vertices. Select a pendent vertex of $\mathcal{T}$, label it as vertex 0, and label the incident edge $e(0)$. Now consider the tree $\hat{\mathcal{T}}$ formed from $\mathcal{T}$ by deleting vertex 0 and edge $e(0)$. Applying the induction hypothesis to $\hat{\mathcal{T}}$, we see that we can label the vertices and edges of $\hat{\mathcal{T}}$ so that i) and ii) hold. Now label the vertices of $\mathcal{T}$ distinct from 0 with the corresponding vertex labels in $\hat{\mathcal{T}}$, and similarly for the edges of $\mathcal{T}$ distinct from $e(0)$. Finally, shift all of the vertex indices and edge labels of $\mathcal{T}$ up by one to produce the desired labelling. This completes the proof of the claim.

With the vertices and edges of $\mathcal{T}$ labelled as in i) and ii), we now construct an $(n-1)\times(n-1)$ matrix N as follows: for each $j = 1, \ldots, n-1$, set $n_{j,j} = \frac{1}{\sqrt{w(e(j))}}$, set $n_{k_j,j} = -\frac{1}{\sqrt{w(e(j))}}$ if k_j is the label of the vertex distinct from j that is incident with $e(j)$, and set $n_{i,j} = 0$ if $i \neq j, k_j$. Observe that from properties i) and ii) of our labelling, the matrix N is lower triangular. Further, the matrix P given by

$$P = \left[\begin{array}{c} N \\ \hline -\mathbf{1}^t N \end{array}\right]$$

is readily seen to satisfy $PP^t = L$. In particular, we find that the matrix N is in fact the Cholesky factor of the leading principal submatrix of L of order $n-1$. Thus we see that for the case of a weighted tree, if we have a labelling satisfying conditions i) and ii) in hand, then the necessary Cholesky factor N can be constructed directly from the labelling, adjacencies, and weights in $(\mathcal{T}, w)$. The remainder of the computation of $L^{\#}$ can then be accomplished in roughly $\frac{2}{3}n^3$ flops, as in Observation 8.1.2.

Next, we outline an approach to finding the group inverse for a singular and irreducible M-matrix; the strategy is essentially due to Golub and Meyer in [45], who considered the case that the matrix is of the form $I-T$ for some irreducible stochastic matrix T. Recall that any real $n \times n$ matrix B has a *QR factorisation* — that is, B can be factored as $B = QR$, where Q is an orthogonal matrix of order n and R is an upper triangular matrix of order n. We refer the reader to [46, section 5.2] for details. The following theorem, which is a modest generalisation of a result in [45], shows that in the setting of irreducible M-matrices, Q and R have some helpful structure.

THEOREM 8.1.4 *Suppose that $A \in \Phi^{n,n}$. Let x be the right Perron vector for A having the form*

$$x = \left[\begin{array}{c} \hat{x} \\ \hline 1 \end{array}\right],$$

and let y denote the left Perron vector for A, normalised so that $y^t x = 1$. Let $r(A)$ be the Perron value of A, and suppose that $r(A)I - A = QR$ is a QR factorisation of $r(A)I - A$. Then we have the following conclusions.
a) The matrix R has the form

$$R = \left[\begin{array}{c|c} U & -U\hat{x} \\ \hline 0^t & 0 \end{array}\right]. \tag{8.2}$$

b) The last column of Q is a scalar multiple of y.
c) We have

$$(r(A)I - A)^{\#} = (I - xy^t)\left[\begin{array}{c|c} U^{-1} & 0 \\ \hline 0^t & 0 \end{array}\right] Q^t (I - xy^t). \tag{8.3}$$

<u>**Proof**</u>: a) As R is upper triangular, the last row of R has a zero in every off-diagonal position. Since x is a null vector for $r(A)I - A$ we have $QRx = 0$, but Q is orthogonal, so it must be the case that $Rx = 0$. It now follows that the n–th diagonal entry of R is zero. Let U denote the leading principal submatrix of R of order $n-1$. Again using the fact that $Rx = 0$, we find that the submatrix of R on rows $1, \ldots, n-1$ and column n must be $-U\hat{x}$. The expression for R in (8.2) now follows.

b) Since $r(A)I - A$ is a singular and irreducible M-matrix, its left null vector is unique up to scalar multiple. Let $w = Qe_n$ denote the last column of Q. Then $w^t(r(A)I - A) = e_n^t Q^t(QR) = e_n^t R = 0^t$. We now conclude that w must be a scalar multiple of y.

c) Let Z denote the matrix on the right-hand side of (8.3), and note that we have $Zx = 0$ and $y^tZ = 0^t$. From b) we find that the last column of Q is either $\frac{1}{\sqrt{y^ty}}y$ or $-\frac{1}{\sqrt{y^ty}}y$, and without loss of generality we assume the former. Partition out the last column of Q, writing Q as $Q = \left[\begin{array}{c|c} Q_1 & \frac{1}{\sqrt{y^ty}}y \end{array}\right]$. Observe that since Q is orthogonal, we have $Q_1Q_1^t = I - \frac{1}{y^ty}yy^t$.

Since $Rx = 0$, we have

$$(r(A)I - A)Z = QR\left(I - xy^t\right)\left[\begin{array}{c|c} U^{-1} & 0 \\ \hline 0^t & 0 \end{array}\right] Q^t\left(I - xy^t\right) =$$

$$Q\left[\begin{array}{c|c} U & -U\hat{x} \\ \hline 0^t & 0 \end{array}\right]\left[\begin{array}{c|c} U^{-1} & 0 \\ \hline 0^t & 0 \end{array}\right] Q^t\left(I - xy^t\right) =$$

$$Q\left[\begin{array}{c|c} I & 0 \\ \hline 0^t & 0 \end{array}\right] Q^t\left(I - xy^t\right) =$$

$$\left[\begin{array}{c|c} Q_1 & \frac{1}{\sqrt{y^ty}}y \end{array}\right]\left[\begin{array}{c|c} I & 0 \\ \hline 0^t & 0 \end{array}\right]\left[\begin{array}{c} Q_1^t \\ \hline \frac{1}{\sqrt{y^ty}}y^t \end{array}\right]\left(I - xy^t\right) =$$

$$Q_1Q_1^t\left(I - xy^t\right) = \left(I - \frac{1}{y^ty}yy^t\right)\left(I - xy^t\right) = I - xy^t.$$

Similarly,

$$Z(r(A)I - A) = \left(I - xy^t\right)\left[\begin{array}{c|c} U^{-1} & 0 \\ \hline 0^t & 0 \end{array}\right] Q^t\left(I - xy^t\right)(r(A)I - A) =$$

$$\left(I - xy^t\right)\left[\begin{array}{c|c} U^{-1} & 0 \\ \hline 0^t & 0 \end{array}\right] Q^t(r(A)I - A) = \left(I - xy^t\right)\left[\begin{array}{c|c} U^{-1} & 0 \\ \hline 0^t & 0 \end{array}\right] Q^tQR =$$

$$\left(I - xy^t\right)\left[\begin{array}{c|c} U^{-1} & 0 \\ \hline 0^t & 0 \end{array}\right] R = \left(I - xy^t\right)\left[\begin{array}{c|c} I & -\hat{x} \\ \hline 0^t & 0 \end{array}\right].$$

A straightforward computation, using the fact $y^tx = 1$, now yields that

$$\left(I - xy^t\right)\left[\begin{array}{c|c} I & -\hat{x} \\ \hline 0^t & 0 \end{array}\right] = I - xy^t.$$

Thus we see that $(r(A)I - A)Z = Z(r(A)I - A) = I - xy^t, y^tZ = 0^t$ and $Zx = 0$. It now follows that Z satisfies the three conditions of (2.1), whence $Z = (r(A)I - A)^{\#}$. □

REMARK 8.1.5 Observe that if we have a QR factorisation of the $n \times n$ singular and irreducible M–matrix $r(A)I - A$ as in Theorem 8.1.4, then necessarily $R^tR = (r(A)I - A)^t(r(A)I - A)$. Let P denote the $(n-1) \times (n-1)$ leading principal submatrix of $(r(A)I - A)^t(r(A)I - A)$, which is necessarily positive definite. It now follows that the matrix U of Theorem 8.1.4 has the property that $U^tU = P$. Recalling that the Cholesky factorisation of any positive definite matrix is unique (see [46, section 4.2]), we find that if there are two upper triangular matrices U_1, U_1 such that $U_1^tU_1 = P = U_2^tU_2$, then there is a diagonal matrix D with diagonal entries in $\{1, -1\}$ such that $U_2 = DU_1$. Thus we find that, up to a signing of its rows, the matrix U of Theorem 8.1.4 is uniquely determined by $r(A)I - A$.

Suppose that we are given a matrix $A \in \Phi^{n,n}$. Theorem 8.1.4 shows us how a QR factorisation of $r(A)I - A$ can be used to produce an expression for $(r(A)I - A)^{\#}$. Next, we consider a strategy suggested by Theorem 8.1.4 for computing $(r(A)I - A)^{\#}$, as well as the amount of work involved in implementing that strategy. Observe that necessarily $r(A)I - A$ has rank $n-1$, and so any submatrix of $r(A)I - A$ formed by deleting a single column will have full column rank. Without loss of generality, we consider the matrix $\tilde{M}$ obtained from $r(A)I - A$ by deleting its last column. Using a sequence of Householder reflections, we can factor $\tilde{M}$ as Q_1U, where Q_1 is $n \times (n-1)$ and has orthonormal columns, and where U is $(n-1) \times (n-1)$ and upper triangular. This factorisation can be computed at a cost of $\frac{8}{3}n^3$ flops; see [46, section 5.2] for details. We note further that the algorithm for a Householder QR factorisation is known to be numerically stable (see [55, section 19.3]).

Let x and y denote right and left null vectors of $r(A)I - A$, respectively, normalised as in Theorem 8.1.4. Since the columns of Q_1 span the column space of $r(A)I - A$, each such column must be orthogonal to y. Consequently, the matrix $Q = \left[\begin{array}{c|c} Q_1 & \frac{1}{\sqrt{y^ty}}y \end{array}\right]$ must be an orthogonal matrix. Since the rows of Q all have Euclidean norm 1, observe that we can compute the entries of $\frac{1}{\sqrt{y^ty}}y$ via the formula $\frac{1}{\sqrt{y^ty}}y_j = \sqrt{1 - e_j^tQ_1Q_1^te_j}, j = 1, \ldots, n$. Thus we may find y from Q_1 at negligible expense (i.e., roughly $2n^2$ flops). Let R be the $n \times n$ triangular matrix given by

$$R = \left[\begin{array}{c|c} U & Q_1^t(r(A)I - A)e_n \\ \hline 0^t & 0 \end{array}\right].$$

Using the facts that $Q_1Q_1^t = I - \frac{1}{y^ty}yy^t$ and $y^t(r(A)I - A) = 0^t$, it is straightforward to determine that $QR = r(A)I - A$, so that we have produced a QR factorisation of $r(A)I - A$. Observe now that from R we can determine $\hat{x}$ (and hence x) by solving the system $-U\hat{x} = Q_1^t(r(A)I - A)e_n$. Since U is a triangular matrix, the cost of finding $\hat{x}$ is roughly n^2 flops, which is negligible compared to the cost of computing Q_1 and U. Having computed Q, R, y and x, we can now find $(r(A)I - A)^{\#}$ via (8.3). Indeed we find from (8.3) that $(r(A)I - A)^{\#}$ can be written as

$$(r(A)I - A)^{\#} = (I - xy^t)\left[\frac{U^{-1}Q_1^t}{0^t}\right](I - xy^t);$$

as U is upper triangular, we can find $U^{-1}Q_1^t$ at a cost of roughly n^3 flops (see [46, section 3.1]), while the remainder of the computation incurs negligible expense. Consequently, we find that the strategy above for finding $(r(A)I - A)^{\#}$ via a QR decomposition of $r(A)I - A$ can be implemented with roughly $\frac{11}{3}n^3$ floating point operations.

Suppose that we are given a matrix $T \in \mathcal{IS}_n$. From Observation 8.1.1, we can compute $(I - T)^{\#}$ in roughly $2n^3$ flops using the approach of Proposition 2.5.1, while from our discussion above, we can use the QR algorithm to compute $(I - T)^{\#}$ in $\frac{11}{3}n^3$ flops. Since the QR approach will be more expensive to compute, one might wonder whether there are circumstances under which the QR approach offers any kind of advantage. The next example attempts to address that issue.

EXAMPLE 8.1.6 Suppose that $0 < \epsilon < 1$, and consider the following irreducible stochastic matrix:

$$T = \begin{bmatrix} 1-\epsilon & \epsilon \\ 1 & 0 \end{bmatrix}.$$

We have

$$I - T = \begin{bmatrix} \epsilon & -\epsilon \\ -1 & 1 \end{bmatrix},$$

and so adopting the notation of Proposition 2.5.1, we see that $(I - T_{1,1})^{-1} = \frac{1}{\epsilon}$. In particular, for values of ϵ close to 0, $(I - T_{1,1})^{-1}$ is quite sensitive to changes in ϵ. Since $(I - T_{1,1})^{-1}$ is the key quantity in each of the expressions for the blocks of $(I - T)^{\#}$ in Proposition 2.5.1, we see that when ϵ is near 0, finding $(I - T)^{\#}$ via Proposition 2.5.1 involves doing computations with quantities that are sensitive to changes in ϵ.

Next we consider a QR factorisation of $(I - T)$. Observe that for the matrices Q and R given by

$$Q = \begin{bmatrix} \frac{\epsilon}{\sqrt{1+\epsilon^2}} & \frac{1}{\sqrt{1+\epsilon^2}} \\ -\frac{1}{\sqrt{1+\epsilon^2}} & \frac{\epsilon}{\sqrt{1+\epsilon^2}} \end{bmatrix}, R = \begin{bmatrix} \sqrt{1+\epsilon^2} & -\sqrt{1+\epsilon^2} \\ 0 & 0 \end{bmatrix},$$

we have $QR = I - T$. Evidently Q is orthogonal and R is upper triangular, so our matrices Q and R furnish a QR factorisation of $I - T$. Further, it is not difficult to see that, up to signing of the columns of the orthogonal factor the rows of the triangular factor, our Q and R are uniquely determined. Observe that none of the entries in Q and R is particularly sensitive to changes in ϵ; indeed it is straightforward to see that for all $\epsilon \in (0,1)$, the entries of Q and R all have derivatives (with respect to ϵ) that are bounded above by 1 in absolute value. Adopting the notation of Theorem 8.1.4, we also find that $x = \mathbf{1}, y^t = \frac{1}{1+\epsilon}\begin{bmatrix} 1 & \epsilon \end{bmatrix}$ and $U^{-1} = \frac{1}{\sqrt{1+\epsilon^2}}$; thus both U^{-1} and the entries of y have derivatives that are bounded above by 1 in absolute value for all $\epsilon \in (0,1)$. Consequently, we find that none of the quantities required to compute $(I-T)^{\#}$ via (8.3) is sensitive to changes in ϵ, even when ϵ is close to 0. Thus, from a numerical standpoint, when ϵ is close to 0, (8.3) offers a computational advantage for this example, as it avoids computations with sensitive quantities.

Finally, we note that part of the reason that the direct application of Proposition 2.5.1 leads to the computationally sensitive quantity $\frac{1}{\epsilon}$ is that we made an unlucky choice of the principal submatrix of $I - T$ to invert. That is, we proceeded by inverting the leading principal submatrix of $I - T$ (i.e., ϵ) rather than the trailing principal submatrix of $I - T$ (i.e., 1). In this example, the latter choice leads to an expression for $(I-T)^{\#}$ involving quantities that are not sensitive to small changes in ϵ. So here we see that there is a way to implement Proposition 2.5.1 in a computationally stable manner, provided that we identify the right principal submatrix of $I - T$ with which to work. For this example, an advantage of the QR factorisation approach is that it provides a computationally stable method for finding $(I-T)^{\#}$ without requiring the user to first make an advantageous choice of a submatrix.

8.2 The shuffle and Hartwig algorithms

In this section, we outline two algorithms for computing the Drazin inverse, and discuss the specific implementation of each to computing the group inverse of a singular and irreducible M-matrix. Both algorithms are based on row operations, and in each case, the algorithm begins with an $n \times n$ matrix A whose index k is not known in advance, and terminates in a finite number of steps, producing both A^D and the index k as output.

We start by describing the so-called *shuffle algorithm*, and follow the presentation given by Anstreicher and Rothblum in [6]. We begin with an initial pair of matrices $A_0 \equiv A, B_0 \equiv I$, which serves as the input to the algorithm. For each $j = 0, 1, 2, \ldots$, consider the $n \times 2n$ partitioned matrix $C_j = \left[\begin{array}{c|c} A_j & B_j \end{array}\right]$, and for concreteness, suppose that A_j has rank m_j. Perform Gauss–Jordan elimination on C_j so as to bring it to the form

$$\left[\begin{array}{c|c} \overline{A}_j & \hat{B}_j \\ \hline 0 & \overline{B}_j \end{array}\right],$$

where $\overline{A}_j$ is $m_j \times n$ with linearly independent rows; evidently $\hat{B}_j$ must also be $m_j \times n$, while both $\overline{B}_j$ and 0 are $(n - m_j) \times n$. Note that the rank m_j does not need to be known in advance, as Gauss–Jordan elimination will simultaneously bring C_j to the desired form and produce m_j as a byproduct. If $m_j = n$, so that $\overline{A}_j$ is square, the algorithm terminates at iteration j. (In particular, note that if A is nonsingular, we have $m_0 = n$, and the algorithm terminates at iteration 0.)

If $m_j < n$, we perform the shuffle step: set

$$A_{j+1} = \left[\begin{array}{c} \overline{A}_j \\ \hline \overline{B}_j \end{array}\right] \text{ and } B_{j+1} = \left[\begin{array}{c} \hat{B}_j \\ \hline 0 \end{array}\right],$$

and let

$$C_{j+1} = \left[\begin{array}{c|c} A_{j+1} & B_{j+1} \end{array}\right].$$

We then continue with the $(j+1)$–st iteration of the algorithm, applied to the matrix C_{j+1}.

Consider the sequence of iterates A_j produced by the shuffle algorithm, and for each $j \geq 1$, partition A_j as

$$A_j = \left[\begin{array}{c} A_j(1) \\ \hline A_j(2) \end{array}\right],$$

where $A_j(1)$ is $m_{j-1} \times n$ and $A_j(2)$ is $(n - m_{j-1}) \times n$. In [6] it is shown that for each $j = 1, \ldots, k$ the rows of the matrix $A_j(2)$ form a basis for the left null space of A^j. Further, it is shown in [6] that the algorithm terminates at step k (corresponding to the index of A), with a matrix of the form

$$\left[\begin{array}{c|c} I & \hat{B} \end{array}\right],$$

and that the Drazin inverse for A can be computed as

$$A^D = (\hat{B})^{k+1} A^k.$$

We remark that, as observed in [6], it is not necessary to perform Gauss–Jordan elimination on C_j at every iteration of the shuffle algorithm. For iterations $j = 0, \ldots, k-1$, we need only perform a sequence of row operations on C_j so as to bring it to the form

$$\left[\begin{array}{c|c} \overline{A}_j & \hat{B}_j \\ \hline 0 & \overline{B}_j \end{array}\right],$$

where $\overline{A}_j$ is $m_j \times n$ and has linearly independent rows. Only on the k-th (i.e., final) iteration is it necessary to apply Gauss–Jordan elimination on C_k to bring it to the form

$$\left[\begin{array}{c|c} I & \hat{B} \end{array}\right].$$

For the situation of interest to us below, namely the application of the shuffle algorithm to singular and irreducible M-matrices, we have $k = 1$ and $m_0 = n-1$. Consequently, in that setting, there is little difference in the number of arithmetic operations incurred by applying Gauss–Jordan at both iterations of the algorithm, and the number of arithmetic operations incurred by applying suitable row operations at iteration 0, then Gauss–Jordan elimination at iteration 1.

We will not present the proofs of the main results in [6] here, but instead we focus on the application of the shuffle algorithm to singular and irreducible M-matrices. To that end, we consider an $n \times n$ singular and irreducible M-matrix Q, and write it in partitioned form as

$$Q = \left[\begin{array}{c|c} \tilde{Q} & -u \\ \hline -v^t & d \end{array}\right], \tag{8.4}$$

where $\tilde{Q}$ is $(n-1) \times (n-1)$. Here we note that $\tilde{Q}$ is a nonsingular M-matrix, u and v are nonnegative vectors, and, since the determinant of the matrix in (8.4) is $\det(\tilde{Q})(d - v^t\tilde{Q}^{-1}u)$, we find that $d - v^t\tilde{Q}^{-1}u = 0$.

Let A_0 be the M-matrix given by (8.4). We start the shuffle algorithm with

$$C_0 = \left[\begin{array}{c|c} A_0 & I \end{array}\right] = \left[\begin{array}{c|c|c|c} \tilde{Q} & -u & I & 0 \\ \hline -v^t & d & 0^t & 1 \end{array}\right];$$

after performing Gauss–Jordan elimination, we arrive at the matrix

$$\left[\begin{array}{c|c|c|c} I & -\tilde{Q}^{-1}u & \tilde{Q}^{-1} & 0 \\ \hline 0^t & 0 & av^t\tilde{Q}^{-1} & a \end{array}\right],$$

where the scalar a is chosen so that the first entry of $av^t\tilde{Q}^{-1}$ is 1.

Next, we perform the shuffle step to produce the matrix

$$C_1 = \left[\begin{array}{c|c|c|c} I & -\tilde{Q}^{-1}u & \tilde{Q}^{-1} & 0 \\ \hline av^t\tilde{Q}^{-1} & a & 0^t & 0 \end{array}\right].$$

(Observe that, in the notation above, $A_1(2)$ is the row vector $\left[\begin{array}{cc} av^t\tilde{Q}^{-1} & a \end{array}\right]$, which is readily seen to span the left null space of Q.) Eliminating under the identity block in C_1 yields

$$\left[\begin{array}{c|c|c|c} I & -\tilde{Q}^{-1}u & \tilde{Q}^{-1} & 0 \\ \hline 0^t & a(1+v^t\tilde{Q}^{-2}u) & -av^t\tilde{Q}^{-2} & 0 \end{array}\right];$$

using the fact that $1 + v^t\tilde{Q}^{-2}u > 0$, we now complete the Gauss–Jordan elimination to produce the matrix

$$\left[\begin{array}{c|c|c|c} I & 0 & \tilde{Q}^{-1} - \frac{1}{1+v^t\tilde{Q}^{-2}u}\tilde{Q}^{-1}uv^t\tilde{Q}^{-2} & 0 \\ \hline 0^t & 1 & -\frac{1}{1+v^t\tilde{Q}^{-2}u}v^t\tilde{Q}^{-2} & 0 \end{array}\right].$$

The algorithm terminates here at the end of iteration 1, which of course corresponds to the index of our original irreducible M-matrix.

Set

$$\hat{B} = \left[\begin{array}{c|c} \tilde{Q}^{-1} - \frac{1}{1+v^t\tilde{Q}^{-2}u}\tilde{Q}^{-1}uv^t\tilde{Q}^{-2} & 0 \\ \hline -\frac{1}{1+v^t\tilde{Q}^{-2}u}v^t\tilde{Q}^{-2} & 0 \end{array}\right]. \tag{8.5}$$

The desired group inverse $Q^{\#}$ is now given by $\hat{B}^2A_0$ (equivalently, $\hat{B}^2Q$) and a computation of the latter yields

$$Q^{\#} = \left[\begin{array}{c|c} Q^{\#}_{1,1} & Q^{\#}_{1,2} \\ \hline Q^{\#}_{2,1} & Q^{\#}_{2,2} \end{array}\right],$$

where the blocks of $Q^{\#}$ are given by

$$Q^{\#}_{1,1} = \tilde{Q}^{-1} - \frac{1}{1+v^t\tilde{Q}^{-2}u}\tilde{Q}^{-2}uv^t\tilde{Q}^{-1} - \frac{1}{1+v^t\tilde{Q}^{-2}u}\tilde{Q}^{-1}uv^t\tilde{Q}^{-2} + \frac{v^t\tilde{Q}^{-3}u}{(1+v^t\tilde{Q}^{-2}u)^2}\tilde{Q}^{-1}uv^t\tilde{Q}^{-1}, \tag{8.6}$$

$$Q^{\#}_{1,2} = \frac{v^t\tilde{Q}^{-3}u}{(1+v^t\tilde{Q}^{-2}u)^2}\tilde{Q}^{-1}u - \frac{1}{1+v^t\tilde{Q}^{-2}u}\tilde{Q}^{-2}u, \tag{8.7}$$

$$Q^{\#}_{2,1} = \frac{v^t\tilde{Q}^{-3}u}{(1+v^t\tilde{Q}^{-2}u)^2}v^t\tilde{Q}^{-1} - \frac{1}{1+v^t\tilde{Q}^{-2}u}v^t\tilde{Q}^{-2}, \text{ and} \tag{8.8}$$

$$Q^{\#}_{2,2} = \frac{v^t\tilde{Q}^{-3}u}{(1+v^t\tilde{Q}^{-2}u)^2}. \tag{8.9}$$

We note that (8.6)–(8.9) coincide with the blocks in the expression for $Q^{\#}$ that is produced when we start with the full rank factorisation

$$Q = \left[\frac{\tilde{Q}}{-v^t}\right]\left[\ I \mid -\tilde{Q}^{-1}u\ \right] \equiv EF$$

and then apply Proposition 2.3.1 to write $Q^{\#}$ as $Q^{\#} = E(FE)^{-2}F$.

While the result of [6] does show that $Q^{\#}$ is given by $\hat{B}^2Q$, an examination of (8.5) and (8.6)–(8.9) shows that in fact $Q^{\#}$ can be computed (more or less) directly from $\hat{B}$ via the following strategy. We start by identifying the nonzero blocks of $\hat{B}$ as

$$\hat{B}_{1,1} = \tilde{Q}^{-1} - \frac{1}{1+v^t\tilde{Q}^{-2}u}\tilde{Q}^{-1}uv^t\tilde{Q}^{-2}, \hat{B}_{2,1} = -\frac{1}{1+v^t\tilde{Q}^{-2}u}v^t\tilde{Q}^{-2}.$$

Next we observe that $\hat{B}_{2,1}u = -\frac{v^t\tilde{Q}^{-2}u}{1+v^t\tilde{Q}^{-2}u}$, so that

$$\frac{1}{1+v^t\tilde{Q}^{-2}u} = 1 + \hat{B}_{2,1}u. \tag{8.10}$$

Also, $\hat{B}_{1,1}u = \frac{1}{1+v^t\tilde{Q}^{-2}u}\tilde{Q}^{-1}u$, so that

$$\frac{v^t\tilde{Q}^{-3}u}{(1+v^t\tilde{Q}^{-2}u)^2} = -\hat{B}_{2,1}\hat{B}_{1,1}u. \tag{8.11}$$

Note that we can now write $\tilde{Q}^{-1}$ as

$$\tilde{Q}^{-1} = \hat{B}_{1,1} - \frac{1}{1+\hat{B}_{2,1}u}\hat{B}_{1,1}u\hat{B}_{2,1}. \tag{8.12}$$

Further, we have

$$\tilde{Q}^{-1}u = \frac{1}{1+\hat{B}_{2,1}u}\hat{B}_{1,1}u, v^t\tilde{Q}^{-1} = v^t\left[\hat{B}_{1,1} - \frac{1}{1+\hat{B}_{2,1}u}\hat{B}_{1,1}u\hat{B}_{2,1}\right], \tag{8.13}$$

$$\tilde{Q}^{-2}u = \frac{1}{1+\hat{B}_{2,1}u}\left[\hat{B}_{1,1} - \frac{1}{1+\hat{B}_{2,1}u}\hat{B}_{1,1}u\hat{B}_{2,1}\right]\hat{B}_{1,1}u \tag{8.14}$$

and

$$v^t\tilde{Q}^{-2} = -\frac{1}{1+\hat{B}_{2,1}u}\hat{B}_{2,1}. \tag{8.15}$$

Observe that the blocks of $Q^{\#}$ in (8.6)–(8.9) are all constructed from the scalars in (8.10) and (8.11), the matrix in (8.12), and the vectors in (8.13), (8.14), and (8.15). Consequently, we find that $Q^{\#}$ can be computed from $\hat{B}_{1,1}, \hat{B}_{2,1}, u$ and v via matrix–vector products, addition of matrices, and outer products of vectors. Those operations are computable in roughly $2n^2, n^2$, and n^2 flops, respectively, while computing $\hat{B}^2Q$, which involves taking two matrix–matrix products, which will cost roughly $4n^3$ flops. Thus, in terms of arithmetic operations, computing $Q^{\#}$ via (8.10)–(8.15) and (8.6)–(8.9) provides a saving over the method of computing $Q^{\#}$ as $\hat{B}^2Q$.

In terms of the total flop count, we find from the above that the application of the shuffle algorithm to Q is essentially the same as the cost of performing Gauss–Jordan elimination to the $n \times 2n$ matrix $\left[\begin{array}{c|c} Q & I \end{array}\right]$, which is roughly $2n^3$. One can then find $Q^{\#}$ from (8.10)–(8.15) and (8.6)–(8.9) at negligible computational expense. For an M-matrix of the form $I - T$ where T is irreducible and stochastic, we see that the shuffle algorithm runs at the same computational cost as the approach to computing $(I-T)^{\#}$ (based on Proposition 2.5.1) that is outlined in Observation 8.1.1. Note however that the shuffle algorithm applies to any irreducible M-matrix, not just those of the form $I - T$ for stochastic T.

Next we turn our attention to the *Hartwig algorithm* for computing the Drazin inverse of a singular matrix given in Hartwig's paper [52]. We note that the Hartwig algorithm is an extension of a method of Robert in [105] for computing the group inverse for a square matrix of index 1. As in our treatment of the shuffle algorithm, we will describe the Hartwig algorithm, then focus on its implementation on singular and irreducible M-matrices.

The Hartwig algorithm proceeds as follows. We begin with a matrix A of order n and rank r_1. As input to the algorithm we set $A_1 = A$, and define $r_0 = n$. For each $j = 1, 2, \ldots,$ we consider the $r_{j-1} \times r_{j-1}$ matrix A_j of rank r_j. We find an invertible matrix R_j such that

$$R_jA_j = \left[\begin{array}{c} C_j \\ \hline 0 \end{array}\right],$$

where C_j is $r_j \times r_j$, then write $R_j A_j R_j^{-1}$ as

$$R_j A_j R_j^{-1} = \left[\begin{array}{c|c} A_{j+1} & B_{j+1} \\ \hline 0 & 0 \end{array}\right],$$

where A_{j+1} is $r_j \times r_j$, with rank r_{j+1}. The algorithm terminates at the end of j-th step in one of the following two circumstances.
i) If, at the end of the j-th step, we find that the matrix A_{j+1} is invertible, then the index of A is j, and we terminate the algorithm.
ii) If, at the end of the j-th step, we find that the matrix A_{j+1} is an $r_j \times r_j$ zero matrix, then A is a nilpotent matrix of index $j+1$, and we terminate the algorithm.

If neither i) nor ii) holds, we proceed to the $(j+1)$-st step of the algorithm, applied to the matrix A_{j+1}.

One of the key results in [52] is that for each $j = 1, 2, \ldots,$ we have

$$A_j^D = R_j^{-1} \left[\begin{array}{c|c} A_{j+1}^D & (A_{j+1}^D)^2 B_{j+1} \\ \hline 0 & 0 \end{array}\right] R_j. \tag{8.16}$$

Consequently, if the algorithm terminates at the end of the k–th step under condition i), then using the fact that A_{k+1} is invertible (so that $A_{k+1}^D = A_{k+1}^{-1}$), we may compute A_k^D from A_{k+1}^{-1}, B_{k+1} and R_k. Then for each $j = k-1, k-2, \ldots, 1$, we can compute A_j^D from A_{j+1}^D, B_{j+1} and R_j, completing that process with the computation of A_1^D (equivalently, A^D). On the other hand, if the algorithm terminates at the end of the k–th step under condition ii), then since A is nilpotent, we have $A^D = 0$.

Next we examine an implementation of the Hartwig algorithm to a singular irreducible M-matrix Q of order n. As in (8.4), we write Q as

$$Q = \left[\begin{array}{c|c} \tilde{Q} & -u \\ \hline -v^t & d \end{array}\right],$$

and recall that the $(n-1) \times (n-1)$ matrix $\tilde{Q}$ is nonsingular. Taking $A_1 = Q$ and choosing

$$R_1 = \left[\begin{array}{c|c} I & 0 \\ \hline v^t \tilde{Q}^{-1} & 1 \end{array}\right], \tag{8.17}$$

we have

$$R_1 A_1 = \left[\begin{array}{c|c} \tilde{Q} & -u \\ \hline 0^t & 0 \end{array}\right],$$

so that

$$R_1 A_1 R_1^{-1} = \left[\begin{array}{c|c} \tilde{Q} & -u \\ \hline 0^t & 0 \end{array}\right] \left[\begin{array}{c|c} I & 0 \\ \hline -v^t\tilde{Q}^{-1} & 1 \end{array}\right] = \left[\begin{array}{c|c} \tilde{Q} + uv^t\tilde{Q}^{-1} & -u \\ \hline 0^t & 0 \end{array}\right].$$

Thus we find that $A_2 = \tilde{Q} + uv^t\tilde{Q}^{-1}$.

Evidently A_2 is invertible, and its inverse is given by

$$A_2^{-1} = \tilde{Q}^{-1} - \frac{1}{1 + v^t\tilde{Q}^{-2}u}\tilde{Q}^{-1}uv^t\tilde{Q}^{-2}. \tag{8.18}$$

Since

$$A_1^{\#} = R_1^{-1} \left[\begin{array}{c|c} A_2^{-1} & -(A_2^{-1})^2 u \\ \hline 0 & 0 \end{array}\right] R_1,$$

a few computations now show that $A_1^{\#}$ (or equivalently, $Q^{\#}$) is given by

$$\left[\begin{array}{c|c} I & 0 \\ \hline -v^t\tilde{Q}^{-1} & 1 \end{array}\right] \times \left[\begin{array}{c|c} A_2^{-1} & \frac{v^t\tilde{Q}^{-3}u}{(1+v^t\tilde{Q}^{-2}u)^2}\tilde{Q}^{-1}u - \frac{1}{1+v^t\tilde{Q}^{-2}u}\tilde{Q}^{-2}u \\ \hline 0^t & 0 \end{array}\right]$$
$$\times \left[\begin{array}{c|c} I & 0 \\ \hline v^t\tilde{Q}^{-1} & 1 \end{array}\right]. \tag{8.19}$$

The blocks arising from (8.19) can be shown to coincide with the corresponding blocks in (8.6)–(8.9).

Finally, we consider the computational cost of applying the Hartwig algorithm to a singular and irreducible M-matrix Q of order n. Observe that any implementation of the Hartwig algorithm to Q will involve the inversion of the $(n-1) \times (n-1)$ matrix A_2, at a cost of roughly $2n^3$ flops. For the specific implementation of the Hartwig algorithm described above, where R_1 is chosen as in (8.17), we see that the key steps of the algorithm involve finding $v^t\tilde{Q}^{-1}$ (so that we can form R_1) and inverting the matrix A_2. The former task can be implemented by solving the linear system $\tilde{Q}x = v$ employing an LU decomposition of $\tilde{Q}$ at a cost of $\frac{2n^3}{3}$ flops; retaining that LU decomposition we may then compute A_2^{-1} via (8.18) at a cost of roughly $\frac{4n^3}{3}$ flops. Consequently, the specific implementation of the Hartwig algorithm to Q described above has a computational cost of $2n^3$ flops, which is comparable to that of the shuffle algorithm.

8.3 A divide and conquer method

Suppose that T is an irreducible stochastic matrix. In section 6.4, we showed how the corresponding mean first passage matrix can be computed in block form, using the mean first passage matrices associated with certain stochastic complements in T. Given that the mean first passage matrix is related to $(I - T)^{\#}$ by (6.3), it is natural to speculate that there is a way of expressing the group inverse of $I - T$ in block form, using the group inverses arising from some stochastic complements in T. In this section we pursue that idea, following the development given by Neumann and Xu in [103]. We note that in fact [103] considers a partitioned version of $(I - T)^{\#}$ as an $m \times m$ block matrix. However for simplicity of exposition, we restrict ourselves to the case $m = 2$ in this section.

We begin by establishing some notation. Denote the order of T by n, let w be the stationary distribution for T, and let M be the corresponding mean first passage matrix. Let $Q = I - T$, and let $W = \text{diag}(w)$. Recall from (6.3) that $M = (I - Q^{\#} + JQ^{\#}_{dg})W^{-1}$. Next we define the matrix U by

$$U = (M - M_{dg})W. \tag{8.20}$$

Evidently U is nonnegative, and can also be expressed as

$$U = -Q^{\#} + JQ^{\#}_{dg}. \tag{8.21}$$

From this last it follows that $U\mathbf{1} = \text{trace}(Q^{\#})\mathbf{1}$. We find immediately from (8.21) that

$$w^t U = \mathbf{1}^t Q^{\#}_{dg}. \tag{8.22}$$

Our expression for U in (8.21) also yields the following.

OBSERVATION 8.3.1 ([103]) Fix k between 1 and $n - 1$, and partition $\{1, \ldots, n\}$ as $S \cup \overline{S}$, where $S = \{1, \ldots, k\}, \overline{S} = \{k + 1, \ldots, n\}$. Then

$$U[S, S] = -Q^{\#}[S, S] + JQ^{\#}[S, S]_{dg}, \tag{8.23}$$

$$U[\overline{S}, S] = -Q^{\#}[\overline{S}, S] + JQ^{\#}[S, S]_{dg}, \tag{8.24}$$

$$U[\overline{S}, \overline{S}] = -Q^{\#}[\overline{S}, \overline{S}] + JQ^{\#}[\overline{S}, \overline{S}]_{dg}, \text{ and} \tag{8.25}$$

$$U[S, \overline{S}] = -Q^{\#}[S, \overline{S}] + JQ^{\#}[\overline{S}, \overline{S}]_{dg}. \tag{8.26}$$

Maintaining the notation above, we now apply Theorem 6.4.2 to derive expressions for the blocks of U.

THEOREM 8.3.2 ([103]) *Let V_S and $V_{\overline{S}}$ be given by (6.37). Then we have*

$$U[S,S] = -(I-\mathcal{P}(T)_S)^{\#} + J(I-\mathcal{P}(T)_S)^{\#}_{dg} + V_S W[S,S], \text{ and} \quad (8.27)$$

$$U[\overline{S},\overline{S}] = -(I-\mathcal{P}(T)_{\overline{S}})^{\#} + J(I-\mathcal{P}(T)_{\overline{S}})^{\#}_{dg} + V_{\overline{S}} W[\overline{S},\overline{S}]. \quad (8.28)$$

Further, we also have

$$U[\overline{S},S] = (I-T[\overline{S},\overline{S}])^{-1}\left[T[\overline{S},S]U[S,S] + JW[S,S]\right], \text{ and} \quad (8.29)$$

$$U[S,\overline{S}] = (I-T[S,S])^{-1}\left[T[S,\overline{S}]U[\overline{S},\overline{S}] + JW[\overline{S},\overline{S}]\right]. \quad (8.30)$$

Proof: Let $M_{\mathcal{P}(T)_S}$ denote the mean first passage matrix associated with the stochastic complement $\mathcal{P}(T)_S$. Let $\gamma_S = \frac{1}{\mathbf{1}^t w[S]}$, where $w[S]$ is the subvector of w corresponding to the indices in S. From Theorem 6.4.2, we see that

$$M[S,S] = \gamma_S M_{\mathcal{P}(T)_S} + V_S.$$

Consequently, we find from (8.20) that

$$U[S,S] = \gamma_S(M_{\mathcal{P}(T)_S} - (M_{\mathcal{P}(T)_S})_{dg})W[S,S] + V_S W[S,S].$$

Observing that $\gamma_S w[S]$ is the stationary vector for $\mathcal{P}(T)_S$, and applying (6.3) to $\mathcal{P}(T)_S$, it follows that

$$\gamma_S(M_{\mathcal{P}(T)_S} - (M_{\mathcal{P}(T)_S})_{dg})W[S,S] = -(I-\mathcal{P}(T)_S)^{\#} + J(I-\mathcal{P}(T)_S)^{\#}_{dg}.$$

Hence $U[S,S] = -(I-\mathcal{P}(T)_S)^{\#} + J(I-\mathcal{P}(T)_S)^{\#}_{dg} + V_S W[S,S]$, so that (8.27) holds. A similar argument applies to $U[\overline{S},\overline{S}]$, establishing (8.28).

Next, we consider $U[\overline{S},S]$. From (8.20), we find that $U[\overline{S},S] = M[\overline{S},S]W[S,S]$, and hence from (6.41) we see that

$$U[\overline{S},S] = (I-T[\overline{S},\overline{S}])^{-1}\left[T[\overline{S},S](M[S,S] - M[S,S]_{dg}) + J\right]W[S,S].$$

Since $(M[S,S] - M[S,S]_{dg})W[S,S] = U[S,S]$, we thus find that

$$U[\overline{S},S] = (I-T[\overline{S},\overline{S}])^{-1}\left[T[\overline{S},S]U[S,S] + JW[S,S]\right]$$

A similar argument applies to $U[S,\overline{S}]$, thus establishing (8.30). □

We continue with the notation above in the following result from [103], which expresses $Q^{\#}$ in block form.

COROLLARY 8.3.3 *We have*

$$Q^{\#}[S,S] = \mathbf{1}(w[S]^tU[S,S] + w[\overline{S}]^tU[\overline{S},S]) - U[S,S], \quad (8.31)$$
$$Q^{\#}[\overline{S},\overline{S}] = \mathbf{1}(w[S]^tU[S,\overline{S}] + w[\overline{S}]^tU[\overline{S},\overline{S}]) - U[\overline{S},\overline{S}], \quad (8.32)$$
$$Q^{\#}[\overline{S},S] = \mathbf{1}(w[S]^tU[S,S] + w[\overline{S}]^tU[\overline{S},S]) - U[\overline{S},S], \textit{ and} \quad (8.33)$$
$$Q^{\#}[S,\overline{S}] = \mathbf{1}(w[S]^tU[S,\overline{S}] + w[\overline{S}]^tU[\overline{S},\overline{S}]) - U[S,\overline{S}]. \quad (8.34)$$

Proof: From (8.22) we have $w^tU = \mathbf{1}^tQ^{\#}_{dg}$. Applying that relation, in conjunction with Observation 8.3.1, now yields (8.31)–(8.34). □

Taken together, the expressions in (8.27)–(8.30) and (8.31)–(8.34) provide a description of $Q^{\#}$ in block form when it is partitioned in conformity with $S \cup \overline{S}$. Evidently the building blocks of those formulas are: the blocks of the partitioned matrix T, the subvectors $w[S]$ and $w[\overline{S}]$ of the stationary distribution vector w, and the matrices $(I - \mathcal{P}(T)_S)^{\#}$ and $(I - \mathcal{P}(T)_{\overline{S}})^{\#}$, which are the group inverses associated with the stochastic complements arising from the partitioning of T. Recalling our discussion from section 5.1, we find that the stationary vector w can be computed from the stationary vectors of the stochastic complements $\mathcal{P}(T)_S$ and $\mathcal{P}(T)_{\overline{S}}$, and the stationary vector of the associated 2×2 coupling matrix.

Observe that once the stochastic complements $\mathcal{P}(T)_S$ and $\mathcal{P}(T)_{\overline{S}}$, have been formed, some of the computations of the blocks of $Q^{\#}$ can be performed independently of the other blocks. That observation suggests the following strategy for finding $Q^{\#}$.

Given an irreducible stochastic matrix T partitioned as

$$T = \left[\begin{array}{c|c} T[S,S] & T[S,\overline{S}] \\ \hline T[\overline{S},S] & T[\overline{S},\overline{S}] \end{array}\right],$$

we proceed as follows.
1. Form the stochastic complements $\mathcal{P}(T)_S$ and $\mathcal{P}(T)_{\overline{S}}$.
2. Compute $(I - \mathcal{P}(T)_S)^{\#}$ and $(I - \mathcal{P}(T)_{\overline{S}})^{\#}$, as well as the stationary distributions vectors for $\mathcal{P}(T)_S$ and $\mathcal{P}(T)_{\overline{S}}$, namely $\gamma_S w[S]$ and $\gamma_{\overline{S}} w[\overline{S}]$, respectively.
3. From the stationary vectors found at step 2, and the coupling matrix, compute the stationary vector w of T.
4. Using the group inverses computed at step 2 and the blocks of T, find V_S and $V_{\overline{S}}$ via (6.37).
5. Use (8.27) and (8.31) to find $U[S,S]$ and $Q^{\#}[S,S]$, and use (8.28) and (8.32) to find $U[\overline{S},\overline{S}]$ and $Q^{\#}[\overline{S},\overline{S}]$.

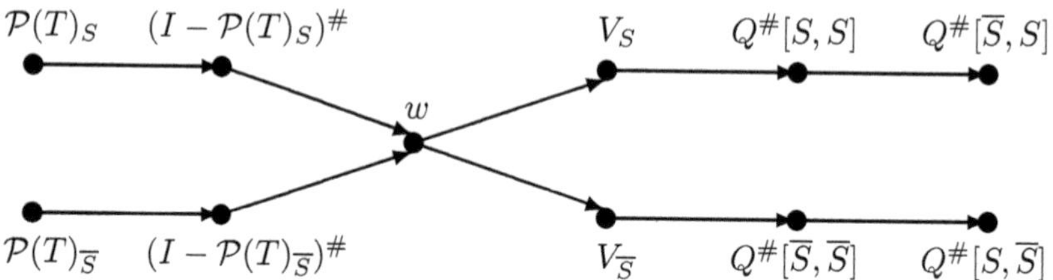

FIGURE 8.1
Computational flow for steps 1–6

6. Using $U[S,S]$ computed at step 5, apply (8.29) and (8.33) to find $U[\overline{S},S]$ and $Q^\#[\overline{S},S]$, and using $U[\overline{S},\overline{S}]$ computed at step 5, apply (8.30) and (8.34) to find $U[S,\overline{S}]$ and $Q^\#[S,\overline{S}]$.

Observe that in step 1, the computations of $\mathcal{P}(T)_S$ and $\mathcal{P}(T)_{\overline{S}}$ can be performed concurrently and in parallel. A similar comment applies to the computations of $(I - \mathcal{P}(T)_S)^\#$ and $(I - \mathcal{P}(T)_{\overline{S}})^\#$ in step 2, the computations of V_S and $V_{\overline{S}}$ in step 4, the computations of $Q^\#[S,S]$, and $Q^\#[\overline{S},\overline{S}]$ in step 5, and the computations of $Q^\#[\overline{S},S]$ and $U[\overline{S},\overline{S}]$ in step 6. Only step 3 requires information from both of the computations performed at step 2, namely the stationary vectors for $\mathcal{P}(T)_S$ and $\mathcal{P}(T)_{\overline{S}}$. Figure 8.1 depicts the computational flow of steps 1–6; it is apparent that this strategy for finding $Q^\#$ is suitable for an implementation in parallel.

Next we consider the number of flops needed to implement steps 1–6 above. For concreteness, we suppose that n is even, and that both S and $\overline{S}$ have cardinality $\frac{n}{2}$. We will provide rough counts of the number of flops needed to complete each of the steps.

• For step 1, in order to find $\mathcal{P}(T)_S$, we need an LU decomposition of $I - T[\overline{S},\overline{S}]$, at a cost of roughly $\frac{2}{3}\left(\frac{n}{2}\right)^3 = \frac{n^3}{12}$ flops. We may then compute $(I - T[\overline{S},\overline{S}])^{-1}T[\overline{S},S]$ column–by–column via two successive solutions of triangular systems, at a total cost of roughly $\frac{n^3}{4}$ flops. We then multiply $(I - T[[\overline{S},\overline{S}])^{-1}T[[\overline{S},S]$ on the left by $T[S,\overline{S}]$, at a rough cost of about $\frac{n^3}{4}$ flops. The cost of adding $T[S,S]$ to $T[S,\overline{S}](I - T[[\overline{S},\overline{S}])^{-1}T[[\overline{S},S]$ is negligible compared to that of the preceding computations, so we conclude that the cost of forming $\mathcal{P}(T)_S$ is roughly $\frac{7n^3}{12}$ flops. The cost of forming $\mathcal{P}(T)_{\overline{S}}$ is also about $\frac{7n^3}{12}$ flops, so that the total cost of implementing step 1 is roughly $\frac{7n^3}{6}$ flops.

• For step 2, in order to compute $(I - \mathcal{P}(T)_S)^\#$, we have a choice of methods. We might compute the desired group inverse either via the approach of Proposition 2.5.1, or via the shuffle algorithm or the Hartwig algorithm; each of these methods will incur a cost of about $2\left(\frac{n}{2}\right)^3 = \frac{n^3}{4}$ flops. Alternatively, we may apply the QR factorisation approach of Theorem 8.1.4 to find $(I-\mathcal{P}(T)_S)^{\#}$, at a rough cost of $\frac{11}{3}\left(\frac{n}{2}\right)^3 = \frac{11n^3}{24}$ flops. Recall also that once $(I-\mathcal{P}(T)_S)^{\#}$ is computed, the stationary vector for $\mathcal{P}(T)_S$ can be found at negligible expense. Evidently similar statements hold for the cost of computing $(I-\mathcal{P}(T)_{\overline{S}})^{\#}$ and the stationary vector for $\mathcal{P}(T)_{\overline{S}}$. Thus we find that the computational cost of step 2 may range from $\frac{n^3}{2}$ flops (if both $(I-\mathcal{P}(T)_S)^{\#}$ and $(I-\mathcal{P}(T)_{\overline{S}})^{\#}$ are found via Proposition 2.5.1 or either the shuffle or Hartwig algorithms) to $\frac{11n^3}{12}$ flops (if both $(I-\mathcal{P}(T)_S)^{\#}$ and $(I-\mathcal{P}(T)_{\overline{S}})^{\#}$ are found via the QR factorisation).

- For step 3, note that the computational cost of finding w is negligible compared to the costs identified above.
- For step 4, in order to find V_S, observe that if the LU factorisation of $I-\mathcal{P}(T)_{\overline{S}}$ is retained from step 1, then $(I-\mathcal{P}(T)_{\overline{S}})^{-1}J$ can be found by computing the matrix–vector product $(I-\mathcal{P}(T)_{\overline{S}})^{-1}\mathbf{1}$. It now follows from (6.37) that V_S can be found with three matrix–vector products and one matrix–matrix addition, at negligible expense compared to the above.
- For step 5, observe from (8.27) and (8.31) that $U[S,S]$ and $Q^{\#}[S,S]$ can be computed from a few matrix additions, matrix–vector multiplications, and multiplication of a matrix by a diagonal matrix. Thus the cost of computing $U[S,S]$ and $Q^{\#}[S,S]$ is negligible compared to those above, as is the cost of finding $U[\overline{S},\overline{S}]$ and $Q^{\#}[\overline{S},\overline{S}]$.
- For step 6, if we retain $(I-T[\overline{S},\overline{S}])^{-1}T[\overline{S},S]$ from step 1, then the cost of computing $(I-T[\overline{S},\overline{S}])^{-1}T[\overline{S},S]U[S,S]$ is about $\frac{n^3}{4}$ flops. Referring to (8.29), it now follows that the cost of finding $U[\overline{S},S]$ is about $\frac{n^3}{4}$ flops, and from (8.33), the cost of finding $Q^{\#}[\overline{S},S]$ is negligible. A similar analysis applies to the cost of computing $U[S,\overline{S}]$ and $Q^{\#}[S,\overline{S}]$. We thus find that the cost of step 6 is roughly $\frac{n^3}{2}$ flops.

Assembling the observations above, and as usual neglecting costs of orders n^2 and lower, we find that the computational cost of implementing steps 1–6 ranges between roughly $\frac{13n^3}{6}$ flops and $\frac{31n^3}{12}$ flops, depending on what methods are used at step 2 to compute the necessary group inverses. In terms of flop counts, we see that the cost of implementing steps 1–6 exceeds $2n^3$ flops, which is the rough cost of finding $Q^{\#}$ via either the shuffle or Hartwig algorithms, or via Proposition 2.5.1.

On the other hand, if steps 1–6 are implemented using the QR method at step 2 to find both $(I-\mathcal{P}(T)_S)^{\#}$ and $(I-\mathcal{P}(T)_{\overline{S}})^{\#}$ then the total computational cost, which is $\frac{31n^3}{12}$ flops, is about 70.5% of the $\frac{11n^3}{3}$ flops incurred by applying the QR method to find $Q^\#$ directly. Thus we see that for a QR–based approach to finding group inverses, the extra computational cost of finding the necessary stochastic complements in step 1 is offset by the savings at step 2, where we work with matrices of smaller order.

In a similar vein, the authors of [103] consider the costs of computing $Q^\#$ from the fundamental matrix:

$$Q^\# = (I - T + \mathbf{1}w^t)^{-1} - \mathbf{1}w^t. \tag{8.35}$$

The strategy outlined in [94] to finding $Q^\#$ via (8.35) is to first find the stationary vector w as a solution to a linear system, at a cost of about $\frac{2n^3}{3}$ flops, then compute $Q^\#$ from (8.35) at a cost of about $2n^3$ flops, for a total computational cost of roughly $\frac{8n^3}{3}$ flops. If steps 1–6 above are implemented and (8.35) is used to find both $(I - \mathcal{P}(T)_S)^\#$ and $(I - \mathcal{P}(T)_{\overline{S}})^\#$ at step 2, then the total computational cost is roughly $\frac{7n^3}{3}$, an improvement of 12.5% over the cost of finding $Q^\#$ directly from (8.35). As we saw with the QR approach above, we see that in the fundamental matrix approach to computing $Q^\#$, the extra computational cost at step 1 is offset by the savings at step 2.

We reiterate a point made in [103] regarding numerical stability. Note that all of the methods for finding $Q^\#$ discussed in sections 8.1 and 8.2 involve the inversion of a matrix of order $n-1$. Observe that if we apply steps 1–6 in the case that $|S| = |\overline{S}| = \frac{n}{2}$ then all of the necessary matrix inversions take place on matrices of order at most $\frac{n}{2}$. Thus, in view of the reduced orders of the matrices to be inverted, we may realise better numerical stability by using steps 1–6.

In Theorem 8.1.4 and section 8.2 we saw several approaches to computing the group inverse of a singular and irreducible M–matrix that is not necessarily of the form $I - T$ where T is stochastic. The remark below shows how steps 1–6 can be adapted to address that problem.

REMARK 8.3.4 Suppose that we have a matrix $A \in \Phi^{n,n}$, partitioned as

$$A = \left[\begin{array}{c|c} A[S,S] & A[S,\overline{S}] \\ \hline A[\overline{S},S] & A[\overline{S},\overline{S}] \end{array}\right].$$

We suppose for simplicity of exposition that the Perron value of A is 1 (if not, then we can multiply A by an appropriate scalar). Our goal in this remark is to show how a variant of steps 1–6 above can be used to compute $(I-A)^{\#}$. Before doing so, we recall a couple of useful facts.

First, suppose that we have an $m \times m$ singular and irreducible M–matrix B, with right and left null vectors u, v respectively, normalised so that $v^t u = 1$. Since $B^{\#}B = BB^{\#} = I - uv^t$, we see that for any indices i, j between 1 and m, we have $u_i v^t = e_i^t - e_i^t B^{\#}B$ and $v_j u = e_j - BB^{\#}e_j$, so that if $B^{\#}$ is known, then u and v can be recovered readily from any column or row, respectively, of $B^{\#}$. This generalises a remark made in Observation 8.1.1 for the case that B is of the form $I - T$ where T is stochastic.

Next, it is straightforward to show that the matrix $\mathcal{P}(A)_S \equiv A[S,S] + A[S,\overline{S}](I - A[\overline{S},\overline{S}])^{-1}A[\overline{S},S]$ has Perron value 1, and that corresponding right and left Perron vectors are scalar multiples of $x[S]$ and $y[S]$, respectively. An analogous statement holds for the matrix $\mathcal{P}(A)_{\overline{S}} \equiv A[\overline{S},\overline{S}] + A[\overline{S},S](I - A[S,S])^{-1}A[S,\overline{S}]$. Let a, b be right and left Perron vectors for $\mathcal{P}(A)_S$, respectively, normalised so that $b^t a = 1$, and let c, d be right and left Perron vectors for $\mathcal{P}(A)_{\overline{S}}$, respectively, normalised so that $d^t c = 1$. Then the 2×2 coupling matrix

$$\begin{bmatrix} b^t A[S,S]a & b^t A[S,\overline{S}]c \\ d^t A[\overline{S},S]a & d^t A[\overline{S},\overline{S}]c \end{bmatrix}$$

has Perron value 1; denoting its right and left Perron vectors by ξ and η, respectively, and normalising them so that $\eta^t \xi = 1$, it follows that the vectors

$$x = \left[\begin{array}{c} \xi_1 a \\ \hline \xi_2 c \end{array}\right], \left[\begin{array}{c} \eta_1 b \\ \hline \eta_2 d \end{array}\right]$$

are right and left Perron vectors for A such that $y^t x = 1$. We remark that all of these observations are slight generalisations of the corresponding statements in section 5.1 for the case that $I - A$ is stochastic.

Finally, recall that if we let $X = \mathrm{diag}(x)$, we find from Remark 2.5.3 that $T = X^{-1}AX$ is an irreducible stochastic matrix. Further, it follows that $w = Xy$ is the stationary distribution vector for T.

With these facts in place, we can now outline a method for finding $(I-A)^{\#}$.
$1'$. Compute $\mathcal{P}(A)_S$ and $\mathcal{P}(A)_{\overline{S}}$.
$2'$. Compute $(I - \mathcal{P}(A)_S)^{\#}$, and from that, find right and left Perron vectors a, b respectively, of $\mathcal{P}(A)_S$, normalised so that $b^t a = 1$. Compute

$(I - \mathcal{P}(A)_{\overline{S}})^{\#}$, and from that, find right and left Perron vectors c, d respectively, of $\mathcal{P}(A)_{\overline{S}}$, normalised so that $d^t c = 1$.
$3'$. Using the coupling matrix and the vectors a, b, c, d, find right and left Perron vectors x, y of A, normalised so that $y^t x = 1$. Also compute the blocks of the stochastic matrix $T = X^{-1}AX$ and its stationary vector $w = Xy$. Using the results computed at steps $1'$ and $2'$, compute the following quantities:
$(I - \mathcal{P}(T)_S)^{\#} = X[S]^{-1}(I - \mathcal{P}(A)_S)^{\#}X[S]$,
$(I - \mathcal{P}(T)_{\overline{S}})^{\#} = X[\overline{S}]^{-1}(I - \mathcal{P}(A)_{\overline{S}})^{\#}X[\overline{S}]$,
$(I - T[S, S])^{-1} = X[S]^{-1}(I - A[S, S])^{-1}X[S]$,
$(I - T[\overline{S}, \overline{S}])^{-1} = X[\overline{S}]^{-1}(I - A[\overline{S}, \overline{S}])^{-1}X[\overline{S}]$,
$(I - T[S, S])^{-1}T[S, \overline{S}] = X[S]^{-1}(I - A[S, S])^{-1}A[S, \overline{S}]X[\overline{S}]$, and
$(I - T[\overline{S}, \overline{S}])^{-1}T[\overline{S}, S] = X[\overline{S}]^{-1}(I - A[\overline{S}, \overline{S}])^{-1}A[\overline{S}, S]X[S]$.
$4'$–$6'$. With the quantities computed at step $3'$, we now compute the blocks of $(I - T)^{\#}$ by proceeding with steps 4–6 in the earlier algorithm for stochastic matrices.
$7'$. Compute $(I - A)^{\#} = X(I - T)^{\#}X^{-1}$.

Observe that at step $2'$, the methods available for computing the group inverse include the shuffle or Hartwig algorithms, or the QR approach. Note that Proposition 2.5.1 will not, in general be applicable here, as the matrices $\mathcal{P}(A)_S$ and $\mathcal{P}(A)_{\overline{S}}$ may not be stochastic. It is straightforward to determine that the computational cost of executing steps $1'$–$7'$ is essentially the same as that of executing steps 1–6 for a stochastic matrix of the same order. As in the stochastic case, the computational cost will be influenced by the choice of method for finding necessary group inverses at step $2'$.

8.4 Stability issues for the group inverse

In this section we address the following issue: given a singular and irreducible M-matrix Q, how can we measure the stability of $Q^{\#}$ under perturbation of Q? We will be concerned with perturbations of Q that yield another singular and irreducible M–matrix, and the results will have implications for the numerical stability of computing $Q^{\#}$. We will also provide some specialised results for group inverses associated with stochastic matrices, and for Laplacian matrices. We note that Wang, Wei and Qiao's book [111] presents perturbation analysis for both the Moore–Penrose inverse ([111, Chapter 7]) and the Drazin inverse ([111, Chapter 8]) in a general setting. Evidently those results may also be applied to the cases of Laplacian matrices, and singular M–matrices, respectively.

We begin our discussion with the following preliminary result. Suppose that we have a matrix $A \in \Phi^{n,n}$. Denote its Perron value by $r(A)$ and let $Q(A) = r(A)I - A$, so that $Q(A)$ is a singular and irreducible M–matrix. We claim that for each pair of indices i, j between 1 and n, $(Q(A))^{\#}$ is a differentiable function of the (i, j)–th entry of A. To be precise, we note that for all sufficiently small values of the scalar t, the spectral abscissa $s(A + tE_{i,j})$ of the matrix $A + tE_{i,j}$ is an algebraically simple eigenvalue; in particular, there is an $\epsilon > 0$ such that $s(A + tE_{i,j})$ is a differentiable function of t for all $t \in (-\epsilon, \epsilon)$. (Note that we have to deal with the spectral abscissa here instead of the Perron value, as $A + tE_{i,j}$ may have a negative entry in the (i, j) position.)

Next, we restrict our attention to the case that $t \in (-\epsilon, \epsilon)$, and we consider the matrix $B(t) = s(A + tE_{i,j})I - (A + tE_{i,j})$. Note that $B(t)$ is singular and has rank $n - 1$. Further, the coefficients of the characteristic polynomial of $B(t)$ are differentiable functions of t. From Theorem 2.3.2, we find that $B(t)^{\#}$ is a polynomial in $B(t)$, and that the coefficients of that polynomial are differentiable functions of the coefficients of the characteristic polynomial of $B(t)$. We thus deduce that $B(t)^{\#}$ is a differentiable function of t on the interval $(-\epsilon, \epsilon)$. In particular, $B(t)^{\#}$ is differentiable at $t = 0$; in keeping with the notation of section 3.2, we henceforth use

$$\frac{\partial (Q(A))^{\#}}{\partial_{i,j}}$$

to denote the partial derivative of $Q(A)^{\#}$ with respect to the (i, j) entry of A. That is,

$$\frac{\partial (Q(A))^{\#}}{\partial_{i,j}} = \lim_{t \to 0} \frac{1}{t} \left(B(t) - B(0) \right).$$

We now present a formula for that derivative.

THEOREM 8.4.1 *Suppose that $A \in \Phi^{n,n}$. Let $x(A)$ and $y(A)$ denote right and left Perron vectors for A respectively, normalised so that $y(A)^t x(A) = 1$. Fix indices i, j between 1 and n. Then we have*

$$\frac{\partial (Q(A))^{\#}}{\partial_{i,j}} = -x_j(A) y_i(A) (Q(A)^{\#})^2 + Q(A)^{\#} e_i e_j^t Q(A)^{\#} \\ - x_j(A)(Q(A)^{\#})^2 e_i y(A)^t - y_i(A) x(A) e_j^t (Q(A)^{\#})^2. \quad (8.36)$$

Proof: In order to ease the notation in the proof, we will suppress the explicit dependence of Q, x and y on A. It follows from our discussion in section 3.1 that there is a neighbourhood of the matrix A such that for each matrix in that neighbourhood, the corresponding right and left Perron vectors x and y can be taken to be differentiable functions of the (i, j) entry, in addition to being normalised so that $y^t x = 1$.

We begin with the equation $QQ^\# = I - xy^t$, and differentiate throughout with respect to $a_{i,j}$ to obtain

$$\frac{\partial Q}{\partial_{i,j}} Q^\# + Q \frac{\partial Q^\#}{\partial_{i,j}} = -\frac{\partial x}{\partial_{i,j}} y^t - x \frac{\partial y^t}{\partial_{i,j}}. \quad (8.37)$$

Observing from (3.2) that

$$\frac{\partial Q}{\partial_{i,j}} = x_j y_i I - e_i e_j^t,$$

and substituting that expression into (8.37), we find that

$$Q \frac{\partial Q^\#}{\partial_{i,j}} = -x_j y_i Q^\# + e_i e_j^t Q^\# - \frac{\partial x}{\partial_{i,j}} y^t - x \frac{\partial y^t}{\partial_{i,j}}. \quad (8.38)$$

Multiplying (8.38) on the left by $Q^\#$, we find that

$$(I - xy^t) \frac{\partial Q^\#}{\partial_{i,j}} = Q^\# Q \frac{\partial Q^\#}{\partial_{i,j}} =$$
$$-x_j y_i (Q^\#)^2 + Q^\# e_i e_j^t Q^\# - Q^\# \frac{\partial x}{\partial_{i,j}} y^t. \quad (8.39)$$

Consequently, we find from (8.39) that

$$\frac{\partial Q^\#}{\partial_{i,j}} = -x_j y_i (Q^\#)^2 + Q^\# e_i e_j^t Q^\# - Q^\# \frac{\partial x}{\partial_{i,j}} y^t + xy^t \frac{\partial Q^\#}{\partial_{i,j}}. \quad (8.40)$$

Differentiating the equation $y^t Q^\# = 0^t$ with respect to $a_{i,j}$ readily yields that $y^t \frac{\partial Q^\#}{\partial_{i,j}} = -\frac{\partial y^t}{\partial_{i,j}} Q^\#$; substituting into (8.40) now yields

$$\frac{\partial Q^\#}{\partial_{i,j}} = -x_j y_i (Q^\#)^2 + Q^\# e_i e_j^t Q^\# - Q^\# \frac{\partial x}{\partial_{i,j}} y^t - x \frac{\partial y^t}{\partial_{i,j}} Q^\#. \quad (8.41)$$

Next, we consider the equation $Qx = 0$. Differentiating it with respect to $a_{i,j}$ now yields

$$(x_j y_i I - e_i e_j^t) x + Q \frac{\partial x}{\partial_{i,j}} = 0. \quad (8.42)$$

Multiplying (8.42) by $(Q^\#)^2$ and recalling that $(Q^\#)^2Q = Q^\#$, it now follows that

$$Q^\# \frac{\partial x}{\partial_{i,j}} = x_j(Q^\#)^2 e_i. \tag{8.43}$$

A similar argument shows that

$$\frac{\partial y^t}{\partial_{i,j}} Q^\# = y_i e_j^t (Q^\#)^2. \tag{8.44}$$

Substituting (8.43) and (8.44) into (8.41) now yields (8.36). □

We observe that once $Q(A)^\#$ has been computed, then as noted in section 8.3, the right and left Perron vectors of A are easily calculated. With $Q(A)^\#$ and the left and right Perron vectors in hand, (8.36) then readily provides the sensitivity of the entries of $Q(A)^\#$ with respect to changes in $a_{i,j}$.

REMARK 8.4.2 Theorem 8.4.1 gives the derivative of $Q(A)^\#$ with respect to a single entry of A. Depending on the context, we might be also be interested in finding a derivative of the form

$$\lim_{t\to 0} \frac{1}{t}\left((s(A+tE)I - (A+tE))^\# - (r(A)I - A)^\# \right) \tag{8.45}$$

for some fixed matrix E. Evidently we can compute any derivative of the type (8.45) by taking appropriate linear combinations of derivatives of the form (8.36).

The following two results are immediate; in both, we retain the notation of Theorem 8.4.1.

COROLLARY 8.4.3 *Fix indices i, j, p, and q between 1 and n. Then we have*

$$\begin{aligned} &e_p^t \left(\frac{\partial (Q(A))^\#}{\partial_{i,j}} \right) e_q = \\ &-x_j(A)y_i(A)e_p^t(Q(A)^\#)^2 e_q + e_p^t Q(A)^\# e_i e_j^t Q(A)^\# e_q \\ &-x_j(A)y_q(A)e_p^t(Q(A)^\#)^2 e_i - y_i(A)x_p(A)e_j^t(Q(A)^\#)^2 e_q. \end{aligned}$$

COROLLARY 8.4.4 *Fix indices i and j between 1 and n. Then we have*

$$e_j^t \left(\frac{\partial (Q(A))^\#}{\partial_{i,j}} \right) e_i = -3x_j(A)y_i(A)e_j^t(Q(A)^\#)^2 e_i + (e_j^t Q(A)^\# e_i)^2.$$

In particular, if $e_j^t(Q(A)^{\#})^2 e_i < 0$, *then at the matrix* A, *the* (j,i) *entry of* $Q(A)^{\#}$ *is increasing in* $a_{i,j}$.

The following example makes use of Theorem 8.4.1.

EXAMPLE 8.4.5 Once again we revisit the population projection matrix A for the desert tortoise in Example 4.2.1. We take A to be given by (4.9), and recall that its Perron value $r(A)$ is approximately 0.9581. Setting $Q(A) = r(A)I - A$, the computed expression for $Q(A)^{\#}$ is

$$\begin{bmatrix} 0.619 & -0.508 & -0.945 & -1.459 & -0.498 & 1.793 & 3.062 & 4.632 \\ 0.930 & 1.355 & -2.443 & -4.541 & -3.730 & -1.702 & 1.060 & 3.163 \\ 0.277 & 0.413 & 1.355 & -2.443 & -2.495 & -2.548 & -1.329 & -0.821 \\ 0.081 & 0.126 & 0.444 & 1.465 & -1.549 & -1.955 & -1.364 & -1.287 \\ 0.029 & 0.049 & 0.196 & 0.691 & 1.336 & -1.618 & -1.228 & -1.254 \\ 0.002 & 0.010 & 0.082 & 0.357 & 0.774 & 1.742 & -1.470 & -1.573 \\ -0.128 & -0.151 & -0.270 & -0.379 & -0.020 & 0.830 & 2.984 & -7.093 \\ -0.044 & -0.056 & -0.126 & -0.277 & -0.328 & -0.438 & -0.036 & 8.429 \end{bmatrix},$$

where, for compactness of presentation, we have rounded the entries to three decimal places. In order to get a sense of how sensitive $Q(A)^{\#}$ is to the positive entries in A, we computed $\frac{\partial Q(A)^{\#}}{\partial_{i,j}}$ for the following (i,j) pairs: $(1,6),(1,7),(1,8)$ (corresponding to the positive entries on the top row of A); $(2,2),\ldots,(8,8)$ (corresponding to the positive entries on the diagonal of A); and $(2,1),\ldots,(8,7)$ (corresponding to the positive entries on the subdiagonal of A). For each such (i,j) pair, we identified the extreme entry of $Q(A)^{\#}$ — i.e, the entry of $Q(A)^{\#}$ having largest absolute value. The results are depicted in Figure 8.2; observe that some of these extreme entries are negative.

Inspecting Figure 8.2, we find that the derivatives of the entries of $Q(A)^{\#}$ with respect to $a_{1,j}$ for $j = 6,7,8$, $a_{j,j}$ for $j = 2,\ldots,6$, and $a_{i,i-1}$ for $i = 2,\ldots,7$, are not especially large in absolute value; we conclude then that at the matrix A, $Q(A)^{\#}$ is not particularly sensitive to small changes in any of those entries. On the other hand, we observe from Figure 8.2 that each of $\frac{\partial Q(A)^{\#}}{\partial_{7,7}}$ and $\frac{\partial Q(A)^{\#}}{\partial_{8,7}}$ has some large positive entries (the extreme values are 117.064 and 141.445, respectively), while $\frac{\partial Q(A)^{\#}}{\partial_{8,8}}$ has a large negative entry (the extreme value is -129.211). It is interesting to note that for the derivatives of $Q(A)^{\#}$ with respect to $a_{7,7}$ and $a_{8,8}$, the extreme entries both appear in the $(2,8)$ position. For the derivative of $Q(A)^{\#}$ with respect to $a_{8,7}$, there are in fact several large entries: a 117.723 in the $(2,8)$ position, a -108.143 in the $(8,8)$ position, and the extreme entry 141.445 in the $(7,8)$ position. In particular, we find that at the matrix A, the $(2, 8)$ entry of $Q(A)^{\#}$ is sensitive to small changes in the values of $a_{7,7}, a_{8,8}$ and $a_{8,7}$.

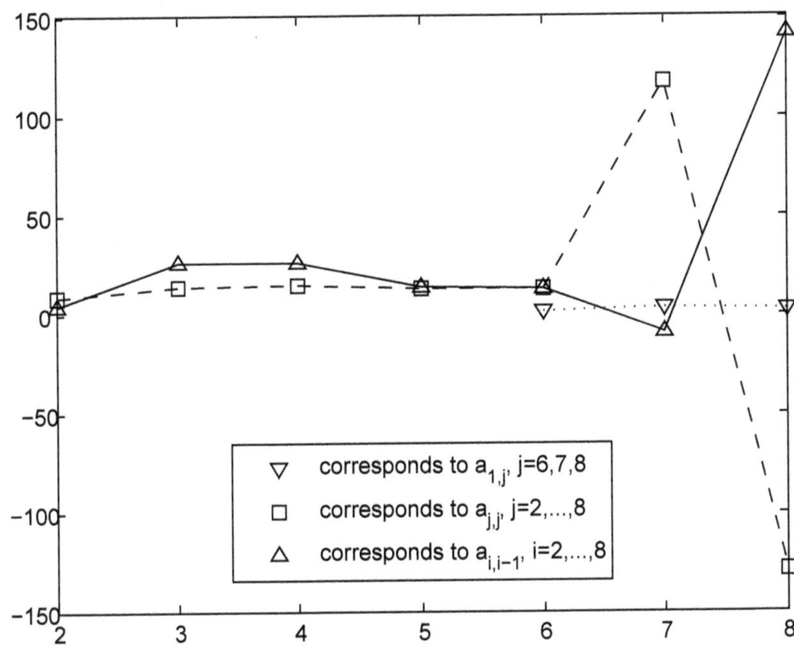

FIGURE 8.2
Extreme entries of $\frac{\partial Q(A)^{\#}}{\partial_{i,j}}$ corresponding to the positive entries on the top row, diagonal, and subdiagonal of the matrix A in Example 8.4.5

Next we consider a scenario analogous to that dealt with in section 5.3. Specifically, we suppose that $T \in \mathcal{IS}_n$. Now consider an irreducible stochastic matrix $\tilde{T} = T + E$. Setting $Q = I - T$ and $\tilde{Q} = I - \tilde{T}$, we would like to get a sense of how close $\tilde{Q}^{\#}$ is to $Q^{\#}$.

We begin with an analysis of the sensitivity of the group inverse with respect to changes in the underlying stochastic matrix. Evidently for any irreducible stochastic matrix T, we have $Q(T) = I - T$. We consider perturbations of T of the type $T + ae_i(e_j - e_k)^t$, where a is a scalar of small absolute value, and where i, j, and k are between 1 and n. We note that the restriction on the type of perturbation under consideration arises from the fact that our perturbing matrices should have row sums zero (so that $T + ae_i(e_j - e_k)^t$ is another stochastic matrix). Strictly speaking, asking for $T + ae_i(e_j - e_k)^t$ to be stochastic also places some restrictions on i, j, k and a, since we would then require $t_{i,j} + a$ and $t_{i,k} - a$

to be nonnegative, however, we will leave that consideration aside, as our discussion below goes through independently of the nonnegativity consideration. As in the preamble to Theorem 8.4.1, the sensitivity of $Q(T)^{\#}$ to perturbations of the type $ae_i(e_j - e_k)^t$ can be measured by the following derivative:

$$\frac{\partial Q(T)^{\#}}{\partial_{i,j}} - \frac{\partial Q(T)^{\#}}{\partial_{i,k}} = \lim_{a\to 0}\frac{1}{a}\left(Q(T + ae_i(e_j - e_k)^t)^{\#} - Q(T)^{\#}\right). \tag{8.46}$$

Theorem 8.4.1 now yields the following result.

THEOREM 8.4.6 *Suppose that $T \in \mathcal{IS}_n$, with stationary distribution vector w. Fix indices i, j, k between 1 and n. Then we have*

$$\frac{\partial Q(T)^{\#}}{\partial_{i,j}} - \frac{\partial Q(T)^{\#}}{\partial_{i,k}} = Q(T)^{\#}e_i(e_j - e_k)^tQ(T)^{\#} - w_i\mathbf{1}(e_j - e_k)^t(Q(T)^{\#})^2. \tag{8.47}$$

Proof: Recalling that $\mathbf{1}$ is a right Perron vector for T and that $w^t\mathbf{1} = 1$, we find from (8.36) that

$$\frac{\partial Q(T)^{\#}}{\partial_{i,j}} =$$
$$-w_i(Q(T)^{\#})^2 + Q(T)^{\#}e_ie_j^tQ(T)^{\#} - (Q(T)^{\#})^2e_iw^t - w_i\mathbf{1}e_j^t(Q(T)^{\#})^2.$$

Similarly, we have

$$\frac{\partial Q(T)^{\#}}{\partial_{i,k}} =$$
$$-w_i(Q(T)^{\#})^2 + Q(T)^{\#}e_ie_k^tQ(T)^{\#} - (Q(T)^{\#})^2e_iw^t - w_i\mathbf{1}e_k^t(Q(T)^{\#})^2,$$

so taking the difference of $\frac{\partial Q(T)^{\#}}{\partial_{i,j}}$ and $\frac{\partial Q(T)^{\#}}{\partial_{i,k}}$ readily yields (8.47). □

We note that any real matrix E with zero row sums can be written as a linear combination of matrices of the form $e_i(e_j - e_k)^t$. Thus, as in Remark 8.45, we find that the result of Theorem 8.4.6 can be used to find derivatives of the type

$$\lim_{a\to 0}\left(Q(T + aE)^{\#} - Q(T)\right)$$

for any real matrix E with zero row sums.

The following example considers (8.47) for stochastic matrices of rank one.

EXAMPLE 8.4.7 Let T be an irreducible $n \times n$ rank one stochastic matrix, i.e., $T = \mathbf{1}w^t$ for some $w \in \mathbb{R}^n$ with $w > 0, w^t\mathbf{1} = 1$. Setting $Q(T) = I - T$, it follows that $Q(T)^\# = I - \mathbf{1}w^t$. Fix indices i, j, k between 1 and n. Observing that $(e_j - e_k)^t Q(T)^\# = (e_j - e_k)^t$, it follows from Theorem 8.4.6 that

$$\frac{\partial Q(T)^\#}{\partial_{i,j}} - \frac{\partial Q(T)^\#}{\partial_{i,k}} = (e_i - 2w_i\mathbf{1})(e_j - e_k)^t.$$

In particular we see that the entries in $\frac{\partial Q(T)^\#}{\partial_{i,j}} - \frac{\partial Q(T)^\#}{\partial_{i,k}}$ are all bounded above by 2 in absolute value, so that $Q(T)^\#$ is not especially sensitive to small perturbations of the type $ae_i(e_j - e_k)^t$.

While Theorem 8.4.6 is stated for the case that our stochastic matrix T is irreducible, it can be verified that the corresponding conclusion also holds in the case that T is stochastic and has 1 as a simple eigenvalue. The next example illustrates (8.47) in that setting.

EXAMPLE 8.4.8 Consider the $n \times n$ stochastic matrix T given by

$$T = \begin{bmatrix} 1 & 0 & 0 & \dots & 0 & 0 \\ 1 & 0 & 0 & \dots & 0 & 0 \\ 0 & 1 & 0 & 0 & \dots & 0 \\ 0 & 0 & 1 & 0 & \dots & 0 \\ \vdots & & & \ddots & & \vdots \\ 0 & 0 & \dots & 0 & 1 & 0 \end{bmatrix},$$

which evidently has 1 as a simple eigenvalue, and stationary distribution given by e_1. A computation reveals that

$$Q(T)^\# = \begin{bmatrix} 0 & 0 & 0 & 0\dots & 0 & 0 & \\ -1 & 1 & 0 & 0\dots & 0 & 0 & \\ -2 & 1 & 1 & 0 & 0 & \dots & 0 \\ -3 & 1 & 1 & 1 & 0 & \dots & 0 \\ \vdots & \vdots & & \ddots & & & \vdots \\ -(n-1) & 1 & 1 & 1 & \dots & 1 & 1 \end{bmatrix}. \tag{8.48}$$

Next, we consider $\frac{\partial Q(T)^\#}{\partial_{1,n}} - \frac{\partial Q(T)^\#}{\partial_{1,1}}$, which measures the sensitivity of the group inverse when a small positive mass is moved from the $(1,1)$ entry of T to the $(1,n)$ entry of T. According to Theorem 8.4.6, we have

$$\frac{\partial Q(T)^\#}{\partial_{1,n}} - \frac{\partial Q(T)^\#}{\partial_{1,1}} = Q(T)^\# e_1(e_n - e_1)^t Q(T)^\# - \mathbf{1}(e_n - e_1)^t (Q(T)^\#)^2. \tag{8.49}$$

Referring to (8.48), we find that

$$(e_n - e_1)^t Q(T)^{\#} = \begin{bmatrix} -(n-1) & 1 & 1 & \dots & 1 \end{bmatrix} \text{ while,}$$

$(e_n - e_1)^t (Q(T)^{\#})^2 = \begin{bmatrix} -\frac{n(n-1)}{2} & n-1 & n-2 & \dots & 2 & 1 \end{bmatrix}$. Substituting into (8.49) and simplifying, we find that

$$\frac{\partial Q(T)^{\#}}{\partial_{1,n}} - \frac{\partial Q(T)^{\#}}{\partial_{1,1}} =$$

$$\begin{bmatrix} \frac{n(n-1)}{2} & -(n-1) & -(n-2) & -(n-3) & \dots & -2 & -1 \\ \frac{(n+2)(n-1)}{2} & -n & -(n-1) & -(n-2) & \dots & -3 & -2 \\ \frac{(n+4)(n-1)}{2} & -(n+1) & -n & -(n-1) & \dots & -4 & -3 \\ \frac{(n+6)(n-1)}{2} & -(n+2) & -(n+1) & -n & \dots & -5 & -4 \\ \vdots & \vdots & & & \ddots & & \vdots \\ \frac{(3n-2)(n-1)}{2} & -(2n-2) & -(2n-1) & -2n & \dots & -(n+1) & -n \end{bmatrix}. \tag{8.50}$$

An inspection of (8.50) reveals that the entries of $\frac{\partial Q(T)^{\#}}{\partial_{1,n}} - \frac{\partial Q(T)^{\#}}{\partial_{1,1}}$ are increasing in absolute value as we move down each column of $\frac{\partial Q(T)^{\#}}{\partial_{1,n}} - \frac{\partial Q(T)^{\#}}{\partial_{1,1}}$. Note also that the entries in the first column of $\frac{\partial Q(T)^{\#}}{\partial_{1,n}} - \frac{\partial Q(T)^{\#}}{\partial_{1,1}}$ range between $\frac{n(n-1)}{2}$ and $\frac{(3n-2)(n-1)}{2}$; when n is large we thus find that the entries in the first column of $Q(T)^{\#}$ will be quite sensitive to small perturbations of T that shift weight from its $(1,1)$ entry to its $(1,n)$ entry.

Our next result maintains the hypotheses and notation of Theorem 8.4.6 (while suppressing the explicit dependence of $Q(T)^{\#}$ on T) and provides a bound on the moduli of the sensitivities of the entries in $Q^{\#}$.

COROLLARY 8.4.9 *For each pair of indices a, b between 1 and n, we have*

$$\left| e_a^t \frac{\partial Q^{\#}}{\partial_{i,j}} e_b - e_a^t \frac{\partial Q^{\#}}{\partial_{i,k}} e_b \right| \leq$$

$$|q^{\#}_{a,i}||q^{\#}_{j,b} - q^{\#}_{k,b}| + \frac{w_i}{2}||(Q^{\#})^t(e_j - e_k)||_1 \left(q^{\#}_{b,b} - \min_{1 \leq l \leq n} q^{\#}_{l,b} \right)$$

$$\leq 2\kappa(T) \left(\max_{1 \leq r,s \leq n} |q^{\#}_{r,s}| + \tau(Q^{\#}) \right). \tag{8.51}$$

Proof: It follows immediately from Theorem 8.4.6 and the triangle inequality that

$$\left| e_a^t \frac{\partial Q^\#}{\partial_{i,j}} e_b - e_a^t \frac{\partial Q^\#}{\partial_{i,k}} e_b \right| \leq |q^\#_{a,i}||q^\#_{j,b} - q^\#_{k,b}| + w_i|(e_j - e_k)^t(Q^\#)^2 e_b|.$$

Next, we write $(e_j - e_k)^t(Q^\#)^2 e_b$ as the inner product of the vector $(e_j - e_k)^t Q^\#$ with the vector $Q^\# e_b$. Since $(e_j - e_k)^t Q^\# \mathbf{1} = 0$, we find from Lemma 5.3.4 that

$$|(e_j - e_k)^t(Q^\#)^2 e_b| \leq \frac{1}{2}||(Q^\#)^t(e_j - e_k)||_1 \max_{1 \leq c,d \leq n}(q^\#_{c,b} - q^\#_{d,b}).$$

Recalling that the maximum entry in the b–th column of $Q^\#$ occurs on the diagonal, it now follows readily that

$$\left| e_a^t \frac{\partial Q^\#}{\partial_{i,j}} e_b - e_a^t \frac{\partial Q^\#}{\partial_{i,k}} e_b \right| \leq$$

$$|q^\#_{a,i}||q^\#_{j,b} - q^\#_{k,b}| + \frac{w_i}{2}||(Q^\#)^t(e_j - e_k)||_1 \left(q^\#_{b,b} - \min_{1 \leq l \leq n} q^\#_{l,b} \right). \quad (8.52)$$

The final inequality of (8.51) follows immediately from the definitions of $\tau(Q^\#)$ and $\kappa(T)$, and the fact that $w_i \leq 1$. □

REMARK 8.4.10 Recall from section 5.3 that the quantities $\kappa(T)$ and $\tau(Q^\#)$ can be used to measure the conditioning of the stationary vector w under perturbation of the underlying stochastic matrix T. A similar comment applies to the quantity $\max_{1 \leq r,s \leq n} |q^\#_{r,s}|$, which has also been used as a condition number for the stationary distribution (see [23]). If none of these three quantities is too large, then the stationary distribution will be well-conditioned with respect to perturbations of T; further, in that case, we see from (8.51) that $Q^\#$ will not be too sensitive to small perturbations that take T to another stochastic matrix.

Theorem 8.4.6 gives us a sense of the sensitivity of $Q(T)^\#$ to small changes in T, but it is also worth discussing how $Q(T)^\#$ changes when T is perturbed to another stochastic that is not necessarily close to T. Specifically, suppose that $\tilde{T} = T + E$, where both T and $\tilde{T}$ have 1 as an algebraically simple eigenvalue. Denoting the stationary distribution for T by w, we recall from Lemma 5.3.1 that the matrix $I - EQ(T)^\#$ is invertible, while from Theorem 5.3.30, we have that

$$\begin{aligned} Q(\tilde{T})^\# =& Q(T)^\#(I - EQ(T)^\#)^{-1} \\ & -\mathbf{1}w^t(I - EQ(T)^\#)^{-1}Q(T)^\#(I - EQ(T)^\#)^{-1}. \quad (8.53) \end{aligned}$$

The following result, which is an immediate consequence of Corollary 5.3.31 considers a very simple perturbation of a stochastic matrix and the ensuing effect on the corresponding group inverse. Again, we will suppress the explicit dependence of $Q(T)^\#$ on T.

PROPOSITION 8.4.11 *Suppose that $T \in \mathcal{IS}_n$ and has stationary distribution w. Fix indices i, j, k between 1 and n, and let $\tilde{T} = T + ae_i(e_j - e_k)^t$ for some scalar a. If $\tilde{T} \in \mathcal{IS}_n$, then*

$$Q(\tilde{T})^\# = Q^\# + \frac{a}{1 - a(q^\#_{j,i} - q^\#_{k,i})} Q^\# e_i (e_j - e_k)^t Q^\#$$
$$- \frac{w_i}{1 - a(q^\#_{j,i} - q^\#_{k,i})} \mathbf{1} \Bigg(a(e_j - e_k)^t (Q^\#)^2 +$$
$$\frac{a^2 (e_j - e_k)^t (Q^\#)^2 e_i}{1 - a(q^\#_{j,i} - q^\#_{k,i})} (e_j - e_k)^t Q^\# \Bigg).$$

Next we consider the norm of the difference between $Q(T)^\#$ and $Q(\tilde{T})^\#$ for two stochastic matrices $T, \tilde{T}$. As we will see in the result below, $Q(T)^\#$ is well–conditioned with respect to small perturbations of T provided that $||Q(T)^\#||$ is not too large.

THEOREM 8.4.12 *Suppose that $T, \tilde{T} \in \mathcal{IS}_n$. Denote the stationary vector of T by w, and let $E = \tilde{T} - T$. Then for any matrix norm $|| \cdot ||$ such that $||E|| ||Q(T)^\#|| < 1$, we have*

$$||Q(\tilde{T})^\# - Q(T)^\#|| \le \frac{||Q(T)^\#||^2 ||E||}{1 - ||E|| ||Q(T)^\#||} \left(1 + \frac{||\mathbf{1}w^t||}{1 - ||E|| ||Q(T)^\#||} \right). \tag{8.54}$$

In particular, if $||E||_\infty ||Q(T)^\#||_\infty < 1$, then

$$||Q(\tilde{T})^\# - Q(T)^\#||_\infty \le \frac{||Q(T)^\#||^2_\infty ||E||_\infty}{1 - ||E||_\infty ||Q(T)^\#||_\infty} \left(\frac{2 - ||E||_\infty ||Q(T)^\#||_\infty}{1 - ||E||_\infty ||Q(T)^\#||_\infty} \right). \tag{8.55}$$

__Proof__: From (8.53), we see that

$$Q(\tilde{T})^\# =$$
$$Q(T)^\#(I - EQ(T)^\#)^{-1} - \mathbf{1}w^t(I - EQ(T)^\#)^{-1} Q(T)^\# (I - EQ(T)^\#)^{-1}.$$

Using the facts that $(I - EQ(T)^\#)^{-1} = I + EQ(T)^\#(I - EQ(T)^\#)^{-1}$ and $w^t Q(T)^\# = 0^t$, it follows that

$$Q(\tilde{T})^\# - Q(T)^\# = Q(T)^\# EQ(T)^\# (I - EQ(T)^\#)^{-1}$$
$$-\mathbf{1}w^t EQ(T)^\# (I - EQ(T)^\#)^{-1} Q(T)^\# (I - EQ(T)^\#)^{-1}.$$

From the hypothesis that $||E||||Q(T)^{\#}|| < 1$, we find that

$$||(I - EQ(T)^{\#})^{-1}|| \leq \frac{1}{1 - ||E||||Q(T)^{\#}||}.$$

An application of the triangle inequality now yields

$$||Q(\tilde{T})^{\#} - Q(T)^{\#}|| \leq \frac{||Q(T)^{\#}EQ(T)^{\#}||}{1 - ||E||||Q(T)^{\#}||} + \frac{||\mathbf{1}w^t EQ(T)^{\#}||||Q(T)^{\#}||}{(1 - ||E||||Q(T)^{\#}||)^2}$$
$$\leq \frac{||Q(T)^{\#}||^2||E||}{1 - ||E||||Q(T)^{\#}||}\left(1 + \frac{||\mathbf{1}w^t||}{1 - ||E||||Q(T)^{\#}||}\right).$$

Observe that (8.55) follows immediately from (8.54) upon observing that $||\mathbf{1}w^t||_\infty = 1$. □

EXAMPLE 8.4.13 Suppose that T is a positive rank one stochastic matrix of order n, say $T = \mathbf{1}w^t$ where $w > 0$ and $w^t\mathbf{1} = 1$. As in Example 8.4.7, $Q(T)^{\#} = I - \mathbf{1}w^t$. In particular, we find that $||Q(T)^{\#}||_\infty = 2(1 - \min_{j=1,\ldots,n} w_j)$. It now follows from Theorem 8.4.12 that if $\tilde{T} \in \mathcal{IS}_n$, and if we set $E = \tilde{T} - T$, then

$$||Q(\tilde{T})^{\#} - Q(T)^{\#}||_\infty \leq \frac{8||E||_\infty}{1 - 2||E||_\infty}\left(\frac{1 - ||E||_\infty}{1 - 2||E||_\infty}\right),$$

provided that $||E||_\infty < \frac{1}{2}$. Thus we find that for a positive rank one stochastic matrix T, the corresponding group inverse is well–conditioned when T is perturbed to yield another irreducible stochastic matrix.

Recall that for a square matrix B, its 2-norm $||B||_2$ coincides with its largest singular value (see [57, section 5.6]). Next, we adapt an idea appearing in [45] to discuss the conditioning of $Q(T)^{\#}$ in the 2-norm.

COROLLARY 8.4.14 *Suppose that $T \in \mathcal{IS}_n$ with stationary distribution vector w. Consider $\tilde{T} \in \mathcal{IS}_n$, and set $E = \tilde{T} - T$. Suppose that we have a factorisation of $Q(T)$ as PR, where P is orthogonal and where R is upper triangular and of the form*

$$R = \left[\begin{array}{c|c} U & -U\mathbf{1} \\ \hline 0^t & 0 \end{array}\right].$$

If $nw^tw||U^{-1}||_2||E||_2 < 1$, Then

$$||Q(\tilde{T})^{\#} - Q(T)^{\#}||_2 \leq$$
$$\frac{\left(nw^tw||U^{-1}||_2\right)^2||E||_2}{1 - nw^tw||U^{-1}||_2||E||_2}\left(1 + \frac{\sqrt{nw^tw}}{1 - nw^tw||U^{-1}||_2||E||_2}\right). \quad (8.56)$$

Proof: First we claim that $||Q(T)^{\#}||_2 \leq nw^tw||U^{-1}||_2$. To see the claim, we begin by noting that from (8.3) it follows that

$$||Q(T)^{\#}||_2 \leq ||I - \mathbf{1}w^t||_2^2||U^{-1}||_2.$$

Thus, in order to establish the claim, it suffices to show that

$$||I - \mathbf{1}w^t||_2^2 = nw^tw. \tag{8.57}$$

It is straightforward to establish (8.57) in the case that $w = \frac{1}{n}\mathbf{1}$. So, henceforth we suppose that $w \neq \frac{1}{n}\mathbf{1}$. Note that $(I - \mathbf{1}w^t)(I - w\mathbf{1}^t) = I - \mathbf{1}w^t - w\mathbf{1}^t + (w^tw)J$, and observe that the eigenvalues of $I - \mathbf{1}w^t - w\mathbf{1}^t + (w^tw)J$ are: 0 (with eigenvector w), nw^tw (with eigenvector $(w^tw)\mathbf{1} - w$) and 1 (with the $n-2$–dimensional eigenspace consisting of the vectors orthogonal to both $\mathbf{1}$ and w). From the Cauchy–Schwarz inequality it follows that $nw^tw \geq (w^t\mathbf{1})^2 = 1$, so that the spectral radius of $I - \mathbf{1}w^t - w\mathbf{1}^t + (w^tw)J$ is equal to nw^tw. This establishes (8.57), from which the claim follows.

From our claim above, we see that if $nw^tw||U^{-1}||_2||E||_2 < 1$, then certainly $||Q(T)^{\#}||_2||E||_2 < 1$. Applying Theorem 8.4.12 with the 2-norm, our upper bound $||Q(T)^{\#}||_2 \leq nw^tw||U^{-1}||_2$, and the readily determined fact that $||\mathbf{1}w^t||_2 = \sqrt{nw^tw}$, (8.56) now follows. □

We now turn our attention to Laplacian matrices for connected weighted graphs, and consider the behaviour of the corresponding group inverse under perturbation of the edge weights. We begin with a result that generalises Lemma 7.4.5.

THEOREM 8.4.15 *Consider two connected weighted graphs $(\mathcal{G}, w)$ and $(\tilde{\mathcal{G}}, \tilde{w})$, both on n vertices. Denote the corresponding Laplacian matrices by L and $\tilde{L}$ respectively, and let E be given by the relation $\tilde{L} = L - E$. Then $(I - EL^{\#})$ is invertible, and*

$$\tilde{L}^{\#} = L^{\#}(I - EL^{\#})^{-1}. \tag{8.58}$$

Proof: The proof that $(I - EL^{\#})$ is invertible proceeds using the technique employed in Lemma 5.3.1; we omit the details here.

Next, we note that since $L^{\#}\mathbf{1} = 0$ and $\mathbf{1}^tL^{\#} = 0^t$, it follows that $L^{\#}(I - EL^{\#})^{-1}\mathbf{1} = 0$, and $\mathbf{1}^tL^{\#}(I - EL^{\#})^{-1} = 0^t$. We now consider the product $\tilde{L}L^{\#}(I - EL^{\#})^{-1} = (L - E)L^{\#}(I - EL^{\#})^{-1}$. It is readily determined that

$$(L - E)L^{\#}(I - EL^{\#})^{-1} = \left(I - \frac{1}{n}J\right)(I - EL^{\#})^{-1} - EL^{\#}(I - EL^{\#})^{-1},$$

where J is the $n \times n$ all ones matrix. Since L and $\tilde{L}$ are both Laplacian matrices, we necessarily have $JE = 0$, from which it follows that

$$\left(I - \frac{1}{n}J\right)(I - EL^{\#})^{-1} - EL^{\#}(I - EL^{\#})^{-1} =$$
$$(I - EL^{\#})^{-1} - \frac{1}{n}J - EL^{\#}(I - EL^{\#})^{-1} = I - \frac{1}{n}J.$$

A similar argument shows that $L^{\#}(I - EL^{\#})^{-1}\tilde{L} = I - \frac{1}{n}J$, and (8.58) follows readily. □

Next, we provide an analogue of Theorem 8.4.12 for Laplacian matrices.

THEOREM 8.4.16 *Suppose that $(\mathcal{G}, w)$ and $(\tilde{\mathcal{G}}, \tilde{w})$, are connected weighted graphs, each on n vertices, with Laplacian matrices L and $\tilde{L}$, respectively. Set $E = L - \tilde{L}$, and let $||\cdot||$ denote a matrix norm. If $||E||||L^{\#}|| < 1$, then*

$$||\tilde{L}^{\#} - L^{\#}|| \leq \frac{||L^{\#}||^2||E||}{1 - ||E||||L^{\#}||}. \tag{8.59}$$

In particular, for the 2-norm, if $||E||_2 < \alpha(\mathcal{G}, w)$ then we have

$$||\tilde{L}^{\#} - L^{\#}||_2 \leq \frac{||E||_2}{\alpha(\mathcal{G}, w)^2 - \alpha(\mathcal{G}, w)||E||_2}. \tag{8.60}$$

Proof: We begin by noting that $(I - EL^{\#})^{-1} = I + EL^{\#}(I - EL^{\#})^{-1}$. From (8.58), it now follows that

$$\tilde{L}^{\#} - L^{\#} = L^{\#}EL^{\#}(I - EL^{\#})^{-1}.$$

Hence we find that

$$||\tilde{L}^{\#} - L^{\#}|| \leq ||L^{\#}EL^{\#}||||(I - EL^{\#})^{-1}|| \leq \frac{||L^{\#}||^2||E||}{1 - ||E||||L^{\#}||},$$

as desired.

Recalling that the nonzero eigenvalues of $L^{\#}$ are the reciprocals of the nonzero eigenvalues of L, we find that

$$||L^{\#}||_2 = \frac{1}{\alpha(\mathcal{G}, w)}. \tag{8.61}$$

Specialising (8.59) to the 2-norm, and using (8.61) now yields (8.60). □

Next, we develop a notion of the sensitivity of $L(\mathcal{G},w)^\#$ to the weight of a particular edge. So, let $(\mathcal{G},w)$ be a connected weighted graph on n vertices with Laplacian matrix L, and fix a pair of indices i,j between 1 and n. We define

$$\frac{\partial L(\mathcal{G},w)^\#}{\partial w_{i,j}}$$

as follows:

$$\frac{\partial L(\mathcal{G},w)^\#}{\partial w_{i,j}} = \lim_{\epsilon\to 0}\frac{1}{\epsilon}\left(\left(L(\mathcal{G},w)+\epsilon(e_i-e_j)(e_i-e_j)^t\right)^\# - L(\mathcal{G},w)^\#\right).$$

It is straightforward to determine, either from Theorem 8.4.1 or from Lemma 7.4.5, that

$$\frac{\partial L(\mathcal{G},w)^\#}{\partial w_{i,j}} = -L(\mathcal{G},w)^\#(e_i-e_j)(e_i-e_j)^t L(\mathcal{G},w)^\#. \tag{8.62}$$

Using (8.62), we obtain the following result.

PROPOSITION 8.4.17 *Let $(\mathcal{G},w)$ be a connected weighted graph on n vertices. Then*

$$\left|\left|\frac{\partial L(\mathcal{G},w)^\#}{\partial w_{i,j}}\right|\right|_2 \le \frac{2}{\alpha(\mathcal{G},w)^2}. \tag{8.63}$$

Proof: From (8.62), we find that

$$\left|\left|\frac{\partial L(\mathcal{G},w)^\#}{\partial w_{i,j}}\right|\right|_2 = ||L(\mathcal{G},w)^\#(e_i-e_j)(e_i-e_j)^t L(\mathcal{G},w)^\#||_2 =$$

$$(e_i-e_j)^t(L(\mathcal{G},w)^\#)^2(e_i-e_j). \tag{8.64}$$

Recalling that the spectral radius of $L(\mathcal{G},w)^\#$ is $\frac{1}{\alpha(\mathcal{G},w)}$, (8.63) now follows. □

Our next example shows that equality is attainable in (8.63).

EXAMPLE 8.4.18 Consider the unweighted star $K_{1,n-1}$ on $n \ge 3$ vertices, labelled so that vertices $1,\ldots,n-1$ are pendent. From Example 7.3.7, it follows that the eigenvalues of the corresponding Laplacian matrix are $0, n$ and 1, the last with multiplicity $n-2$. Consequently, $\alpha(K_{1,n-1}) = 1$. Further, we find from Example 7.2.11 that

$$L^\# = \sum_{j=1}^{n-1}\left(e_j - \frac{1}{n}\mathbf{1}\right)\left(e_j - \frac{1}{n}\mathbf{1}\right)^t.$$

Observe that

$$L^{\#}(e_1 - e_2) = \left(e_1 - \frac{1}{n}\mathbf{1}\right) - \left(e_2 - \frac{1}{n}\mathbf{1}\right) = e_1 - e_2.$$

From (8.62), we thus find that

$$\frac{\partial L^{\#}}{\partial w_{1,2}} = -L^{\#}(e_1 - e_2)(e_1 - e_2)^t L^{\#} = -(e_1 - e_2)(e_1 - e_2)^t. \quad (8.65)$$

Hence, $||\frac{\partial L^{\#}}{\partial w_{1,2}}||_2 = 2 = \frac{2}{\alpha(K_{1,n-1})}$, so that equality holds in (8.63). Indeed it is readily seen that for any pair of distinct indices i, j between 1 and $n-1$, $||\frac{\partial L^{\#}}{\partial w_{i,j}}||_2 = \frac{2}{\alpha(K_{1,n-1})}$.

Taken together, (8.60) and (8.63) suggest that when the algebraic connectivity of a weighted graph is not too small, then the group inverse of the corresponding Laplacian matrix will be reasonably well–conditioned. Our final example considers the conditioning of the Laplacian matrix in case that the algebraic connectivity is small.

EXAMPLE 8.4.19 Consider P_n, the unweighted path on n vertices, and label the vertices so that vertices 1 and n are pendent, and where vertex j is adjacent to vertex $j+1, j = 1, \ldots, n-1$. Recall from Example 7.3.3 that $\alpha(P_n) = 2\left(1 - \cos\left(\frac{\pi}{n}\right)\right)$, which is asymptotic to $\frac{\pi^2}{n^2}$ as $n \to \infty$ (in the sense that the ratio of the two quantities approaches 1 as $n \to \infty$). Letting L denote the corresponding Laplacian matrix, it follows from Example 7.2.8 that for each $i = 1, \ldots, n$, we have $l^{\#}_{i,1} - l^{\#}_{i,n} = \frac{n+1}{2} - i$.

Next, we consider $\frac{\partial L^{\#}}{\partial w_{1,n}}$. From (8.62) we find readily that for each $i, j = 1, \ldots, n$,

$$\frac{\partial l^{\#}_{i,j}}{\partial w_{1,n}} = -\frac{(n+1-2i)(n+1-2j)}{4}.$$

In particular, we observe that

$$\min_{i,j=1,\ldots,n} \frac{\partial l^{\#}_{i,j}}{\partial w_{1,n}} = -\frac{(n-1)^2}{4} \text{ while } \max_{i,j=1,\ldots,n} \frac{\partial l^{\#}_{i,j}}{\partial w_{1,n}} = \frac{(n-1)^2}{4}.$$

Thus we see that if n is large, some of the entries of $L^{\#}$ are sensitive to small changes in the weight of the edge between vertices 1 and n. From (8.64), we find that

$$\left|\left|\frac{\partial L^{\#}}{\partial w_{1,n}}\right|\right|_2 = \sum_{j=1}^{n} \left(\frac{n+1}{2} - j\right)^2 = \frac{n(n^2-1)}{12}.$$

Again, if n is large, the 2-norm of $\frac{\partial L^{\#}}{\partial w_{1,n}}$ is also large, indicating that $L^{\#}$ is sensitive to changes in the weight of the edge between 1 and n.

Symbol Description

Symbol	Description
$\mathcal{AS}_n$	set of $n \times n$ stochastic matrices having 1 as an algebraically simple eigenvalue
$\alpha(\mathcal{G}, w)$	algebraic connectivity for the weighted graph $(\mathcal{G}, w)$
$b_{i,j}$	(i,j) entry of B
B^D	Drazin inverse of the square matrix B
$B_{(j)}$	matrix obtained from the square matrix B by deleting its j-th row and column
B^t	transpose of the matrix B
B^*	conjugate transpose of the matrix B
$B^\dagger$	Moore-Penrose inverse of the matrix B
B_{dg}	diagonal matrix formed from the square matrix B by setting all off-diagonal entries equal to 0
$B[S_1, S_2]$	submatrix of B on rows indexed by the set S_1 and columns indexed by the set S_2
$\mathbb{C}^{n,n}$	$n \times n$ complex matrices
$C^{(\infty,\infty)}_{\mathcal{AS}_n}(T)$	structured condition number for the stationary distribution vector of the stochastic matrix T, with respect to the infinity norms
$\mathcal{D}(B)$	directed graph of the matrix B
$\mathrm{diag}(u)$	diagonal matrix whose diagonal entries are the corresponding entries of the vector u
$\frac{\partial r(A)}{\partial_{i,j}}$, $\frac{\partial^2 r(A)}{\partial^2_{i,j}}$	first and second derivatives of the Perron value with respect to (i,j) entry, at A
$\frac{\partial x(A)}{\partial_{i,j}}$	derivative of the Perron vector x with respect to the (i,j) entry, at A
e_i	i-th standard unit basis vector
$e_{i,j}$	elasticity of the Perron value with respect to the (i,j) entry
$E_{i,j}$	matrix with a 1 in the (i,j) position, and zeros elsewhere
E_λ	eigenprojection matrix corresponding to λ
$\Phi^{n,n}$	set of all $n \times n$ irreducible nonnegative matrices
$\tilde{\Phi}^{n,n}$	set of all $n \times n$ irreducible essentially nonnegative matrices
$(\mathcal{G}, w)$	a weighted graph
$\gamma(T)$	absolute value of a subdominant eigenvalue of the stochastic matrix T
$\mathcal{IS}_n$	set of $n \times n$ irreducible stochastic matrices
J	all ones matrix
$\mathrm{Kf}(\mathcal{G}, w)$	Kirchhoff index of the weighted graph $(\mathcal{G}, w)$
$\underline{\mathrm{Kf}}(\mathcal{G})$	minimum Kirchhoff index for the graph $\mathcal{G}$
$\mathcal{K}(T)$	Kemeny constant associated with the stochastic matrix T

$\kappa(T)$ condition number for the stochastic matrix T

$L(\mathcal{G}, w)$ Laplacian matrix for the weighted graph $(\mathcal{G}, w)$

M mean first passage matrix

$\mu(\Delta)$ infimum of $\mathcal{K}(T)$, taken over $T \in \mathcal{S}(\Delta)$

$\mathcal{N}(B)$ null space of the matrix B

$\mathbf{1}$ all ones vector

$\mathcal{P}^n$ set of all $A \in \mathbb{R}^{n,n}$ having all principal minors positive

$\mathcal{P}_0^n$ set of all singular $A \in \mathbb{R}^{n,n}$ having all proper principal minors positive

$\mathcal{P}(T)_S$ stochastic complement of $T(S)$ in T

$P_{\mathcal{R}(B),\mathcal{N}(B)}$ oblique projector with range $\mathcal{R}(B)$ and null space $\mathcal{N}(B)$

$Q(A)$ $r(A)I - A$ for the irreducible nonnegative matrix A

$Q^\#$ group inverse of the matrix Q

$q_{i,j}^\#$ (i, j) entry of $Q^\#$

rank(B) rank of the matrix B

$r(B)$ Perron value of the essentially nonnegative matrix B

$\mathcal{R}(B)$ range of the matrix B

$r(i, j)$ resistance distance between vertices i and j in a graph

$\mathbb{R}^n$ real vectors of order n

$\mathbb{R}^{n,n}$ $n \times n$ real matrices

$s(B)$ spectral abscissa of the square matrix B

$\mathcal{S}(\Delta)$ set of stochastic matrices T such that $\mathcal{D}(T)$ is a spanning subgraph of Δ

$\mathsf{si}(T)$ scrambling index of the primitive stochastic matrix T

trace(B) trace of the matrix B

$\tau(B)$ $\frac{1}{2}\max_{i,j}\sum_{k=1}^{n}|b_{i,k} - b_{j,k}|$, for the matrix $B \in \mathbb{R}^{n,n}$ having constant row sums

$\mathcal{W}(\mathcal{G})$ Weiner index of $\mathcal{G}$

$w[S]$ subvector of w on the entries indexed by the set S

$x_i(t)$, $x_i(B)$ i–th component of the vector x, which depends explicitly on the parameter t and the matrix B, respectively

0 the zero matrix, or the zero vector

Bibliography

[1] N.M.M. Abreu. Old and new results on algebraic connectivity of graphs. *Linear Algebra and its Applications*, 423:53–73, 2007.

[2] M. Akelbek. *A Joint Neighbour Bound for Primitive Digraphs.* Ph.D. Thesis, University of Regina, 2008.

[3] M. Akelbek and S. Kirkland. Coefficients of ergodicity and the scrambling index. *Linear Algebra and its Applications*, 430:1099–1110, 2009.

[4] Y.A. Alpin and N.Z. Gabassov. A remark on the problem of localization of the eigenvalues of real matrices. *Izvestija Vysših Učebnyh Zavedeniĭ Matematika*, 11:98–100, 1976.

[5] A.L. Andrew, K.-W.E. Chu, and P. Lancaster. Derivatives of eigenvalues and eigenvectors of matrix functions. *SIAM Journal on Matrix Analysis and Applications*, 14:903–926, 1993.

[6] K.M. Anstreicher and U.G. Rothblum. Using Gauss–Jordan elimination to compute the index, generalized nullspaces, and Drazin inverse. *Linear Algebra and its Applications*, 85:221–239, 1987.

[7] H. Bai. The Grone–Merris conjecture. *Transactions of the American Mathematical Society*, 363:4463–4474, 2011.

[8] R.B. Bapat. *Graphs and Matrices.* Springer, London; Hindustan Book Agency, New Delhi, 2010.

[9] F.L. Bauer, E. Deutsch, and J. Stoer. Abschätzungen für die Eigenwerte positiver linearer Operatoren. *Linear Algebra and its Applications*, 2:275–301, 1969.

[10] A. Ben-Israel and T.N.E. Greville. *Generalized Inverses: Theory and Applications, 2nd Ed.*. Springer–Canadian Mathematical Society Books in Mathematics, New York, 2003.

[11] E. Bendito, A. Carmona, A.M. Encinas, and M. Mitjana. Distance-regular graphs having the M–property. *Linear and Multilinear Algebra*, 60:225–240, 2012.

[12] A. Berman and R.J. Plemmons. *Nonnegative Matrices in the Mathematical Sciences.* Society for Industrial and Applied Mathematics, Philadelphia, 1994.

[13] J.A. Bondy and U.S.R. Murty. *Graph Theory with Applications.* American Elsevier Publishing Co., New York, 1976.

[14] A. Brauer. Limits for the characteristic roots of a matrix IV: applications to stochastic matrices. *Duke Mathematical Journal*, 19:75–91, 1952.

[15] R.A. Brualdi and H.J. Ryser. *Combinatorial Matrix Theory.* Cambridge University Press, Cambridge, 1991.

[16] S.L. Campbell and C.D. Meyer. *Generalized Inverses of Linear Transformations.* Dover Publications, New York, 1991.

[17] H. Caswell. *Matrix Population Models: Construction Analysis and Interpretation, 2nd Ed.*. Sinauer, Sunderland, 2001.

[18] M. Catral, M. Neumann, and J. Xu. Proximity in group inverses of M-matrices and inverses of diagonally dominant M-matrices. *Linear Algebra and its Applications*, 409:32–50, 2005.

[19] S. Chaiken. A combinatorial proof of the all minors matrix tree theorem. *SIAM Journal on Algebraic and Discrete Methods*, 3:319–329, 1982.

[20] P. Chebotarev and E. Shamis. The matrix–forest theorem and measuring relations in small social groups. *Automation and Remote Control*, 58:1505–1514, 1997.

[21] Y. Chen, S. Kirkland, and M. Neumann. Nonnegative alternating circulants leading to M–matrix group inverses. *Linear Algebra and its Applications*, 233:81–97, 1996.

[22] G.E. Cho and C.D. Meyer. Markov chain sensitivity measured by mean first passage times. *Linear Algebra and its Applications*, 316:21–28, 2000.

[23] G.E. Cho and C.D. Meyer. Comparison of perturbation bounds for the stationary distribution of a Markov chain. *Linear Algebra and its Applications*, 335:137–150, 2001.

[24] J.E. Cohen. Derivatives of the spectral radius as a function of non–negative matrix elements. *Mathematical Proceedings of the Cambridge Philosophical Society*, 83:183–190, 1978.

[25] P. Courrieu. Fast computation of Moore–Penrose inverse matrices. *Neural Information Processing Letters and Reviews*, 8:25–29, 2005.

[26] E. Crisostomi, S. Kirkland, and R. Shorten. A Google-like model of road network dynamics and its application to regulation and control. *International Journal of Control*, 84:633–651, 2011.

[27] D.M. Cvetkovic, M. Doob, I. Gutman, and A. Torgasev. *Recent Results in the Theory of Graph Spectra*. North-Holland, Amsterdam, 1988.

[28] D.M. Cvetkovic, M. Doob, and H. Sachs. *Spectra of Graphs. Theory and Applications, 3rd Ed.*. Johann Ambrosius Barth, Heidelberg, 1995.

[29] H. de Kroon, A. Plaisier, J. van Groenendael, and H. Caswell. Elasticity: The relative contribution of demographic parameters to population growth rate. *Ecology*, 67:1427–1431, 1986.

[30] E. Deutsch and M. Neumann. Derivatives of the Perron root at an essentially nonnegative matrix and the group inverse of an M–matrix. *Journal of Mathematical Analysis and Applications*, 102:1–29, 1984.

[31] E. Deutsch and M. Neumann. On the first and second order derivatives of the Perron vector. *Linear Algebra and its Applications*, 71:57–76, 1985.

[32] N. Dmitriev and E. Dynkin. On the characteristic numbers of a stochastic matrix. *Comptes Rendus (Doklady) de l'Académie des Sciences de l'URSS, Nouvelle Série*, 49:159–162, 1945.

[33] D. Doak, P. Kareiva, and B. Klepetka. Modeling population viability for the desert tortoise in the western Mojave desert. *Ecological Applications*, 4:446–460, 1994.

[34] P. Doyle and J.L. Snell. *Random Walks and Electrical Networks*. Mathematical Association of America, Washington, 1984.

[35] L. Elsner. On convexity properties of the spectral radius of nonnegative matrices. *Linear Algebra and its Applications*, 61:31–35, 1984.

[36] L. Elsner, C.R. Johnson, and M. Neumann. On the effect of the perturbation of a nonnegative matrix on its Perron eigenvector. *Czechoslovak Mathematical Journal*, 32:99–109, 1982.

[37] M. Fiedler. Algebraic connectivity of graphs . *Czechoslovak Mathematical Journal*, 23:298–305, 1973.

[38] M. Fiedler. A property of eigenvectors of nonnegative symmetric matrices and its application to graph theory. *Czechoslovak Mathematical Journal*, 25:619–633, 1975.

[39] M. Fiedler. Eigenvectors of acyclic matrices. *Czechoslovak Mathematical Journal*, 25:607–618, 1975.

[40] M. Fiedler. Absolute algebraic connectivity of trees. *Linear and Multilinear Algebra*, 26:85–106, 1990.

[41] M. Fiedler and V. Pták. On matrices with nonpositive off–diagonal elements and positive principal minors. *Czechoslovak Mathematical Journal*, 12:382–400, 1962.

[42] M. Fiedler and V. Pták. Some generalizations of positive definiteness and monotonicity. *Numerische Mathematik*, 9:163–172, 1966.

[43] S. Friedland. Convex spectral functions. *Linear and Multilinear Algebra*, 9:299–316, 1980/81.

[44] A. Ghosh, S. Boyd, and A. Saberi. Minimizing effective resistance of a graph. *SIAM Review*, 50:37–66, 2008.

[45] G.H. Golub and C.D. Meyer. Using the QR factorization and group inversion to compute, differentiate, and estimate the sensitivity of stationary probabilities for Markov chains. *SIAM Journal on Algebraic and Discrete Methods*, 7:273–281, 1986.

[46] G.H. Golub and C.F. Van Loan. *Matrix Computations, 3rd Ed.* Johns Hopkins University Press, Baltimore, 1996.

[47] R.T. Gregory and D.L. Karney. *A Collection of Matrices for Testing Computational Algorithms*. Wiley–Interscience, New York, 1969.

[48] T.N.E. Greville. The Souriau–Frame algorithm and the Drazin pseudoinverse. *Linear Algebra and its Applications*, 6:205–208, 1973.

[49] I. Gutman and B. Mohar. The quasi–Weiner and the Kirchhoff indices coincide. *Journal of Chemical Information and Computer Sciences*, 36:982–985, 1996.

[50] F. Harary. Status and contrastatus. *Sociometry*, 22:23–43, 1959.

[51] F. Harary. *Graph Theory*. Addison–Wesley, Reading, 1969.

[52] R.E. Hartwig. A method for calculating A^d. *Mathematica Japonica*, 26:37–43, 1981.

[53] M. Haviv and L. Van der Heyden. Perturbation bounds for the stationary probabilities of a finite Markov chain. *Advances in Applied Probability*, 16:804–818, 1984.

[54] D.J. Higham and N.J. Higham. Structured backward error and condition of generalized eigenvalue problems. *SIAM Journal on Matrix Analysis and Applications*, 20:493–512, 1998.

[55] N.J. Higham. *Accuracy and Stability of Numerical Algorithms, 2nd Ed.* Society for Industrial and Applied Mathematics, Philadelphia, 2002.

[56] E. Hopf. An inequality for positive linear integral operators. *Journal of Mathematics and Mechanics*, 12:683–692, 1963.

[57] R.A. Horn and C.R. Johnson. *Matrix Analysis*. Cambridge University Press, Cambridge, 1985.

[58] H. Hosoya. Topological index. A newly proposed quantity characterizing the topological nature of structural isomers of saturated hydrocarbons. *Bulletin of the Chemical Society of Japan*, 4:2332–2339, 1971.

[59] J.J. Hunter. Stationary distributions and mean first passage times of perturbed Markov chains. *Linear Algebra and its Applications*, 410:217–243, 2005.

[60] J.J. Hunter. Mixing times with applications to perturbed Markov chains. *Linear Algebra and its Applications*, 417:108–123, 2006.

[61] J.J. Hunter. Variances of first passage times in a Markov chain with applications to mixing times. *Research Letters in the Information and Mathematical Sciences*, 10:17–48, 2006.

[62] I.C.F. Ipsen and S. Kirkland. Convergence analysis of a Pagerank updating algorithm by Langville and Meyer. *SIAM Journal on Matrix Analysis and Applications*, 27:952–967, 2006.

[63] C.R. Johnson. Inverse M-matrices. *Linear Algebra and its Applications*, 47:195–216, 1982.

[64] C.R. Johnson and R.L. Smith. Inverse M-matrices, II. *Linear Algebra and its Applications*, 435:953–983, 2011.

[65] C.R. Johnson and C. Xenophontos. Irreducibility and primitivity of Perron complements: Application of the compressed directed graph. In *Graph Theory and Sparse Matrix Computation, IMA Vol. Math. Appl. 56, A. George, J. R. Gilbert, and J. W. H. Liu, (Eds).*, 101–106. Springer, New York, 1993.

[66] J.G. Kemeny and J.L. Snell. *Finite Markov Chains.* Van Nostrand, Princeton, 1960.

[67] N. Keyfitz and W. Flieger. *Population: Facts and Methods of Demography.* W.H. Freeman, San Francisco, 1971.

[68] S. Kirkland. The group inverse associated with an irreducible periodic nonnegative matrix. *SIAM Journal on Matrix Analysis and Applications*, 16:1127–1134, 1995.

[69] S. Kirkland. On a question concerning condition numbers for Markov chains. *SIAM Journal on Matrix Analysis and Applications*, 23:1109–1119, 2002.

[70] S. Kirkland. Conditioning properties of the stationary distribution for a Markov chain. *Electronic Journal of Linear Algebra*, 10:1–15, 2003.

[71] S. Kirkland. Girth and subdominant eigenvalues for stochastic matrices. *Electronic Journal of Linear Algebra*, 12:25–41, 2004/05.

[72] S. Kirkland. Sensitivity of the stable distribution vector for a size–classified population model. *Linear Algebra and its Applications*, 398:3–23, 2005.

[73] S. Kirkland. Algebraic connectivity. In *Handbook of Linear Algebra, (Ed.) Hogben L.* . Chapman and Hall/CRC, Boca Raton, 36-1 to 36-12, 2007.

[74] S. Kirkland. Fastest expected time to mixing for a Markov chain on a directed graph. *Linear Algebra and its Applications*, 433:1988–1996, 2010.

[75] S. Kirkland, J. Molitierno, and M. Neumann. The sharpness of a lower bound on algebraic connectivity for maximal graphs. *Linear and Multilinear Algebra*, 48:237–246, 2001.

[76] S. Kirkland, J.J. Molitierno, M. Neumann, and B.L. Shader. On graphs with equal algebraic and vertex connectivity. *Linear Algebra and its Applications*, 341:45–56, 2002.

[77] S. Kirkland, M. Neumann, N. Ormes, and J. Xu. On the elasticity of the Perron root of a nonnegative matrix. *SIAM Journal on Matrix Analysis and Applications*, 24:454–464, 2002.

[78] S. Kirkland, M. Neumann, and B.L. Shader. Distances in weighted trees and group inverses of Laplacian matrices. *SIAM Journal on Matrix Analysis and Applications*, 18:827–841, 1997.

[79] S. Kirkland, M. Neumann, and B.L. Shader. Bounds on the subdominant eigenvalue involving group inverses with applications to graphs. *Czechoslovak Mathematical Journal*, 47:1–20, 1998.

[80] S. Kirkland and M. Neumann. Convexity and concavity of the Perron root and vector of Leslie matrices with applications to population models. *SIAM Journal on Matrix Analysis and Applications*, 15:1092–1107, 1994.

[81] S. Kirkland and M. Neumann. Group inverses of M-matrices associated with nonnegative matrices having few eigenvalues. *Linear Algebra and its Applications*, 220:181–213, 1995.

[82] S. Kirkland and M. Neumann. The M-matrix group generalized inverse problem for weighted trees. *SIAM Journal on Matrix Analysis and Applications*, 19:226–234, 1998.

[83] S. Kirkland and M. Neumann. Cutpoint decoupling and first passage times for random walks on graphs. *SIAM Journal on Matrix Analysis and Applications*, 20:860–870, 1999.

[84] S. Kirkland and M. Neumann. The case of equality in the Dobrushin–Deutsch–Zenger bound. *Linear Algebra and its Applications*, 431:2373–2394, 2009.

[85] S. Kirkland, M. Neumann, and J. Xu. A divide and conquer approach to computing the mean first passage matrix for Markov chains via Perron complement reductions. *Numerical Linear Algebra with Applications*, 8:287–295, 2001.

[86] S. Kirkland, M. Neumann, and J. Xu. Transition matrices for well–conditioned Markov chains. *Linear Algebra and its Applications*, 424:118–131, 2007.

[87] D.J. Klein and M. Randic. Resistance distance. *Journal of Mathematical Chemistry*, 12:81–95, 1993.

[88] O. Krafft and M. Schaefer. Mean passage times for tridiagonal transition matrices and a two-parameter Ehrenfest urn model. *Journal of Applied Probability*, 30:964–970, 1993.

[89] M. Levene and G. Loizou. Kemeny's constant and the random surfer. *American Mathematical Monthly*, 109:741–745, 2002.

[90] N.V.R. Mahadev and U.N. Peled. *Threshold Graphs and Related Topics, Annals of Discrete Mathematics 56.* North-Holland, Amsterdam, 1995.

[91] M. Marcus and H. Minc. *A Survey of Matrix Theory and Matrix Inequalities.* Dover Publications, New York, 1992.

[92] R. Merris. Degree maximal graphs are Laplacian integral. *Linear Algebra and its Applications*, 199:381–389, 1994.

[93] L.A. Metzler. A multiple–county theory of income transfer. *Journal of Political Economy*, 59:14–29, 1951.

[94] C.D. Meyer. The role of the group generalized inverse in the theory of finite Markov chains. *SIAM Review*, 17:443–464, 1975.

[95] C.D. Meyer. The condition of a finite Markov chain and perturbation bounds for the limiting probabilities. *SIAM Journal on Algebraic and Discrete Methods*, 1:273–283, 1980.

[96] C.D. Meyer. Uncoupling the Perron eigenvector problem. *Linear Algebra and its Applications*, 114/115:69–94, 1989.

[97] C.D. Meyer. Sensitivity of the stationary distribution of a Markov chain. *SIAM Journal on Matrix Analysis and Applications*, 15:715–728, 1994.

[98] S.R. Mohan, M. Neumann, and K.G. Ramamurthy. Nonnegativity of principal minors of generalized inverses of M-matrices. *Linear Algebra and its Applications*, 58:247–259, 1984.

[99] B. Mohar. Eigenvalues, diameter, and mean distance in graphs. *Graphs and Combinatorics*, 7:53–64, 1991.

[100] B. Mohar. Laplace eigenvalues of graphs—a survey. *Discrete Mathematics*, 109:171–183, 1992.

[101] J.J. Molitierno. *Applications of Combinatorial Matrix Theory to Laplacian Matrices of Graphs.* Chapman and Hall/CRC, Boca Raton, 2012.

[102] M. Neumann and N-S. Sze. On the inverse mean first passage matrix problem and the inverse M-matrix problem. *Linear Algebra and its Applications*, 434:1620–1630, 2011.

[103] M. Neumann and J. Xu. A parallel algorithm for computing the group inverse via Perron complementation. *Electronic Journal of Linear Algebra*, 13:131–145, 2005.

[104] P. Pokarowski. Uncoupling measures and eigenvalues of stochastic matrices. *Journal of Applied Analysis*, 4:259–267, 1998.

[105] P. Robert. On the group-inverse of a linear transformation. *Journal of Mathematical Analysis and Applications*, 22:658–669, 1968.

[106] N.J. Rose. A note computing the Drazin inverse. *Linear Algebra and its Applications*, 15:95–98, 1976.

[107] E. Seneta. *Non-negative Matrices and Markov Chains, 2nd Ed.* Springer–Verlag, New York, 1981.

[108] E. Seneta. Sensitivity analysis, ergodicity coefficients, and rank–one updates for finite Markov chains. In *Numerical Solution of Markov Chains, (Ed.) Stewart W.J.* Marcel Dekker, New York, 121–129, 1991.

[109] G.W. Stewart. *Introduction to Matrix Computations.* Academic Press, New York, 1973.

[110] R. Vahrenkamp. Derivatives of dominant root. *Applied Mathematics and Computation*, 2:29–39, 1976.

[111] G. Wang, Y. Wei, and S. Qiao. *Generalized Inverses: Theory and Computations.* Science Press, Beijing, 2004.

[112] J.H. Wilkinson. *The Algebraic Eigenvalue Problem.* Oxford University Press, London, 1965.

[113] J.H. Wilkinson. Note on the practical significance of the Drazin inverse. *Technical report STAN-CS-79-736, Stanford University,* 1979.

[114] F. Zhang. *Matrix Theory, Basic Results and Techniques, 2nd Ed.* Springer, New York, 2011.

Index

Milton Keynes UK
Ingram Content Group UK Ltd.
UKHW021625071024
449327UK00020BA/1185

9 780367 380489